DICTIONARY OF
MECHANICAL ENGINEERING

DICTIONARY
OF MECHANICAL
ENGINEERING

J. L. NAYLER,
M.A., C.Eng., F.R.Ae.S., F.A.I.A.A.

G. H. F. NAYLER,
M.Sc., C.Eng., M.I.Mech.E., A.F.R.Ae.S.

BUTTERWORTHS

London – Boston – Sydney – Wellington – Durban – Toronto

First published by George Newnes Ltd, 1967
Second edition by Newnes–Butterworths, 1975
   Reprinted 1978
   Reprinted by Butterworths, 1981

ISBN 0 408 00175 5

Reproduced and printed by photolithography and bound in Great Britain at
The Pitman Press, Bath

PREFACE

THIS dictionary has been compiled to cover the very broad subject of mechanical engineering in such a way that terms in common use can be found in a book of handy size like many others already published in Newnes' series of technical dictionaries. We have interpreted mechanical engineering as mainly the production of, the means for, and the utilisation of, mechanical power in engines, transport and mechanisms. We have had in mind that tools, and the making of them, are of first importance, but in order to keep within certain limits we have omitted—with rare exceptions—those tools that are used by hand. As regards power, its production involves the design and construction of many types of devices to enable energy to be developed from fundamental sources and then on to prime movers. As far as possible, terms likely to be found in other dictionaries of the series have been omitted, except where the application of a term is clearly common to more than one branch of engineering. Consequently we have largely neglected the many fields allied to the mechanical engineering industry, such as foundry practice, metallurgy, metrology and welding, all of which are vital to the industry but are not in themselves mechanical.

In preparing a work of this kind it is necessary to consult many sources, since the choice of clear and concise definitions is always a difficult task. The selection of terms has been based mainly on the reading of current literature, including the foremost engineering journals. Thus many well known but little used terms may not be found. We have specialised to some extent on machine tools and mechanisms that have wide application, although intrinsically the latter were first used in clocks and watches.

Terms printed in italics in the text indicate other entries that will provide the reader with supplementary information.

Our illustrations are intended to help the less expert and are spread over the large field of mechanical engineering, whilst avoiding intricate subjects which are too complicated for simple line drawings.

The authors are much indebted to Miss E. E. Metcalfe for her valuable assistance in the preparation of the line drawings, and to the Publisher's staff for their helpful co-operation at all times during the passage of the dictionary through the press.

J. L. N.
G. H. F. N.

A

a. The symbol for acceleration. Also *f.*

α. The symbol for angular acceleration.

abs. Absolute.

acc. Acceleration.

Abradant. A material, such as emery and generally in powder form, used for grinding.

Abrasion. The wearing or rubbing away of a surface. See also **Dresser.**

Abrasion Test. A test by means of a scratch on a smooth surface of the material. See **Hardness Tests.**

Absorber. An auxiliary vibratory system, for modifying the vibration characteristics of the main system, either damped or undamped.

Absorption Dynamometer. A dynamometer which measures the work done by absorbing or dissipating the power, e.g. by the friction of a brake or as in a *Froude brake.* Cf. **Transmission Dynamometer.**

Abutment Engine. See **Engine (Servomotor Types).**

Acceleration. The rate of change of *velocity* (speed) or the average increase of velocity in a unit of time, usually expressed in feet (or centimetres) per second per second. Similarly, ' angular acceleration ' is the rate of increase of rotational (angular) velocity, usually expressed in radians per second per second.

Acceleration due to Gravity. The acceleration of a freely falling body in a vacuum, varying with the distance from the earth's centre and having a mean value of $32 \cdot 2 \, \text{ft/sec}^2$ or $9 \cdot 806 \, \text{m/s}^2$ at sea level.

Acceleration Impedance. See **Effective Inertia, Effective Mass.**

Accelerator (Accelerator Pedal). (*a*) A pedal in a motor-vehicle which acts on the throttle valve and thus controls the power and speed of the engine.

(*b*) A pedal which controls the fuel injection into an oil engine.

Accelerometer. An instrument for measuring acceleration (*a*) by the movement of a mass supported by a spring, (*b*) by a simple pendulum in which its period is inversely proportional to the square

1

root of the acceleration being measured, or (*c*) by the precession rate of a pendulous gyroscope.

Accessory Gear-box. A gear-box, driven by and remote from, an engine for mounting accessories such as a hydraulic pump on an aero-engine.

Acid Pump. A pump with the barrel and valves made of glass so as to be resistant to acids.

Ackermann Steering. The arrangement on an automobile whereby the inner axle moves through a greater angle than the outer axle during cornering so as to give approximate true rolling of the respective wheels.

Acme Thread. A standard American screw thread, with the flanks having an inclined angle of 29°, which is used extensively for feed screws. It has the same depth as the square thread but is stronger as the bottom of the thread is wider than the square thread. (See Fig. 183(*b*).)

Actuator. (*a*) An electric, hydraulic, mechanical or pneumatic device, or combinations of these, to effect some predetermined linear or rotating movement. One example used in automatic train conpressed air usually operated at a pressure of from 400 to 500 kPa. If steam is used, it is usually at a pressure from 525–700 kPa. pipe pressure reduction.

(*b*) A *servomotor* producing a limited output motion.

(*c*) A complete self-contained *servo-mechanism* producing a limited output motion.

Adapter. An accessory appliance so that objects of different sizes can be interchanged such as on a spindle or other fitting.

Addendum. (*a*) The radial distance from the *reference cylinder* (reference addendum) or the *pitch cylinder* (working addendum) to the *tip cylinder*.

(*b*) Half the difference between the diameter of the *pitch circle* of a worm wheel and the *throat diameter*.

(*c*) The distance from the *pitch circle* to *tip cone* measured on the *back cone* of a bevel gear.

(*d*) In a gear-wheel, the radial distance between the *pitch circle* and the *addendum circle*. (Figs. 84, 93, 172, 173.) In a British Standard form of tooth it equals 0·3183 times the circular pitch.

 Addendum Circle. The circle passing through the tips of the teeth and equal in diameter to the blank or disc from which the wheel is to be cut. Cf. **Tip Cylinder.**

 Chordal Addendum. The radial distance measured as in (*a*) above but from the chord of the *pitch circle*.

Addendum Angle. The difference between the *tip angle* and the *pitch angle* of a bevel gear.

Addendum Circle. *Tip circle.* See also **Tip Cylinder.**

Addendum (Screw Threads). The radial distance between the major and pitch cylinders (or cones) of an external screw thread; the radial distance between the pitch and minor cylinders (or cones) of an internal thread. See **Pitch Cylinder (Cones).** (Fig. 156.)

Adhesion (Adhesive Force). The frictional grip between two surfaces in contact, e.g. between the locomotive driving wheel and the rail where it is the product of the weight on the wheel and the friction coefficient (0·1 to 0·2 depending on the condition of the rail surface).

Adjustable Pitch Propeller. A propeller, the *pitch* of whose blades can be altered on the ground but not in flight.

Adjusting Rod. A rod with an adjustable clamp for attaching to a *fusee* or barrel arbor and provided with sliding weights for balancing the pull exerted by the main spring and thus testing its pull.

Adjusting Screw. A screw, usually with a very fine thread, in an instrument or tool by which one part is moved relative to another, to give adjustment in focus, level, tension, etc.

Adjustment Strips. Metal strips for the accurate adjustment of the exact bearing loads on sliding surfaces. The precise amount of contact is effected by pressure imparted to the strips from set or adjusting screws.

Admission. The instant in the working cycle of an internal-combustion or steam-engine when the inlet valve allows entry of the working fluid into the cylinder.

Admission Corner. The corner on an *indicator diagram* which corresponds with the entry of the working fluid into a cylinder.

Admission Line. The side of the *indicator diagram* which shows the actual condition while steam is entering an engine cylinder.

Admission Port. The passage by which steam or the combustible fluid enters an engine cylinder.

Admittance. See **Mechanical Admittance.**

Advance. (*a*) Altering the time of ignition of an internal-combustion engine to cause the spark in the cylinder to pass at an earlier point.

(*b*) See **Angular Advance.**

Advance of Spark. In ignition, turning the *contact breaker* so that the spark will ignite the charge earlier during the compression stroke.

Advance of Spark

An internal-combustion engine has 'advanced ignition' if the spark has been advanced.

Advanced Ignition. See **Advance of Spark.**

Aerial (Cable) Railway (Aerial Ropeway). A system of overhead cables with small cars or containers for conveying persons or loads, usually over mountainous country.

Aero-engine. The power unit of an aircraft, including originally piston-engine types and later the gas-turbine types after these had been invented.

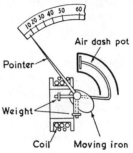

FIG. 1. AIR DASHPOT (MOVING-IRON INSTRUMENT).

Aero-thread Insert. See **Thread Insert.**

Agitators. Mechanical stirrers or reverberators for settling concrete, sorting coal or sand, or for mixing molten metals or for rocking fluid paper-pulp on a frame to make the wood fibres interlace.

Air Bearing. A shaft bearing maintained wholly by compressed air with no contact between fixed and moving surfaces.

Air Brake. (*a*) A mechanical brake on a railway train which is operated by air pressure acting on a piston.

(*b*) A *flap* on an aeroplane wing or fuselage which protrudes from the surface and thus reduces the speed of the aeroplane.

Air-chuck. A pneumatically-operated chuck for gripping machine tools with a mechanical safety device against the failure of the air pressure.

Air-compressor. A machine which compresses air after it has been drawn in at one pressure and delivers it at a higher pressure, as a source of power or for ventilation. It may be of a reciprocating, rotary or fan type.

Air-cushion. A cushion of air compressed on the far side of the piston of a high-speed single-acting steam-engine, both to absorb

4

shocks at the release end of the cylinder and to assist the engine past the dead centre.

Air-cushion Vehicle. A craft which rides on a cushion of air over land or water, the air pressure being maintained beneath the vehicle by power-driven rotors or fans.

Air Dashpot. (See Fig. 1 and **Moving-iron Instrument.**) An instrument having a loosely fitting piston in a cylinder which allows the slightly compressed air to by-pass the piston and thus to slow down the motion of the indicating pointer.

Air Drill. A *rock drill* operated by compressed air. See also **Pneumatic Drill.**

Air-ejector. An *air-pump* which maintains a partial vacuum in a vessel by a high-velocity steam jet to entrain the air and exhaust it against atmospheric pressure.

Air Engine. (*a*) A *heat engine* in which air is used as the working substance. It is impracticable except for very small powers. Cf. **Hot-air Engine.**

(*b*) A small reciprocating engine driven by compressed air.

Air Hammer. (*a*) A double-acting power hammer used in drop forging for roughing out heavy forgings in foundry work. The power of the hammer is based only on the weight of the tup and no account is taken of the additional power provided by the compressed air usually operated at a pressure of from 400 to 500 kPa. If steam is used, it is usually at a pressure from 525–700 kPa.

(*b*) A pneumatic hammer in the form of a pistol and used for rivetting.

Air-lift Pump. (*a*) A pump which forces air from an air-compressor down a small pipe into the water in a well or borehole, and admits the air into the lower end of an immersed pipe so that alternate plugs of water and air are forced up the pipe by the hydrostatic pressure of the water in the open borehole.

(*b*) An air-operated displacement pump.

Air-meter. An apparatus for metering the flow rate of a gas.

Air-pressure Reducing Valve. A mechanical device that takes air at high pressure on the inlet side and delivers it with a lower pressure at the outlet side.

Air-pump. (*a*) A pump or other device used for transferring air from one place to another, for exhausting or for compressing air.

(*b*) A reciprocating pump fitted to condensing steam-engines to draw water from the condenser together with any vapour or air liberated in the process.

Air-pump

(*c*) A vacuum pump to reduce the pressure on the low-pressure side of a system.

(*d*) A blower to obtain a rapidly moving air blast.

Airscrew. See **Propeller.**

Air-speed Indicator (A.S.I.). An instrument which gives the speed of the aircraft through the air; its reading is usually subject to corrections for instrument error, position error and compressibility error.

Fig. 2 shows the principle of its construction. A Pitot-static tube mounted on the aircraft gives the Pitot pressure from an open-ended forward-facing tube and the atmospheric static pressure is from a hole in the side of another tube, usually concentric. These pressures reach the instrument along tubes marked Pitot line and Static line, the former to a flexible capsule and magnifying mecha-

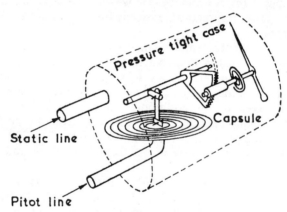

FIG. 2. AIR-SPEED INDICATOR.

nism mounted in an airtight case and the pressure in this case is that of the (pressure) Static line. Thus the instrument is measuring the difference between the two pressures which is proportional to the square of the air speed.

Air Standard Cycle. A standard cycle of reference for comparing the performances and calculating the relative efficiences of different internal-combustion engines. See also **Relative Efficiency.**

Air Standard Efficiency. The thermal efficiency of an internal-combustion engine working on the appropriate *air standard cycle.*

Air Starting Valve. A small piston valve in a diesel engine actuated by the camshaft and operating the main valve to admit

6

starting air to the working cylinders.

Air Valve. (*a*) Any valve controlling the passage of air.

(*b*) A valve located at the highest point in a pipe-line to let out air which has accumulated there.

Air Vane. See **Vane.**

Air-vessel. A vessel containing air which is fitted (*a*) to the delivery side of a reciprocating water pump to smooth out the pulsating discharge, or (*b*) to promote an even flow in long pipe-lines.

Alarum (Clock). A clock which rings on a bell, gong or clock case at a pre-set time.

Alighting Gear. The part of any aircraft which supports it on land or water and absorbs the shock of landing, but excluding the hull of a flying boat. It includes all undercarriage units of landplanes and the main and wing-tip floats of seaplanes.

Alignment. A setting in line such as the centres of a lathe, the centres of the bearings of an engine crankshaft and the axial continuity of shafting and shaft-bearings.

Allan Valve. *Trick valve.*

All-or-nothing Piece (Stop Slide). A piece of the mechanism of a *repeater* which either allows striking or entirely prevents it.

Allowance. A prescribed difference in dimensions in order to allow of some quality of fit between two pieces when mated together such as a hole and a mating shaft. (See also **Fit.**) It is positive for a *clearance fit* or a *transition fit*. (For screw threads, see **Fitting Allowance** and **Wrenching Allowance.**)

Alternate Cones. Two equal cones arranged on parallel shafts with their bases facing in opposite directions. Their mutual function is to provide speed variation by means of a shifting belt that can travel from end to end.

Ambulator. *Perambulator.*

Amplifier. A device in which an input controls by hydraulic, pneumatic or electrical means a local source of power to produce an output greater than, and bearing a definite relationship to, the input.

Torque Amplifier (*Capstan Amplifier*). A mechanism, with input and output shafts rotating at the same speed on the principle of a capstan, to give an amplified output torque when an input torque is applied; the additional energy is supplied by the rotating capstan drum.

Amplitude. The peak (maximum) value of a periodically varying quantity. Cf. **Maximum Value.**

Amplitude Spectrum. The values of the amplitudes of the components of a vibration at each frequency, arranged in the order of the frequencies. It is called a ' line spectrum ' if the amplitude is zero except at discrete frequencies and the vibration is sustained, and a ' continuous spectrum ' for a vibration which is not sustained.

Anchor. See **C.Q.R. Anchor, Stockless Anchor.**

Anchor Escapement. *Recoil escapement.*

Anemometer. An instrument for measuring and registering the velocity and direction of the wind or the rate of flow of a gas, usually by mechanical or electrical methods.

Aneroid Barometer. An instrument, usually portable, for recording changes in atmospheric pressure and for the determination of altitude. It is constructed on the principle of a *vacuum chamber unit* with a train of levers to magnify the amount of the expansion and contraction of a bellows type of unit. The zero must be set to the correct sea-level atmospheric pressure at the time in order that its reading should indicate the correct altitude.

Angle Bearing. A shaft-bearing in which the joint between the base and the *cap* is set at an angle, and thus not perpendicular to the direction of the load.

Angle Bending Machines. Machines either for bending bars to various curvatures by means of rolls or for squeezing bars into angular forms in presses.

Angle Cutter. A milling cutter for making flutes on taps, reamers, etc.

Angle Gear. An arrangement of bevel gearing to drive a shaft, at other than a right angle with the driving shaft, by the inter-position of a third mitre wheel.

Angle Motion. Canting motion. See **Canting.**

Angle of Advance. *Angular advance.*

Angle of Flexure. The angle through which torsion deflects a shaft.

Angle of Friction. *Friction angle.*

Angle of Inclination. (*a*) The angle which the thread of a screw makes with its axis.

(*b*) The angle made with the horizontal plane.

Angle of Obliquity. The deviation of the direction of the force between two gear-teeth in contact, from that of their common tangent.

Angle of Relief (Clearance Angle). The angle between the back face or the lower part of a cutting tool and the surface of the material which is being cut.   (See Figs. 3, 17, 46.)

Angle of Thread. *Included angle.*

Angle of Twist.   The angle through which one section of a shaft is twisted by a torque, relative to some other section.

Angle of Upset.   The angle at which a portable-type balance crane will upset or overturn with the weight of its load.

Angle Valve.   A screw-down stop valve with the casing or body of a spherical shape.   The axis of the stem is in line with one body end and at right angles to the other.   Cf. **Globe and Oblique Valves.**

Angledozer. See **Bulldozer.**

Angles of Cutting Tools (Cutting Angles).   The angles between the surfaces of the materials being cut and the cutting faces of the tools. The chip thickness varies with the approach angle of the tool.   See **Angle of Relief, Cutting Tools,** Figs. 3, 14, 47.

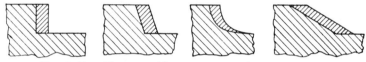

FIG. 3. CHIP THICKNESS VARIATION WITH APPROACH ANGLE.

Angular Acceleration.   See **Acceleration.**

Angular Advance (Angle of Advance).   In steam-engine valve gear the angle which the centre of an eccentric *sheave* makes with a line set at 90° in advance of the crank pin.   Its magnitude depends on the *lead* and the outside *lap*.   See **Lead of Valve** and **Lap.**

Angular Cutter.   A milling cutter with the cutting face at an angle to the axis of the cutter.

Angular Displacement.   (*a*) The angle turned through by a body about a given axis, usually measured in degrees.

(*b*) The angle turned through by a line joining a fixed point to a moving point.

Angular Momentum.   See **Momentum.**

Angular Thread. *Vee-thread.*

Angular Velocity.   The rate of change of *angular displacement,* usually expressed in radians per second or revolutions per minute. Cf. **Linear Velocity.**

Annular Gear.   An annular ring with gear-teeth cut on it.

Annular Seating.   A ring-shaped seating for a valve as found in pumps.

Annular Valve (Circular Disc Valve).   A valve consisting of a circular disc seating on a concentric hole.

9

Annular Wheel. A cog-wheel with the teeth fixed to its internal diameter; also called an internal wheel. It always revolves in the same direction as its *pinion*.

Anti-backlash Gear. See **Gear**.

Anti-friction Bearing. A bearing in which special means are taken to reduce friction such as rollers to support a rotating shaft. Special metals, plastics, polyurethane rubbers and other complex compounds are also often used by themselves or impregnated in the material of the main bearing bush.

Anti-friction Metals. A term originally used to describe *white metal*, a tin-base alloy containing over 50% tin, and now applied to a wide range of metals specially suitable for bearings, especially tin-lead alloys.

Anti-friction Rollers. Live *rollers* which sustain the pressure of a rotating spindle or shaft. Cf. **Anti-friction Bearing**.

Antinode. See **Nodes**.

Antinous Release. A flexible release cable in a camera to operate the shutter.

Antiphase. When the difference in *phase angle* is π.

Anti-resonance. When a small change in the frequency of an externally-applied excitation causes an increase in the amplitude of a specified response of a mechanical system. (Cf. **Resonance**.) See also **Frequency**.

Anti-static Belting. Belting with an extremely high coefficient of friction on a highly conductive traction face which does not clog with fluff, etc. See also **Belt**.

Anti-torque Rotor. *Tail rotor*.

Anti-vibration. See **Vibration** and **Mounting**.

Anvil (Anvil Block). (*a*) A massive block of cast- or wrought-iron, sometimes steel-faced, on which work is supported during forging.

(*b*) The jaws of a *micrometer* are also called anvils. (See Fig.112.)

Aperiodic. Having no natural frequency; not resonant at any one frequency.

Apex. (*a*) The common intersection of the two axes of a pair of *bevel gears* and the instantaneous axis of relative motion of either gear with respect to the other, called the *pitch element*, which all lie in the axial plane. (Figs. 11, 12, 131.)

(*b*) The corner of the *fundamental triangle* opposite to its base in the geometry of a screw thread. (Fig. 157.)

Apron. A plate or fixing bolted to the front of the saddle of a lathe, which encloses the gear operated by the *lead screw*. (See Fig. 102.)

Apron Conveyor. A travelling belt composed of a number of linked sections, usually metal or wood slats, for transport horizontally or on a gentle gradient; also called ' slat conveyor '. It is used on large machines to protect the *bed* from swarf. Aircraft wing spar milling machines often incorporate this type of apron.

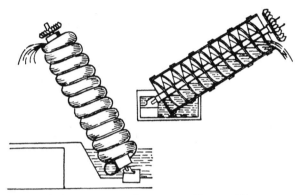

FIG. 4. ARCHIMEDEAN SCREWS FOR RAISING WATER.

Arbor. (*a*) A rotating shaft, *spindle* or bar which forms part of an instrument or machine or machine tool. (See Fig. 167.)

(*b*) A spindle of a wheel as in a watch or clock. See **Barrel Arbor, Fusee Arbor.**

Arbor Chuck. A chuck used in a lathe for turning the outside diameters of cylindrical work after the hole has been first bored, the hole fitting over an arbor, mandrel or spindle to ensure concentricity of outer and inner diameters. See also **Mandrel.**

Arbor Press. An appliance used for forcing arbors or mandrels into or out of work by a screw-press or hydraulic power.

Arboring. Cutting back a flat bearing face to receive the washers and nuts of attachment bolts by means of a broad-faced cutter wedged transversely on an arbor.

Arc of Approach. The arc on the pitch circle of a gear-wheel over which two teeth are in contact and approaching the *pitch point*.

Arc of Contact. The arc on the *pitch circle* of a gear-wheel over which two teeth are in contact.

Arc of Recess. The arc on the pitch circle of a gear-wheel over which two teeth are in contact while receding from the *pitch point*.

11

Arc Thickness. The difference between the chordal addendum and the *addendum* measured along the gear-tooth centre line. (See Figs. 93, 173(*b*).)

Archimedean Drill (Persian Drill). A drill with a quick multiple thread over which a nut works in a to-and-fro axial movement to give an alternating rotary motion of a *bit*.

Archimedean Screw. A hollow inclined screw, or a pipe forming a helix around an inclined axis, with its lower end in water. When it is rotated the water is lifted to a higher level. (Fig. 4.) Cf. **Worm (Screw) Conveyor.**

Archimedean Spiral. (*a*) The locus of a point moving with uniform velocity along the radius vector while the radius vector also moves about the pole with constant angular velocity. In polar co-ordinates $r = a\theta$. See **Pinwheel Gear.**

(*b*) A device for raising water by rotation of a spiral. (Fig. 4.)

Articulated Blade. A rotorcraft blade mounted on one or more hinges to permit flapping and fore-and-aft movement during flight.

Articulated Connecting-rods. The auxiliary connecting-rods of a *radial engine* working on pins carried by the master-rod instead of on the main *crank pin.*

Artificial Horizon (Gyro Horizon). An instrument embodying a *gyroscope* which simulates the natural horizon.

Assembly. The putting together of a machine, or mechanism, from its component parts; also the final product after putting the parts together.

Atmospheric Engine. (*a*) An engine of the piston type which is not supercharged.

(*b*) An early form of steam-engine, in which a partial vacuum created by condensation of the steam allowed atmospheric pressure to drive down the piston; also called a *single-acting engine.*

Atmospheric Line. A datum line on an *indicator diagram* drawn by allowing atmospheric pressure to act on the indicator piston or diaphragm; it divides the steam area above from the vacuum area below.

Atmospheric Pump. *Suction pump.*

Atwood Machine. A device consisting of a pulley over which is passed an inextensible cord connecting two weights. It can be used to determine the acceleration of gravity.

Auger. A tool used for boring holes consisting of a long steel shank with a cutting edge at one end and a cross-piece for handle at the other. A 'shell- ' or ' pod-auger ' has a straight channel or groove. A ' screw auger ' has a twisted blade, the chips being discharged by the spiral groove.

Auger Stem. The heavy bar to which the drill bit is attached when boring a well.

Auto-collimator. See **Optical Tooling.**

Autogiro (Autogyro). A trade name for an aircraft with a main lifting rotor that draws energy from the air stream and is not driven by power. A type of *gyroplane.*

Automatic Expansion. The control by governors and their gear of the expansion of steam in a steam-engine. See also **Governor, Governor Valve Gear.**

Automatic Lathe. A lathe with an unmanned repetitive action. Fig. 103 shows an automatic machine action with five chucks, five cross-slides and three independant axial slides, including the centre block.

Automatic Pilot. See **Autopilot.**

Automatic Stoker. *Mechanical stoker.*

Automation. A technique for controlling the whole or a part of a manufacturing process, including inspection and rejection. Part or all of the technique is automatically under electronic control.

Automobile. *Motor-vehicle.*

Autopilot. The mechanism and its associated controls for controlling automatically the flight of an aircraft or a missile along a given path. The autopilot is usually set by the pilot, but it is sometimes set by radio control as it must be in a missile when in flight. In an aircraft it is called, colloquially, ' George '.

Auxiliary Rotor. A small rotor mounted on the tail of a helicopter to provide directional control and to counteract the torque of the main rotor, usually called *tail rotor.*

Axial (Axial-flow) Compressor. A compressor with alternate rows of fixed and rotating blades, radially mounted, with the flow through the compressor in·the direction of the axis. (See Figs. 192, 194.) There are two main types (*a*) with tapered casing, and (*b*) with tapered rotary drum. Some jet engines have two-spool compressors, dividing the compression into two stages called Low Pressure (L.P.) and High Pressure (H.P.) compressors (see **Spool**). The compression ratio is less for the higher flight speeds of aeroplanes due to the greater external compression.

Axial compressors are often used instead of fans for driving high-speed wind tunnels when the pressure ratio required is too great for a fan design. Cf. **Centrifugal Compressor.**

Axial Engine. (*a*) A turbojet engine with an axial-flow compressor.

(*b*) A piston engine with cylinders parallel with the driving shaft.

Axial Pitch. The *pitch* of a screw or gear measured in a direction parallel to the axis.

Axial Plane. A plane containing the axis of a symmetrical body.

Axial Pressure Angle. See **Pressure Angle (Axial)**.

Axial Run-out. See **Run-out**.

Axial Section. A section in a plane containing the axis of a screw or gear.

Axial Thickness. The distance measured along a line parallel to the axis between the traces of a gear-tooth (see **Tooth Trace**); that is, the distance measured across the reference cylinder of a helical, spur or worm gear-wheel in a direction parallel to the axis.

Axial Turbine. A *turbine* in which the water passes through the wheel in the axial direction.

Axis (Axis of Rotation). A straight line about which a body, a screw or a gear rotates.

Axis of Symmetry. An imaginary straight line around which a symmetrically developed body is formed. The centre of gravity is on this axis when the density is uniform.

Axle. A cross-shaft which carries the driving or freely-mounted wheels of any vehicle. In the ' dead axle ' type the wheel turns on the axle which is inserted in the hub and forms the axis of rotation. In the ' live axle ' type the wheel is rigidly fixed to the axle which turns in bearings.

Stub Axle. A short dead axle carrying the wheels and swivel pins of a vehicle for steering purposes and capable of limited angular movement about the swivel pins.

' Live axles ' can be full-floating, three-quarter floating or semi-floating. The shaft in the full-floating type transmits torque only and the wheel is supported entirely on the axle housing, the shaft being positively connected to the wheel. The wheel in the three-quarter floating type is supported on a set of bearings on the axle housing, the shaft being rigidly fixed in the wheel hub to transmit torque only except when rounding bends. In the semi-floating type the axle shaft rotates on bearings in the axle housing, the axle being rigidly fixed in the wheel. In this last case the axle shaft supports all the weight which is transferred to the wheel and thus experiences bending as well as torque.

' Two-speed axles ' have an epicyclic mechanism interposed between the driven bevel gear and the differential gear.

Axle-box. The complete bearing arrangements for the axles of railway rolling-stock with the upper half of the bearing containing a box-shaped housing to hold the lubricant.

Axle-grinding Machine. A machine for grinding railway axles with the wheels on the axle.

Axle Guard. *Horn plate.*

Axle Housing. An enclosing structure for the axle of railway rolling-stock.

Axle Lathe. A lathe with a loose *poppet* at each end of the bed between which the axle is centred and rotated through a central headstock with double change-speed gears.

Azimuth Control. *Cyclic pitch control.*

B

B.A. *British Association screw thread.*

B.G. Birmingham gauge.

B.H.N. Brinell Hardness Number. **See Brinell Hardness Test.**

b.h.p. *Brake horse-power.*

B.M. Bench mark; bending moment.

b.m.e.p. *Brake mean effective pressure.*

B.o.T. Board of Trade.

B.S. British Standard.

B.S.F. *British Standard fine thread.*

B.S.P. *British Standard pipe screw thread.*

B.S.S. British Standard Specification of the *British Standards Institution.*

B.S.W. *British Standard Whitworth screw thread.*

B.Th.U. British Thermal Unit. Its mean value is the 180th part of the quantity of heat required to raise one pound of water from 32°F to 212°F. Cf. **Mechanical Equivalent of Heat.**

B.W.G. *Birmingham Wire Gauge.* **See Gauges Commonly Used.**

Babbitting. The process of lining bearings with *Babbitt's metal* or with *white metal.*

Babbitt's Metal. An alloy used for bearings containing tin alloyed with copper and antimony plus varying amounts of lead. A common formula is tin 40, copper 1·5, antimony 10 and lead 48·5 as percentages.

Back Centre. A pointed spindle on the *loose headstock* of a lathe for supporting the end of the work remote from the chuck.

Back Cock. The bracket on the back plate of a clock from which the pendulum is suspended.

Back Cone. The cone whose generator is perpendicular to the *pitch cone* generator at the *pitch circle* of a bevel gear.

Back Cone Angle. The angle between the axis and the *back cone* generator of a bevel gear, being the complement of the *pitch angle*.

Back Cone Pressure Angle. The acute angle between the normal to the intersection of the tooth flank of a bevel gear and the *back cone* at the *pitch circle* and the tangent to the *pitch circle* at that point.

Back Cut-off Valve. A sliding and adjustable plate on the back face of the main *slide valve* of a steam-engine to regulate the point of cut-off for the steam and worked independently from a separate eccentric.

Back-firing. (*a*) A premature ignition in an internal-combustion engine before the end of the compression stroke, with the consequent reversal of the direction of rotation during starting.

(*b*) The ignition of gases while the exhaust valve is still open.

Back Gear. A train of gear-wheels fitted to the *headstock* of a lathe or other machine tool for the reduction of the speed of the *mandrel* below that of the cone pulley, thus increasing the power of the machine. Cf. **Eccentric Throw-out.**

Backlash. (*a*) The amount an element of a mechanism has to move before communicating its motion to a second element. See also under **Gear.**

(*b*) For two *gear-wheels*, backlash is the minimum distance between the tooth flanks which are in *mesh*. (See Fig. 172.)

Back Pressure. (*a*) The pressure opposing the motion of a piston during the exhaust stroke, or working stroke, in an internal-combustion engine or steam-engine.

(*b*) The exhaust pressure of a turbine.

Back-pressure Engine. A steam-engine in which the steam is exhausted for heating purposes at a pressure greater than the normal terminal pressure.

Back-pressure Turbine. A steam-turbine from which the whole of the exhaust steam is taken at a suitable pressure for heating purposes.

Back-pressure Valve. A valve to prevent the return flow of fluids in a pipe. Cf. **Check Valve (Non-return Valve).**

Back Rest. (*a*) A guide attached to the *slide rest* of a lathe and placed in contact with the work to steady it when turning.

(*b*) The roller or oscillating bar at the back of a *loom* over which the warp threads pass from *beam* to *healds*.

Back Shaft. The shaft which runs along the whole length of the rear of a self-acting lathe and through which motion is transmitted from the headstock to the slide rest for sliding and surfacing only; it is capable of reversal for traversing the *saddle*.

Back Stay. *Steady*.

Back Steady Rest (Back Stay). See **Steady.**

Backward Eccentric. The eccentric which opens the slide valve to the steam supply when the engine is required to run backward. Sometimes called *backward gear*.

Backward Gear. The relative arrangement of eccentrics, etc., in a steam-engine whereby the engine will, on the admission of steam, run backward. See also **Backward Eccentric.**

Back Washer. A machine for scouring, drying and opening out carded *slivers* in worsted manufacture.

Backing-off. (*a*) The operation of relieving or bevelling off the backs of the teeth of milling cutters.

(*b*) The operation of bevelling the hinder or leaving edge of the threads of a *tap*.

(*c*) The reversal of the spindles of the *mule* in cotton spinning, to unwind the yarn after the completion of twisting and drawing out.

Backing-off Lathe. A lathe in which the teeth of milling cutters and taps are bevelled off by a to-and-fro movement of a cutting tool on the slide rest.

Balance. (*a*) An instrument used for weighing. The ' chemical balance ' gives the weight of small amounts of material to a very high degree of accuracy. See **Weighing Machines.**

(*b*) The vibrating member of a watch or clock with platform escapement forming, with the balance spring, the time-controlling element. (Fig. 39.)

Balance Arc. The portion of the vibration of the *balance* of a watch or clock during which it is in contact with the *escapement*.

Balance Arm. The portion of the *balance* connecting the rim to the *staff*.

Balance Ball. A spherical weight which is attached to a crane chain just above the hook for pulling it down when there is no load being lifted. Larger weights are usually in the shape of a pear—hence ' pear weight '.

Balance Box. A box for a cantilever-type crane containing a heavy load to counterbalance the weight of the jib and the crane's load.

Balance Bridge. See **Bascule Bridge.**

Balance Cock. The detachable bracket carrying the upper pivot of the *balance staff*.

Balance Crane. See **Crane**.

Balance Cylinder (Balancing Cylinder). A small auxiliary steam-cylinder sometimes fitted to large vertical steam-engines to reduce the load on the valve gear by admitting steam to the underside of the *balance* (*dummy*) *piston* which is connected to the engine slide valve.

Balance Gear (U.S.). The *differential gear* of a motor-vehicle.

Balance Piston. *Dummy piston*.

Balance Rim. The circular rim of the *balance* of a watch or clock. See **Compensation Balance**.

Balance Spring. A very fine metal ribbon in the form of a flat spiral or cylinder, or of a helical spring, fixed around the *balance staff* of a watch or small clock. See also **Balance Wheel**.

Balance Staff. The staff which carries the *balance wheel* in a watch or clock and the collet to which the *balance spring* is attached.

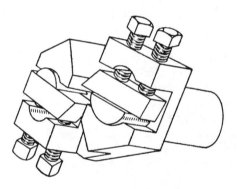

Fig. 5. Balance
Turning Tool.

Balance Turning Tool. A very rugged tool for making roughing cuts. It comprises two tool bits designed to cut tangentially on opposite sides of the workpiece that is being turned. (See Fig. 5.)

Balance Weight. (*a*) A weight used to counterbalance a moving part of a machine, engine or mechanism (e.g. Fig. 45).

(*b*) A weight placed in the driving wheel of a locomotive.

(*c*) Any weight used as a counterpoise.

(*d*) A small weight used on a chemical balance.

(*e*) The small lead weight clipped to the wheel-rim of a motor-vehicle to balance the wheel and tyre on the axle.

Balance Wheel. (*a*) A *flywheel*.

(*b*) A spring-controlled and dynamically-balanced wheel which regulates the beats of a watch or chronometer by its oscillations; a *balance*. See **Balance Spring**.

Balanced Crank. *Disc crank*.

Balanced Pulley. A pulley in balance both statically and dynamically. See **Balancing (Dynamic)** and **Balancing (Static)**.

Balanced Valve. *Equilibrium valve*.

Balanced Wheel. (*a*) A rapidly rotating wheel with a truly-turned rim and holes drilled in or near the rim so that the wheel turns freely and comes to rest in any position. See **Balancing (Dynamic)**.

(*b*) A wheel having dynamic balance.

Balancing Cylinder. *Balance cylinder*.

Balancing (Dynamic). The balancing of the centrifugal forces in rotating machines so that there is no residual unbalance to initiate vibration in any plane. Cf. **Balancing (Static)**.

Balancing Machine. A machine for testing *static balancing* and determining the weight and position of the masses to be added to obtain balance.

Balancing (Static). The act of balancing a piece of rotating mechanism whilst it is stationary. This form of balancing is not sufficient to ensure balance during rotation, since it does not take into account the balancing forces at all stations along the whole length of the axis of rotation.

Baler. A machine for compressing loose bulky material and securing it in a convenient form for transport, such as hay or cotton. See also **Combine Baler**.

Ball-and-disc Integrator. See **Integrator** and Fig. 111.

Ball-and-socket Joint. A joint in which a spherical end is placed within a socket that has been recessed to fit it, thus permitting free motion within a given cone or cut-out in the socket. It is the same as ' ball joint '.

Ball-bearing. A bearing on a shaft composed of a number of hardened-steel balls rolling between an inner race forced on to the shaft and an outer race carried in a housing. The balls are equally spaced by a light metal cage and run in shallow grooves called ball-tracks. The cage or ball retainer is crimped or riveted into place after the balls have been inserted (see Fig. 6). An alternate method utilizes a filling slot in the races which allows assembly with a greater number of balls but reduces the thrust

Ball-bearing

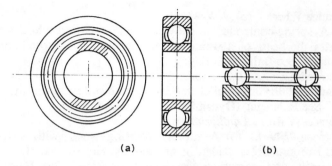

(a)   (b)

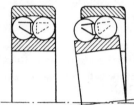

FIG. 6 (*above*). BALL-BEARING, SINGLE-ROW: (*a*) radial or journal, (*b*) thrust.

FIG. 7 (*left*). BALL-BEARING, DOUBLE-ROW SELF-ALIGNING-RADIAL.

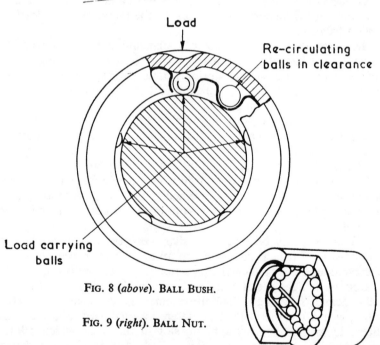

Load

Re-circulating balls in clearance

Load carrying balls

FIG. 8 (*above*). BALL BUSH.

FIG. 9 (*right*). BALL NUT.

20

load carrying capacity of the bearings; that is the load in the axial direction. A self-aligning ball-bearing has the balls running in a spherical housing which enables the inner and outer races to be at an angle to each other as shown in Fig. 7.

Ball Bush. An outer cylindrical sleeve running on balls along a shaft, each row of balls taking the load in turn and then recirculating as shown in Fig. 8.

Ball Catch. A spring-controlled ball, projecting through a smaller hole, which engages with a hole in a striking plate, as for a door fastening.

Ball-cock. A self-regulating valve which, through a linkage system, turns the flow of water (or a liquid) on and off by the falling and the rising of a partly submerged sphere, usually a hollow ball.

Ball Cutter. A spherical cutting tool or a cutter with a rounded edge.

Ball Joint. See **Ball-and-socket Joint.**

Ball Mill. A fine grinder or ore crusher which is a slightly inclined or horizontal rotating cylinder containing balls, usually ceramic, or steel, to grind the material to the necessary fineness by the rubbing and impact of the tumbling balls. Wet ball-milling is usually a batch process, but dry ball-milling may be continuous with the ' fines ' removed by an air current.

Ball Nut. A nut having a semi-circular helical groove on the inside which fits over a shaft with a mating groove. The load is transmitted by balls running in the grooves and returning through a non-load-carrying section. (See Figs. 9, 134.)

Ball Race. (*a*) A steel ring forming part of the ball-track of a *ball-bearing*.

(*b*) A complete *ball-bearing*.

Ball (Spherical) Resolver. A sphere with a fixed centre, two degrees of rotational freedom and two output rollers in the equatorial plane with their axes at right angles to each other. The driving-roller axis is in a plane parallel to the equatorial plane and its projection on the equatorial plane makes an angle θ with the cosine output roller. For a rotation α of the driving-roller the output rollers rotate through angles equal to $\alpha \cos \theta$ and $\alpha \sin \theta$.

Ball-track. See **Ball-bearing.**

Ball Turning. The production of spherical objects by means of a special rest moved by worm gears in a circular path or by means of special curved tools, the tool post swinging through a circular arc.

Ball Valve. *Cage valve (a), cock valve (b).*

(a) A non-return valve consisting of a globe or ball working on a cup-shaped seat, usually within a suitable cage. It is used in small water-and-air-pumps and for small check valves. (See Fig. 30.)

(b) A spherical ball with a cylindrical hole through its centre allowing fluid to flow. When turned through 90° the face of the ball stops the flow. (See Fig. 197.)

Ballistic Pendulum. A heavy block, usually of wood, suspended by string so that it swings only in one plane. When a bullet is fired into the block, its velocity can be calculated from a measurement of the angle of swing of the pendulum.

Band. *Belt.*

Band Brake. A flexible band, with one end anchored and a force applied to the other, which is wrapped partially round the periphery of a wheel or drum.

Band Clutch. See **Clutch.**

Band Conveyor (Belt Conveyor). A travelling endless belt for conveying materials, small articles, etc., from one place to another and passing over, and being driven by, horizontal drums. (Fig. 40.) Cf. **Chain Conveyor.**

Band Mill. A wide *bandsaw.*

Band Resaw. *Band mill.*

Bandsaw (Bandsawing Machine). An endless band of steel with saw-teeth upon one edge which passes over and is driven by two pulleys with horizontal axes, the pulleys keeping the band in tension. In one type the band runs vertically, and in another type the machine cuts horizontally.

Banjo Axle. Rear-axle casings in an automobile with a differential casing in the centre, thus resembling a banjo with two necks.

Banjo Bolt. A *bolt* having a *blind hole* drilled up the centre and several radial holes drilled into it from a peripheral groove. The bolt is used as an easy method of attaching a pipe flush with a machined surface. Frequently used for gas, lubricant feed and bleed pipes.

Banjo Frame. *Bow connecting-rod.*

Banking Pins. Vertical pins, limiting the motion of the *lever,* which are located in the bottom plate of a watch.

Bar. (a) A piece of material of uniform cross-section which may be either rolled or extruded.

(b) A steel bar, mounted parallel to a running rail, which, when depressed by the wheels of a train, is capable of holding points or giving information of the position of a train.

(c) A narrow detachable plate in a clock or watch.

Bar Gauge. A substitute for a *plug gauge* for checking the dimensions of large-diameter plain holes.

Bar Lathe. A small lathe, the bed of which is made in a single piece of circular, triangular or rectangular section.

Bar Movement. A watch *movement* with the upper pivots carried in bars.

Bar Saw. A large and robust mechanical hacksaw for cutting iron or steel in bar form; it has a very slow action.

Barker's Mill. A mill, driven by an excess of hydrostatic pressure, in which water is admitted to a central vertical container and passes out under pressure at the bottom through side holes in hollow radiating horizontal arms, thus driving the mill. See also **Scotch Turbine.**

Barostat. (a) A device, similar to an *aneroid barometer*, for regulating the pressure supply to, or delivery from, the fuel meter of an *aero-engine*.

(b) A device for initiating movement or controlling a mechanism when a given pressure is reached.

Barrel. (a) A cylindrical part of a machine such as the body of a pump in which the piston moves. See also **Pump Barrel.**

(b) The cylindrical portion of a locomotive boiler or a portable engine boiler.

(c) The drum around which a chain is wound in hoisting machinery.

(d) The main piece of a *capstan*.

(e) The hollow cylindrical part of a key.

(f) The cylindrical container for housing the mainspring of a clock or watch.

(g) The central part of a propeller hub which takes the centrifugal force on the blades.

Barrel Arbor. The *arbor* in a clock or watch carrying the *barrel* on which is coiled the mainspring during winding.

Barrel Cap. The detachable cover of a *barrel* in a clock or watch.

Barrel Cover. The cover which snaps into a grooved recess at one end of the *barrel* in a clock or watch.

Barrel Elevator. An *elevator* using parallel travelling chains with projecting curved arms. The chains pass over sprocket wheels at the top and the bottom and lift barrels in the curved arms from a loading platform to a runway.

23

Barrel Hook. The means by which the mainspring is attached to the *barrel* (*f*).

Barrel-type Crankcase. A petrol-engine crankcase from which the crankshaft must be removed from one end. Cf. **Split Crankcase.**

Barrel Wheel. The last, a large, wheel in the train of gearing of a crane, keyed upon the same shaft as the lifting barrel.

Bascule Bridge. A counterpoise bridge which can be rotated in a vertical plane about axes at one or both ends, the roadway rising and the counterpoise descending into a pit. Also called ' balance bridge '.

Base Circle. (*a*) Any transverse section of the *base cylinder.*

(*b*) The circle used in setting out the profiles of gear-wheel teeth which are of *involute* form. (See Figs. 84, 93, 172.)

Base Cylinder (of a Gear). The cylinder to which the generators of an involute *helicoid* are tangent.

Base Diameter. The diameter of the *base circle* of a helical spur or worm gear.

Base Helix. The curve traced by a point of contact between the generator of an involute *helicoid* and the *base cylinder.*

Base Helix Angle. The *helix angle* of a *base helix.*

Base Lead Angle. The *lead angle* of a *base helix* of an involute *helicoid* gear.

Base Pitch (Normal). The distance between similar flanks of two adjacent teeth of a gear measured along a common normal (Fig. 84). See also **Pitch.**

Base Pitch (Transverse). The distance between similar profiles of two adjacent teeth of an involute *helicoid* gear measured in a transverse plane along a common normal.

Base Plate. *Bed plate.*

Basic Angle. The angle size on which the design size of a gear, thread, etc., is based.

Basic Form. That form of a surface or profile on which the *design forms* are based.

Basic Form of Screw Thread. That form on which the design forms for both the external and internal threads are based. (See Fig. 156 and **Thread.**)

Basic Member. A mating part whose design is equal to the *basic size.*

Basic Size. The size of a dimension, or part, on which both the limits of size and the design sizes are based. See also **Tolerance.**

Basic Taper. The taper size on which the design size is based.

Basic Truncation (Major or Minor). The distance, measured perpendicular to the axis, between the appropriate cylinder, or cone and the adjacent *apex* of the *fundamental triangle*. See also **Crest Truncation.**

Basil. The bevelled edge of a drill or chisel.

Bastard Thread. A *screw thread* which is not a standard.

Batten. The swinging frame of a loom which controls the *reed*, carries the *race board* and beats up each pick of weft to the fabric already formed.

Bayonet Engine. A horizontal engine with the *bed plate* curved round to one side of the crank, the curved portion carrying the bearing for the crankshaft.

Beading. Making a rounded edge on sheet metal.

Beading Machine. A sheet-metal tool used to make flanges, beads (rounded edges) and miscellaneous odd curves and angles.

Beam. (*a*) A girder supported at its ends and loaded transversely.

(*b*) A flanged roller or steel tube carrying the warp threads in the *loom* when they are in position.

(*c*) The beam on which the sheet of threads is wound in beam-warping.

(*d*) A hollow metal cylinder carrying a warp thread in lace manufacture.

Beam-engine. An early well-known but obsolete type of *steam-engine* in which the connection between the piston of the inverted steam-cylinder and the flywheel or pump-cylinder was made through a beam whose point of oscillation was set midway between the centres of the two rods.

Bearded Needle. *Spring needle.*

Bearing Materials. Bronzes, white metal, *Babbitt's metal*, copper, nylon and similar materials, polyurethane, metals impregnated with P.T.F.E. (polytetrafluoroethylene).

Bearing Neck. The portion of a rotating shaft in contact with a bearing.

Bearing Plate. The part of a lathe bed over which the *cross-slide* slides.

Bearing Pressure. The pressure of a rotating shaft on its bearing, usually measured in kPa of projected area.

Bearing, Roller. See **Roller-bearing.**

Bearing Spring. The spring which carries the weight of a vehicle and lessens the effect of jars and shocks.

Bearing Surface. (*a*) The area of the surface upon which a shaft rotates.

(*b*) In machinery, the surfaces of bearing parts in mutual contact.

Bearings. The supports for holding a revolving shaft in its correct position. See **Ball-bearing, Cage, Journal, Plummer Block, Roller-bearings, Spherical Roller-bearing, Split Bearing, Taper Roller-bearing.**

Beat. (*a*) A measured sequence of strokes or sounds.

(*b*) A sound of regularly varying intensity due to the combination of two sounds of slightly different frequencies.

(*c*) The blow given by a tooth of an *escape wheel* as it strikes the *pallet* in a clock. When the blow is uniform on both pallets, the *escapement* is said to be ' in beat '.

Beat Pins. The pins projecting from the ends of the gravity arms of a *gravity escapement*.

Beat Screws. Screws used to adjust the relative position of the *crutch* and pendulum so that the *escapement* can be brought into *beat*.

Beater. (*a*) A revolving shaft with blades which break up and loosen matted lumps of cotton during the cotton-spinning processes of opening and beating.

(*b*) A trough containing a cylinder, both fitted with knives, for reducing paper pulp to the right consistency.

Beating. A term describing the regular thudding sound of a locomotive or a steam vessel.

Beating Engine. *Hollander.*

Beating-up. The movements of the reed to push each thread of weft against the edge of the woven fabric.

Bed. A generic term of general application, implying that upon which something is laid or placed, such as *lathe* (Fig. 102). See **Bed Plate, Lathe Bed,** etc.

Bed Plate (Base Plate). A heavy cast-iron or fabricated steel base used as a foundation for an engine or other machine.

Bedding (Bedding-in). (*a*) The adjustment and fitting of the journals of a shaft and its bearings to each other.

(*b*) A seating or a *bed*.

(*c*) The laying of a piece of machinery on its foundation.

Beetle. A machine in which a row of wooden hammers fall on a roll of cloth as it revolves and impart a soft glossy finish to it.

Beetle-head. The *monkey* of a *pile-driver*.

Bell Chuck. *Cup chuck.*

26

Bell Crank. A triangular frame which oscillates about a pivot and converts horizontal reciprocal motion into vertical motion or vice versa. A bell crank is often simply two levers at right angles keyed on the same shaft which can be mounted in any plane.

Bell-crank Lever. A lever with two arms at right angles and a common fulcrum at their junction.

Bell Hopper. The hopper which releases and spreads out by its bell shape the charge of iron ore, coke and limestone into a blast-furnace.

Bellows. (*a*) A portable or fixed contrivance for producing a jet of air from a flexible-ended or -sided box which is alternately expanded and contracted, drawing in air through a non-return valve and expelling it through a nozzle.

(*b*) The flexible folding light-tight part of some cameras, uniting the back and front portions.

(*c*) The convoluted portion of a pipe to allow for thermal elongation or misalignment (Figs. 42 and 87). See also **Sylphon Bellows.**

Belt (Belting; Driving Band). An endless band of leather or other flexible material for transmitting power from one shaft to another by running over flat, convex or grooved rim pulleys. Belts may be flat, vee-shaped or ribbed to fit on to appropriately shaped pulleys. Belt velocities may be as high as 800 m/s. See also **Anti-static Belting, Link Belting, Open Belt,** etc.

Belt Compressor. An air-compressor driven by a belt and pulleys from an independent engine or from shafting.

Belt Conveyor. *Band conveyor.*

Belt Coupling. The union of the ends of a belt. Cf. **Belt Fastener.**

Belt Drive. A method of power transmission from one shaft to another by means of an endless belt passing around a pulley on each shaft.

Belt Fastener. A connecting piece that joins together the ends of a belt. Cf. **Belt Coupling.**

Belt Fork (Strap Fork, Belt Striker). *Belt shifter.*

Belt Perch. A bar or rest placed alongside a belt pulley on which to rest the belt when unshipped for repairs.

Belt Polisher. A polishing machine consisting of a belt covered with abrasive or polishing material and passing around pulleys. See also **Linear Sander.**

Belt Punch. A cutting tool shaped like pliers with an annular edge used for cutting holes in leather or similar material.

Belt Sander

Belt Sander. A machine in which a belt covered with abrasive material is moved rapidly by rotating pulleys over woodwork to finish its surface. See also **Linear Sander**.

Belt Shifter (Belt Fork). A forked device used for shifting a belt from one pulley to an adjacent pulley or from a fast to a loose pulley and vice versa. Also called ' belt striker '.

Belt Shipping. The placing of a belt on its pulley or moving it from one pulley to another.

Belt Slip. The slipping of a driving belt when there is insufficient frictional grip between the pulley and the belt.

Belt Speeder (U.S.). A pair of cone pulleys carrying a belt which is moved to alter the rate of motion; e.g. some spinning machines vary the rate of rotation of the spool as the cap increases in size.

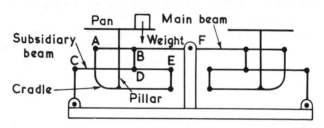

FIG. 10. BERANGER BALANCE.

Belt Tightener. (a) A contrivance for pulling the ends of belts together for coupling up.

(b) A device to maintain a uniform tension upon driving belts or to cause them to conform more nearly to the circumference of the pulleys (U.S.). See also **Binder Pulley**.

Belt Tripper. A contrivance for tilting sideways, at a convenient point, either a *belt* or *apron conveyor*.

Bench Drilling Machine. A small drilling machine which can be bolted to a work bench and actuated by hand or by an electric motor via a suitable speed reduction mechanism.

Bench Lathe. A small lathe mounted on a work bench.

Bench Work. Work carried on at a bench, or vice, with hand tools or small machines in contrast to work carried on with a lathe or other machine with its own stand or pedestal.

Bending. The curvature of a beam about its axis under load. See also **Flexural Axis** and **Angle of Flexure**.

Bending Machine. *Straightening machine.*

28

Bending Rolls. Heavy rollers of cast-iron or steel set in strong standards to straighten crooked plates, to bend them into arcs, or to bend them into complete cylinders.

Bent-tail Carrier. A *lathe carrier* with a bent shank projecting into, and engaged by, a slot in the driving plate or chuck.

Beranger Balance. A balance in which the motion of each scale pan is such that the pan remains horizontal for any vertical displacement. The position of the load on the pan does not affect the equilibrium of the balance and the subsidiary beams and links ensure that there are no lateral thrusts on the knife edges. The total equivalent downward force at A is equal to the weight on the pan if $CE/CD = FA/FB$ as in Fig. 10. Cf. **Roberval Balance.**

Between Centres. In lathe work, this term signifies chucking the work between the centres of the *headstock* and the *tailstock*. The workpiece is then rotated by means of a *lathe carrier*. Other methods involve the use of a *chuck, collet* or *face plate*.

Bevel. An angle, which is not a right angle, between two surfaces.

Bevel Gear (Bevel Gearing). An arrangement of *bevel wheels* for the transmission of motion from one shaft to another on intersecting axes. See also **Straight Bevel Crown Gear** and **Spiral Bevel Crown Gear.**

Fig. 11 shows the axial and transverse planes of a bevel gear. Fig. 12 shows the reference planes of a bevel gear. Straight bevel gears in mesh are illustrated in Fig. 13.

Bevel Gear Shaper (Planer). A machine tool for shaping *bevel wheels*.

Bevel Mortise Wheel. One of a pair of bevel wheels which has been fitted with inserted wooden teeth to secure a silent drive.

Bevel-turning Slide. See **Profile-turning Slide.**

Bevel Wheels. Toothed wheels shaped like the frustrum of a cone, which are used in pairs to transmit motion between two shafts whose axes intersect at an angle to each other. See also **Bevel Gear.**

Beveloid* Gearing. An *involute gear* with tapered tooth thickness, root and outside diameter. [* Registered trade mark.]

Bezel. (*a*) The sloped cutting-edge of a chisel or other cutting tool.

(*b*) The groove and projecting flange or lip by which the crystal of a watch, or the stone of a jewel, is held in its setting.

(*c*) The grooved ring in which the glass cover of a watch or of an instrument is held in its setting.

Bevel Gears

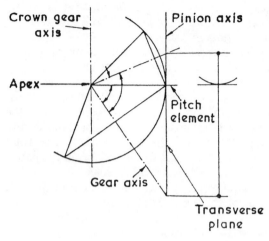

FIG. 11. BEVEL GEARS—AXIAL AND TRANSVERSE PLANES.

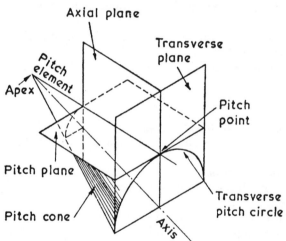

FIG. 12 (*above*). BEVEL GEARS—REFERENCE PLANES.

FIG. 13 (*left*). STRAIGHT BEVEL GEARS IN MESH.

Bib-valve. A draw-off tap closed by screwing down a washered disc on to a seating.

Bicycle. (*a*) A vehicle having only two wheels, one behind the other.

(*b*) An aeroplane's undercarriage unit with two wheels or wheel units, one behind the other.

Bifilar Suspension. A suspension by two parallel vertical wires (or threads) to provide a controlling torque on a body, as in some instruments.

' Big Ben '. See **Gravity Escapement, Zinc and Steel Pendulum.**

Big-end Bolts. *Connecting-rod bolts.*

Bight. A hanging loop of a chain or rope which is held by its ends, such as the operating chain of manually-worked pulleys in lifting tackle.

Bilateral. (*a*) Having, or pertaining to, two sides.

(*b*) Measured in two directions, especially applied to *tolerances.*

Bilateral Limit System. See **Limit System.**

Bilateral Tolerance. See **Tolerance.**

Billet Mill. A rolling mill for reducing steel ingots to billets. Cf. **Blooming Mill** and **Bloom.**

Billeting Rolls. *Roughing rolls.*

Binary Vapour-engine. A heat-engine using two separate working fluids for the high- and low-temperature portions of the cycle respectively.

Binder (Harvester). A harvesting machine which, in addition to cutting corn, gathers it, forms it into sheaves and ties the sheaves. It is also called a ' harvester '.

Binder Pulley. An adjustable pulley which tightens a belt or cord on its driving and driven pulleys to its correct tension. Cf. **Jockey Pulley,** Fig. 97.

Binding Head Screw. A screw head which is *undercut* round the stem so that when it is screwed down tight the peripheral material beds into the *mating* part.

Binding Screw. A *set screw* used for clamping two parts together.

Birmingham Wire Gauge (B.W.G.). See **Gauges Commonly Used.**

Bit. (*a*) A boring tool which is rotated in a vice or machine.

(*b*) The cutting edge of the borer used in rock drilling. In America, the entire length of the borer, including shank, steel and bit.

(*c*) The cutting iron of a wood-worker's plane.

Bit Stop. An attachment to a *bit* which limits drilling or boring to a given depth. Also called ' bit gauge '.

Bite. (*a*) The hold which one part of a machine has on another.

(*b*) The grip of the jaws of lathe-chucks, vices, etc.

(*c*) In general, a gripping action.

Black Diamond. *Carbonado.*

Black Work. Work that has not been machined or polished.

Blacking Mill. A large revolving closed cylinder containing heavy rollers rotating freely on its internal diameter or spherical balls for grinding graphite and carbonaceous materials for the preparation of blacking for painting the inside of casting moulds.

Blade (*a*) The cutting part of some edge tools.

(*b*) A vane of a turbine or fan.

(*c*) A radial arm of a propeller.

(*d*) The movable part of a knife-switch.

Blade Angle. The acute angle between the chord of a section of a propeller, or of a rotor blade, and a plane perpendicular to the axis of rotation.

Blank. A specially prepared piece of metal ready for machining, grinding or pressing, to a particular shape. (See Figs. 63, 71.)

Blanking. A shearing process in which sheet-metal is cut or punched to form a flat blank, which is later bent or formed in a desired shape (cf. **Piercing**). A combined blanking and forming tool is illustrated in Fig. 185.

Blast. Air under pressure impelled by mechanical means.

Blast-engine. An engine for creating a *blast* of air, usually to aid combustion in a furnace; now largely driven by the waste gases (after cleaning) from the furnace.

Blast-pipe. The exhaust pipe of a locomotive terminating in a blast nozzle. The steam through the nozzle provides a draught to entrain and exhaust the flue gases.

Bleed. (*a*) Air under pressure taken from an axial compressor to run an auxiliary service.

(*b*) Liquid taken from a hydraulic system to release trapped air. See also **Bleeding**.

Bleed Choke. See **Choke**.

Bleeding. The tapping of air from a compressor (*a*) to prevent surging in a turbojet or turboprop engine; (*b*) to feed air to auxiliary equipment; (*c*) to provide a source of power for blowing away the stagnant boundary layer in the bends in intake ducts; (*d*) withdrawing some of the steam from the high-pressure stages of a turbine to heat the feed-water and thus improve the thermal efficiency. See also **Bleed**.

Blind Hole. A hole drilled part of the way into a piece of material to any required depth. See **Thread Insert** and Fig. 181.

Block. The housing holding the *sheaves* or pulleys over which a rope or chain passes, as in a lifting tackle. The term commonly includes the sheaves. A block has a hook eye or strap for attaching it to an object and it can be used for changing the direction of a running rope.

Block Brake. A vehicle brake in which a block of metal or hardened material is forced against the rim of a revolving wheel by hand power or a mechanism.

Block Carriage. The travelling frame, carrying the chain *sheaves* upon the horizontal *jib* of a crane, which is traversed along the jib by *racking gear*.

Block Chain. A power-transmitting chain in which alternate links are steel blocks.

Block Clutch. See **Clutch**.

Block Gauge. A distance gauge made of hardened steel with its opposite faces parallel and accurately ground flat, the faces being separated by a definite distance—the ' gauge distance '. A block gauge is used for checking the accuracy of other gauges. See also **Slip Gauge**.

Block System. A method of railway working with the line sub-divided into short sections or blocks and arranged so that only one train is in a given block at one time. This system of signalling is often automatically controlled by the movement of the train.

Block and Tackle. A *block* with the rope woven through it for hoisting or obtaining a purchase.

Blocking-down. Fitting sheet-metal into an intricate mould or die by beating on it with a hammer with a block of lead interposed.

Blocking-up. Raising and supporting machinery or other con-structions with the aid of cranes, jacks, levers and wooden blocking.

Blocking Girders. Girders attached to the underside of the truck frames of a travelling crane; fitted back and front, they are wider than the frames. Jacks fitted at the girder ends are screwed down to the ground to prevent overturning when lifting a load crossways.

Blondin. *Cable-way*.

Bloom. A product in the rolling of steel with a cross-section greater than 36 in.^{2} Smaller sizes are known as billets.

Blooming down. The rolling down of steel ingots into *blooms*.

33

Blooming Mill. A rolling mill for reducing steel ingots to *blooms*; also known as cogging mills. Cf. **Billet Mill.**

Blow Back. The return of some of the induced mixture through the carburetter of a piston engine when running at slow speeds due to the late closing of the inlet valve during compression.

Blow-by. The gas which leaks past the piston of a piston engine during the period of maximum pressure.

Blower. (*a*) A rotary compressor for supplying a large volume of air at low pressure.

(*b*) A ventilating fan or venturi tube for supplying air.

(*c*) A ring-shaped perforated pipe encircling the top of the *blast-pipe* of a locomotive.

Blower and Spreader. A machine combining the action of beaters and blower for spreading cotton into a *lap*.

Blowing Cylinders. Double-acting cylinders employed for pumping the air under pressure into the blast-main of a blast-furnace. See also **Blowing Engine.**

Blowing Engine. A combined steam-or-gas engine coupled to a large reciprocating air-blower for supplying air to a blast-furnace.

Blowing Through. The sending of a jet of steam through the cylinders and valves to warm a steam-engine before starting.

Blueing. (*a*) Thermal treatment of watch springs to obtain the desired elastic properties as indicated by the colour.

(*b*) Forming a protective coating of a blue oxide film on polished steel by heating in contact with saltpetre or wood ash or incidental to annealing.

(*c*) Applying a blue dye to metal objects before scratching the dimensions on to the work.

Blunger (Blunging Machine). A pottery machine in the form of a vertical cylinder containing a rotating shaft armed with fixed horizontal knives and used for amalgamating clay with water in making *slip*, which is clay reduced to the consistency of cream.

Blunt Start. The condition resulting from the removal of the partial thread at the end of a screwed member to facilitate the entry of the threads without damage, when repeatedly assembled.

Bob. (*a*) The suspended weight at the end of a plumb-line.

(*b*) The weight at the end of a pendulum.

(*c*) A small buff wheel, perforated and nearly spherical, mounted on a spindle and used for polishing the inside of spoons.

(*d*) A working beam of a steam-engine.

Bobbin. (*a*) A spool on which yarn is wound; the spool has flanges for holding warp yarn and no flanges for holding the weft.

(b) A small spool adapted to receive thread and installed within the shuttle of a sewing machine so that the thread produces the lower half of each stitch. See also **Cop, Pirn.**

Bob-weight. A counter-balancing weight on a machine.

Body (of a **Valve**). The main part of a valve where the flow of fluid is controlled.

Body Boss (of a **Valve**). See **Boss.**

Body Plug. A plug for sealing a tapped hole in a body boss or drain boss. See **Boss.**

Body Seat Facing. A deposit on the body seat ring (or body) of different material on which the body seat is machined.

Body Seat Ring. The part of a renewable seated valve on which the body seat is machined. It is made separate from the body and secured in it.

Bogie (Bogie Truck). A truck of short wheelbase resting on two or more pairs of wheels, which forms a pivoted support at one or both ends of a long vehicle such as a locomotive or a railway coach. Its use is to enable a long vehicle to run round sharp curves. (See Fig. 107.)

Bogie Engine. A locomotive with a *bogie*; in some locomotives the leading wheels are on a bogie and in the double-bogie type both the leading and trailing wheels are on bogies.

Bogie Frame. The frame upon which the axles of a *bogie* are mounted.

Boiler. A steam generator of one of two main types, (a) and (b).

(a) A shell boiler, in which the water is contained within more or less cylindrical vessels traversed by tubes, through which the heated gases of combustion pass to impart their heat to the water, such as in a locomotive and a bathroom geyser.

(b) A water-tube boiler in which the water is contained in the tubes and the heated gases circulate round the tubes on courses directed by baffles.

(c) A small water heater, heated directly by electricity or gas.

Boiler Capacity. The weight of steam in pounds per hour which a boiler can evaporate when steaming at maximum output.

Boiler Pressure. The pressure of the steam in a boiler varying according to the type of boiler from a little over atmospheric pressure to more than 10 MPa for use with high-pressure turbines.

Boiler (Stay) Tap.   A threading tap to tap holes for the reception of boiler stays.   It may be in one piece or consist of two shell taps secured by keys to a long shaft.

Bolster.   (*a*) A steel block supporting the pad or die in a pressing or punching machine.

(*b*) The rocking steel frame by which the *bogie* of a locomotive supports the weight of the engine.

(*c*) The bearings that fit within the housings in forge and mill rolls and sustain these rolls.

Bolt.   Any piece of material used to connect parts together which has a thread on one end and a head on the other.   The head can be hexagonal, square, slotted, etc.

Bolt Cutter.   A machine for cutting the heads of *bolts*.

Bolt-making Machine (Bolt Machine).   A machine for forging bolts by forming a head on a round bar.

Bond.   The matrix of non-abrasive material for binding together the abrasive grains, such as rubber, resin, shellac, ceramics, etc., in grinding wheels.   ' Vitrified bond ' refers to glasses resulting from the fusion of ceramic materials during firing in kilns at high temperatures.

Bonnet (Cover). (*a*) That part of a stop or gate valve which is attached to the *body* and carries the operating mechanism; in general, a movable protecting cover and hence ' bonneted safety valve '.   See also **Bridge** (*a*).

(*b*) A cover over an engine as in a motor-vehicle.

Book-folding Machine (or Folding Machine).   A machine for folding sheets, and gathering, sewing and binding them, which is adapted to fold sheets of various sizes from folio downwards.   The machine can also be adapted to fold two separate sheets together, pasting the separate pages at the back, or for cutting sheets into a number of pieces and folding them separately.

Boom.   (*a*) The *jib* of a *derrick crane*.

(*b*) The upper or lower flange of a girder.

(*c*) The main spar of a lifting tackle.

(*d*) The spar holding the lower part of a fore-and-aft sail or a spar for extending its foot.

(*e*) In general, any long beam or line of timbers.

Boost.   The increase over atmospheric pressure of the induction pressure in a supercharged *piston engine* measured in Pa.

Boost Gauge.   An instrument for measuring the *manifold pressure* of a supercharged piston engine either in relation to ambient pressure or in absolute terms.

Booster Pump. (*a*) A pump for increasing the pressure of a liquid in a closed pipe system.

(*b*) A pump for maintaining a positive pressure between the fuel tank and an aero-engine or liquid propellant rocket engine.

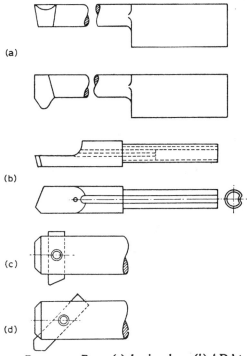

(a)

(b)

(c)

(d)

FIG. 14. BORING BARS AND BITS: (*a*) boring bar, (*b*) ' D '-type boring bit, (*c*) and (*d*) single-point boring bar at 90° and 45° for blind holes.

Bore. (*a*) The internal diameter of a pipe or a cylinder.

(*b*) The interior of a small-arm or piece of ordinance, including both the chamber and the rifled portion.   See **Rifling.**

(*c*) The internal wall of an engine or pump cylinder.   See also **Cylinder Bore.**

Boring. (*a*) The operation of machining a cylindrical hole in or through any piece of work, performed in a lathe or boring mill. Boring usually presupposes an existing hole which has to be made true and enlarged to the proper size.   Cf. **Drilling Machine.**

Boring

(*b*) The process of drilling holes into the ground or rock for the insertion of blasting charges or to obtain information about the ground or rock. See also **Boring Machine.**

Boring Bar. A stiff cylindrical bar supported at the machine table, carrying the boring tool and driven by the spindle of a *boring machine* or held in the tool-post of a lathe. Various types are shown in Fig. 14. (See also Fig. 23.)

Boring Collar. See **Cone Plate.**

Boring Flange. *Drill plate.*

Boring Head (Cutter Head). The ring carrying the cutters of a *boring bar.*

Boring Machine. A machine tool used in boring cylinders, the work being clamped on a bed and the tool carried by a driving spindle and rotating in fixed supports. The 'horizontal' boring machine is designed either (*a*) with a spindle which only rotates, or (*b*) with a spindle which both rotates and has a horizontal movement. The 'vertical' boring machine has a vertical spindle, is similar to a radial driller and for boring cylinder-blocks has the requisite number of spindles to finish the operation in one pass of the tools.

Boring Mill. A vertical *boring machine* in which the work is carried by the rotating table and the *boring bar* is fixed.

Boring Stem. A heavy bar on which the drill bit is mounted in boring artesian wells and which adds force to the blow by its weight. See also **Boring Tool.**

Boring Tool. A tool used for internal turning and held on a *boring bar.* It has usually only a single cutting edge whereas a *drilling tool* has two edges placed on opposite sides of the axis of the tool.

Boss. (*a*) The centre or hub of a wheel.

(*b*) The cylindrical projecting part of a machine casting or forging to support a shaft, pin or studs.

(*c*) The large part of a shaft on which a wheel is keyed or at the end where it is coupled to another shaft. See also **Key Boss.**

(*d*) Any protuberant part.

Body Boss. A boss formed on the exterior of the body of a valve to provide sufficient metal to permit a tapped connection.

Drain Boss. A *body boss* to provide for a tapped connection for drainage purposes. See also **Propeller.**

Bossed Up. A forging is said to be bossed up when (*a*) a circular disc of sensible thickness is formed upon the face of a forging; (*b*) the turned-up edges of such bosses are formed in a lathe.

Bossing Machine. A steam-hammer, when used for welding the bosses on the wheels of rolling stock at the appropriate temperature.

Bottle Jack. A light-weight *screw jack* with the lower part shaped like a bottle, usually provided with a handle at the side for carrying.

Bottom. (*a*) *Cogs* are said to bottom when their tops touch the periphery of the co-acting wheel in machinery.

(*b*) A piston which strikes or touches the end of its cylinder is said to bottom.

(*c*) A *tap* is said to bottom when it reaches the end of a *blind hole*. A tap designed to produce threads far down into a blind hole is known as a *bottoming tap* or *plug tap*.

Bottom Clearance. The shortest distance between the tooth crest of one gear-wheel and the bottom of the tooth space of another gear-wheel in mesh.

Bottom Dead-centre. *Outer dead-centre.*

Bottom End. The *crank-pin* end of a marine engine connecting-rod.

Bottom Plate (Dial Plate). The plate in a watch to which the *pillars* are fixed.

Bottom Rake. The *angle of relief* in cutting tools.

Bottom Tool. The lower half of a *fullering tool*.

Bottoming. Impinging against the bottom so as to impede free mechanical movement as when a cog strikes the bottom of a space between two cogs. See **Bottom.**

Bottoming Tap. *Plug tap.*

Bouncing Pin Detonation Meter. An apparatus for fuel testing by determining quantitatively the degree of detonation in a piston engine.

Boundary Lubrication. Lubrication by a very thin closely adherent film of oil between two surfaces, the oil, such as a long-chain fatty acid, being one, or at the most, only a few molecules thick.

Bourdon Gauge (Pressure Gauge). A metal tube of a flattened oval section, which is bent to a curve, with the free end closed and the fixed end open to the pressure. The pressure tends to straighten the tube and the movement is recorded on a dial. Hence sometimes called ' dial gauge '. (See Figs. 15, 188.)

Bow. (*a*) The ring of a pocket-watch case to which the watch chain is attached.

Bow

(*b*) A flexible strip of whalebone or cane with the ends drawn together to give tension to a line which is given a single or pair of turns round a pulley to form a sensitive drive for a drill or mandrel. Hence ' bow drill '.

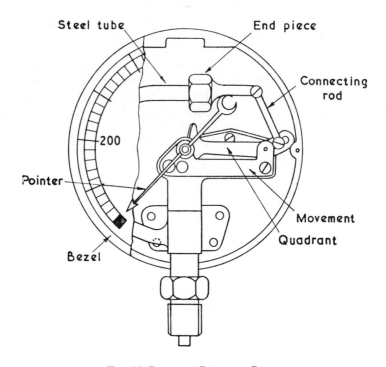

FIG. 15. BOURDON PRESSURE GAUGE.

Bow Connecting-rod (Banjo Frame, Kite Connecting-rod). A triangular connecting-rod used in steam-pumps with the crank driving the flywheel enclosed by the bow.

Bow Drill. A small drill rotated by hand via the frictional grip of a string held in tension by a bow.

Bowden Cable. A movable wire inside a concentric cable, the latter fixed at both ends; the wire is pulled or released to alter gearing such as on a bicycle.

Bow-saw. A thin-bladed saw held in tension by a special frame or sometimes by a bow.

Box. (*a*) A portion of a mechanism resembling a box such as a valve box.

(*b*) A bearing for a shaft.

Box Angle Plate. A metal box or cube with slots machined in the face for the accurate mounting of work.

Box Coupling (Muff Coupling). A cylindrical coupling, split longitudinally, for uniting two lengths of coupling or connecting two shafts, the halves being bolted together and keyed to the shafts. Also known as ' sleeve coupling '. See also **Butt Coupling.**

Box End. A *connecting-rod* with no loose *strap end* but with the *brasses* thrust into a slot from one side, slid along to their seatings and tightened with a *cotter.*

Box Jig. A *jig* made in the form of a box into which the job to be drilled is inserted and located by suitable pins and faces. (See Fig. 96.)

Box Link. A *slot link* with the internal faces recessed so that the edges act as additional bearing surfaces. See also **Link.**

Box Loom. A *loom* provided with several shuttle boxes.

Box of Tricks (Escape Motion). The building motion of a *fly frame* which regulates the speed of the bobbin, reverses the traverse of the lifting rail and reduces this as each layer of *roving* is laid on the bobbin.

Box (Boxed) Standard. The main framework of a machine or engine, hollowed internally to provide the maximum of strength with the minimum of material.

Box Tools. Combinations of separate tools secured in a box for attachment to the faces of a lathe turret.

Bracket Chain Wheel. *Sprocket.*

Brake. A device or mechanism for applying frictional resistance to the motion of a body and thereby absorbing mechanical energy by transferring it into heat (*a*) to retard a vehicle, or (*b*) to measure the power developed by an engine or motor. See **Air Brake, Band Brake, Block Brake, Brake Shoes, Disc Brake, Froude Brake, Hydraulic Brake, Rope Brake.**

Brake Band. The strap or band encircling a *brake drum.*

Brake Block. Blocks of material for applying a frictional force in a brake. See also **Brake Lining.**

Brake Drum. (*a*) A drum or pulley attached to a wheel (or shaft), to which is applied an external band or internal brake shoes. See **Band Brake, Internal Expanding Brake,** Fig. 16.

Brake Drum

(b) A large drum used for winding the rope which lifts and lowers the cage in a colliery or pulls the trucks in quarries or operates the lifts in buildings.   Also called a ' winding drum '.

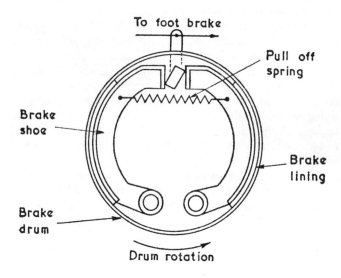

FIG. 16. INTERNAL EXPANDING BRAKE.

Brake Horse-power (b.h.p.).   The *horse-power* developed by an engine as measured by a brake or dynamometer applied to the driving shaft.

Brake Lining.   Asbestos-based fabric riveted or bonded to the shoes of *internal expanding brakes* to increase the friction between them and the drum and, at the same time, to provide a renewable surface.   Cf. **Brake Shoes.**

Brake Mean Effective Pressure (b.m.e.p.).   That part of the *indicated mean effective pressure* (i.m.e.p.) which would give an output equal to the *brake horse-power* of an engine or engine cylinder; the product of the i.m.e.p. and the *mechanical efficiency*.

Brake Power.   The frictional resistance developed by a brake.

Brake Press.   A press in which the energy of a large flywheel is suddenly applied for forming and blanking.

Brake Shoes.   The internal expanding members in a brake drum on which the renewable friction linings are mounted.   The two

42

shoes in the *brake drum* of an automobile may be arranged as shown in Fig. 16. In this arrangement one shoe is a leading shoe and the other is a trailing shoe.

Brake Strap. The encircling band to which the brake blocks are screwed in the friction brake of a crane or dynamometer.

Brake Thermal Efficiency. The ratio of the heat equivalent of the dynamometer brake output to the heat supplied to the engine in the fuel or steam. See also **Thermal Efficiency.**

Brake Wheel. (*a*) A wheel which receives the friction pressure of a *brake strap* or *brake blocks*.

(*b*) A wheel on the platform of a railway coach (or car) by which the brakes are operated.

Braking Propeller. A propeller, the pitch of which can be altered so that the propeller gives a reverse (negative) thrust and acts as an airbrake or waterbrake.

Bramah's Press. *Hydrostatic press.*

Branch Chuck. A chuck having four branches turned up at the ends and each furnished with a screw to grip the work.

Brass(es). The bearing(s) for the journal(s) of a shafting, a half bearing being termed a brass, the two bearings a pair of brasses and the bearings with their seatings a *plummer block*.

Brass Finisher's Lathe. A lathe specially designed with attachments to machine brasswork in quantities. Special hand-operated rests include (*a*) a six-holed circular turret to carry tools for drilling, boring, tapping or turning, mounted on compound slides with longitudinal and transverse movements; the whole saddle can be traversed along the bed and clamped quickly in position; (*b*) a second saddle, easily removable carrying tool-post with similar compound slides and transverse hand adjustment; (*c*) a third rest for cutting screws fitted on a bar at the back of the lathe, which can be thrown over out of the way, and supported at the front by a roller which runs on a guide bar. The lead screw and nut provided with the machine cover most of the screw-cutting requirements and there is provision for replacement for any other pitch required. See also **Brass Tool.**

Brass Tool. A tool used for machining purposes on brass or bronze, usually with no top rake, the front rake and side clearance being about $6°$ with a side rake of 0 to $3°$; drills have straight *flutes*. See also **Brass Finisher's Lathe.**

Brass-winding. Filling brass bobbins with lace thread collectively from a jack of wood bobbins.

Brazing

Brazing. The process of joining two pieces of metal by fusing a layer of spelter or of a brass alloy between the adjoining surfaces.

Brazing Wire. A soft brass wire of small gauge used for binding round joints which are to be brazed. Sprinkled with borax the wire melts on heating and runs in. See also **Welding.**

Break Lathe. *Gap lathe.*

Break Rolls. The first series of rollers in a flour mill which break up the grain.

Breakdown Crane (Accident Crane). A *balance crane* of the portable *jib crane* type mounted on a motor lorry or railway truck. See **Crane.**

Breaking Joint. A stepping of consecutive joints so that they are not in line, such as the joints of piston rings, plates in shipbuilding, etc. Cf. **Labyrinth Seal.**

Breaking Pieces. (*a*) Easily replaceable members of a machine, which, being made weaker than the remainder, break first under overload and so protect the machine from extensive damage.

(*b*) Short lengths of shafting used for coupling an engine with the bottom rolls of a rolling mill, or the rolls from one another, which break first under overload. See **Wabblers.**

(*c*) The weak link connection between an aero-engine and its propeller in case of a crash.

Breaking-down Rolls. *Roughing rolls.*

Breast Wheel. A water wheel in which the water enters the buckets at about the height of the wheel centre, above being called ' high breast ' and below called ' low breast '.

Breech Block. The steel block which closes the rear of the bore of a breech-loading fire-arm against the force of the charge.

Breech-lock Thread. See **Buttress Screw Thread.**

Breech Mechanism. The mechanism by which the breech of a gun or a fire-arm is closed before firing. See also **Breech Block, Buttress Screw Thread.**

Bréguet Spring. A balance spring of a watch having the outer coil raised above the plane of the spiral and the end of the spring bent to a special form before entering the *stud.* Hence ' Bréguet-sprung '.

Bridge. (*a*) The exterior part of an outside screw valve, connected by pillars to the *bonnet,* in which the actuating thread of the stem engages either directly or through a bush or through a sleeve. Cf. **Yoke.**

(*b*) An arched guide casting attached to the cover of a lift or force pump. The free end of the piston or plunger rod travels through the central boss of the casting.

(*c*) A raised platform or support in a watch or clock, usually with two feet, forming a bearing for one or more pivots.

Bridge Gauge. A measuring device used to detect the relative movement of two machine parts due to wear.

Bridge Piece. The loose piece of the bed in a *gap lathe* which fits and bridges the gap.

Bridge Tree. (*a*) The cross-bar of a turbine frame above the casing which affords a bearing and central support for the spindle in millwrighting.

(*b*) A lever for sustaining the footsteps of a millstone spindle, which, when raised or lowered, alters the distance between the faces of the two stones.

Bridle. (*a*) The flange of a slide valve of a steam-engine to hold the rod in position.

(*b*) The loop forged on a slide-valve rod to embrace the back of the valve.

(*c*) A loop or clip for holding test-pieces in a testing machine.

Bridle Rod. *Radius rod.*

Brier-tooth Saw. *Gullet saw.*

Brinell Hardness Test. The measurement of the hardness of a material by the area of the indentation, after equilibrium has been reached (about 20 seconds), produced by a hard steel ball under specified conditions of loading (see **Hardness Tests**). The ball diameter is usually 10 mm, the load 500 kg for soft metals and 3000 kg for hard metals; for geometric similarity the load divided by the square of the diameter must be kept constant. The hardness number is the ratio of the load to the curved area of indentation.

British Association (B.A.) Screw Thread. A symmetrical vee-thread of $47\frac{1}{2}°$ included angle, having its crests and roots rounded with equal radii such that the basic depth of thread is 0·6000 of the pitch. It is designated by numbers from 0 to 25, ranging from 6 mm to 0·25 mm in diameter and from 1 mm to 0·072 mm pitch.

British Standard Brass Thread. A screw thread of Whitworth profile used for thin-walled tubing with 26 threads per inch, irrespective of diameter. See **Whitworth Screw Thread.**

British Standard Channel. A rolled steel channel conforming to the standard dimensions laid down by the British Standards Institution.

British Standard Cycle Screw Thread

British Standard Cycle Screw Thread. A vee-form screw thread with a 60° flank angle and a top and bottom radius equal to one-sixth of the pitch. The actual depth of the thread is 0·5327 pitch.

British Standard Fine (B.S.F.) Thread. A screw thread of Whitworth profile, but of fine pitch for a given diameter. It is largely used in automobiles. See **Whitworth Screw Thread.**

British Standard Pipe (B.S.P.) Thread (British Standard Gas Thread). A *Whitworth screw thread* but designated by the bore and not the full diameter, of the pipe on which it is cut.

British Standard Whitworth Thread. See **Whitworth Screw Thread.**

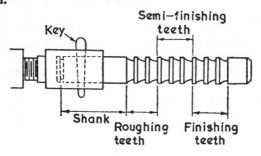

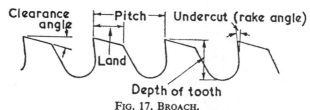

FIG. 17. BROACH.

British Standard Wire Gauge (Imperial) (S.W.G.). An arbitrary series of numbers for expressing the diameter of wires ranging from 0 000 000 (12·7 mm) to 50 (0·254 mm). See also **Browne and Sharpe Wire Gauge** and **Gauges Commonly Used.**

Brittle Lacquer Technique. A qualitative method in which the part to be tested is coated by a brittle lacquer which cracks when the part is subjected to stress.

Brittleness. The fracture of a material under low stress and without appreciable deformation.

Broach. (*a*) A tapered steel shaft with numerous transverse cutting edges increasing in height along its length, used as a tool for smoothing and enlarging holes in metal, or as a smooth tool without

46

cutting edges for burnishing pivot holes in watches. Cf. **Reamer.**

(*b*) A straight tool with file teeth made of steel to be pressed through irregular-shaped holes in metal, that cannot be dressed by revolving tools, and for splined shaft fittings.

(*c*) The pin in a lock which enters the *barrel* of a key.

A circular pull broach is shown in Fig. 17.

Broaching. The enlarging, smoothing and truing of holes with a *broach* or *reamer.* Broaching is done on manually-operated presses, on pull-screw machines, on drilling machines, on lathes, and on hydraulically-actuated broaching machines or presses. Push broaching is performed on vertical machines with short stiff broaches.

Broaching Machine. A machine used for finishing square and polygonal holes.

Broad Gauge. (*a*) A railway gauge of 7 ft as laid down by Brunel.

(*b*) A railway gauge greater than the standard width of 4 ft $8\frac{1}{2}$ in. adopted by Stephenson. (7 ft = 2·13 m approx.; 4 ft $8\frac{1}{2}$ in = 1·44 m approx.).

Brocot Suspension. In clocks, a suspension in which the length of the pendulum can be adjusted from the front of the dial.

Brown and Sharpe Wire Gauge (B.W.G.). The American standard wire gauge. The range of diameters is from 48 (0·0315 mm approx.) to 0000 (11·684 mm), the numbers descending from 48 to 1 and thence 0 to 0000. See also **British Standard Wire Gauge.**

Bucket. (*a*) The piston of a reciprocating pump.

(*b*) A cup-shaped vane divided midway by a ridge and attached to the periphery of a *Pelton wheel.*

(*c*) The cup-shaped receptacle on the impulse wheel of a turbine.

(*d*) The receptacles for the water in *overshot* and *breast water wheels.*

(*e*) The outlet for water in turbines.

(*f*) A dredging scoop.

(*g*) A receptacle on an elevator for lifting loose material.

(*h*) A water-cooled steel jet deflector mounted under a vertical *rocket engine* or *rocket motor.*

Bucket Air Pump. A marine engine air pump with piston, foot and head valves.

Bucket and Plunger Pump. A double-action pump with the bucket and plunger combined on a single rod, the plunger being uppermost. By its combined action half the contents of the barrel are discharged on the up and half during the down stroke.

Bucket Conveyor. A pair of endless chains running over toothed wheels and carrying a series of buckets which are automatically tipped at the delivery end.

Bucket Dredger (Bucket-ladder Dredger). A small-draught vessel with an endless chain of buckets reaching down into the material to be dredged and lifting it for discharge into the vessel or an attendant barge.

Bucket Ladder Excavator. A mechanical excavator with an endless chain of buckets adapted for excavating on land and discharging into a vehicle for removal.

Bucket Pump. A pump with a bucket or piston having valves through it for the passage of the fluid lifted.

Bucket Valve. A flap, non-return, valve fitted in the bucket or piston of some types of reciprocating lift pumps.

Buff. A revolving disc made of layers of cloth charged with abrasive powder for polishing, especially metals.

Buffer Box. The casing which encloses the *buffer spring* and *buffer rod* on rail-mounted vehicles.

Buffer Disc. The spheroidal disc against which the buffers of rail-mounted vehicles make contact.

Buffer Rod. The rod which carries the *buffer disc*.

Buffer Spring. The spring enclosed in the *buffer box* which deadens the impact of collision. It is a type of *helical spring*.

Buffers. (*a*) Spring-loaded contrivances at the ends of railway vehicles to minimize the shock of collision.

(*b*) Any resilient pad or mechanism used for a similar purpose.

Buffing. The process of polishing as with a *buff*. Hence ' buffing lathe '.

Building Motion. A mechanism in *fly frames* and spinning machines which guides the roving or yarn and builds it into a package.

Building Mover (U.S.). A heavy truck on rollers or wide-track wheels. The building rests on a cross-bolster which is supported by two trucks with at least three rollers each.

Bulger Ram. A round-ended ram for forcing metal plates into apertures, in experiments on bulging stress.

Bull Rope. An endless rope which drives the *bull wheel* of a cable-drilling rig, the rope being slipped off the grooved pulley when not actually raising or lowering the ' string of tools '.

Bull Wheel. (*a*) A large wheel engaging with and driving the rack of a planing machine.

(*b*) The driving pulley for the camshaft of a *stamp*.

(*c*) The large wheel at the base of a revolving *derrick*.

(*d*) The driving wheel of a cable-drilling rig.

Bulldozer. (*a*) A heavy motor-driven vehicle mounted on caterpillar tracks and pushing a broad steel blade in front to remove obstacles, to level uneven surfaces, etc. Also called ' Angledozer '.

(*b*) A heavy power-press, having a horizontal reciprocating ram for shaping angle irons, etc., with suitable dies and used in railway and wagon shops.

Bull-headed Rail. A rail with a dumb-bell cross-section but with unequal heads, the larger head being the upper part.

Bullock (or Horse) Gear. A mechanism for utilizing animal power by means of a lever attached to gears, the animal walking in a circle.

Bump Stop. The final stop to prevent further motion of a mechanism, frequently made of hard rubber.

Bumpers. Fenders on motor-vehicles, ships, etc., for mitigating collisions. They are sometimes sprung when fitted on vehicles.

Bundling Machine. A machine for grasping a number of articles into a bundle ready for tying.

Burnishing. (*a*) *Spinning*.

(*b*) The operation of producing a brilliant finish on metal parts, the edges of books and the surfaces of pottery ware.

Burr. (*a*) A roughness left on a metal by a cutting tool.

(*b*) The turned-up edge of metal after punching or drilling.

(*c*) A blank punched from sheet-metal.

(*d*) A small milling cutter used for engraving and dental work.

(*e*) A toothed drum used on a mandrel between lathe centres.

Bursting. (*a*) The breaking of a rotating part of machinery due to centrifugal forces.

(*b*) The bursting of a vessel due to an excessive pressure difference between inside and outside, the inside being the greater. In the reverse case it is called collapsing or bursting inwards.

Bus. *Omnibus.*

Bush. A cylindrical sleeve forming a bearing surface for a shaft or pin; usually as a lining. See also **Ball Bush, Drill Bush.**

Bushing. The fitting or driving in of a *bush* into its seating.

Butt Coupling. A box coupling for connecting shafting and keyed around the two shafts which butt against each other and are co-axial. See also **Box Coupling.**

Butt Joint. A joint between two plates in end contact with a narrow strip riveted or welded to them. See also **Double Strap Butt Joint.**

Butterfly Throttle. (*a*) An elliptical plate pivoted on its centre which throttles the steam passing into a cylinder or closes the induction pipe completely.

(*b*) A circular plate used to control the volume of air, and therefore the air/petrol mixture, entering a petrol engine. See also **Butterfly Valve.**

Butterfly Valve. (*a*) A pair of semi-circular plates hinged axially to a common diametral spindle in a pipe so that the plates permit flow in one direction only.

(*b*) A disc acting as a throttle when turned on its diametral axis in a pipe. (See Fig. 24.)

Button-headed Screws (Half-round Screws). Screws with hemispherical heads, slotted for a screwdriver. (Fig. 155.)

Buttress Screw Thread (Leaning Thread). A screw thread designed to resist heavy axial loads with the front or thrust face perpendicular to the axis and the back of the thread sloping at 45°. Some such threads have their thrust faces inclined to facilitate thread milling and thread grinding. Quick-acting vices and the breech mechanisms of guns use buttress threads; hence the name *breech-lock thread.* (See Fig. 183(*c*).)

By-pass. A passage through which a gas or liquid is allowed to flow instead of, or additional to, its ordinary channel, or any device for arranging this.

By-pass Engine. A turbojet engine which has a relatively large low-pressure compressor and by-passes some of the air from it round both the high-pressure compressor and the turbine into the tail pipe. (See Fig. 194.)

By-pass Valve. A valve for directing flow through a *by-pass.*

C

cal. Calorie.
C.E.I. Cycle Engineers' Institute screw thread.
c.f. Centrifugal force.
c.g. Centre of gravity.
C.G.S. System. Centimetre-gramme-second system of units
cm.p.s. Centimetres per second.
comp.r. Compression ratio.
C.P. Cycle. Constant pressure cycle.

C.Q.R. Anchor. An anchor designed like a plough. (See Fig. 18.) Cf. **Stockless Anchor.**

C. to C. Centre to centre.

c/s, c.p.s. Cycles per second.

C.V. Cycle. Constant volume cycle.

Cabinet Leg. The cupboard under a lathe or machine tool for holding gear, tools, oil tank, etc.

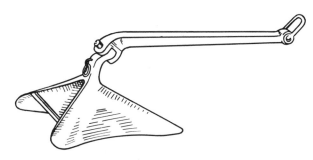

FIG. 18. C.Q.R. ANCHOR.

Cable-drilling Rig. A large earth-drilling machine in which the drill bit at the bottom of a long tube is suspended on a cable from a vertical derrick and rotated at ground level by a prime mover; the tube is lengthened by additions at ground level.

Cable Grip. A flexible cone of wire put on the end of a cable to enable the cable to be pulled into a duct as tension tightens the cone on the cable.

Cable Ploughing. The ploughing of a field by using two *traction engines,* connected by a wire rope attached to a plough, which alternately supply the necessary power to pull the plough across the field.

Cable Railway. A railway with the motive power coming from a continuous moving cable, overhead or underground, to which the car can be rigidly connected by a clutch device at any point on the cable.

Cable Tramcars. Tramcars operated from an underground cable in the same manner as a *cable railway.*

Cable-way (Blondin). A construction for transporting material by skips suspended from cables which are slung over and between a series of towers, the skips being raised, lowered or moved to any position along the cables. Cf. **Cable Railway.**

Cage.　(*a*) A platform, with or without framework, for lowering or raising goods, etc.　A ' cage ' in a mine lowers and raises men or wagons in a shaft.

(*b*) The frame in a travelling crane within which a man sits and controls its movements.

(*c*) The holder for the separate balls in a ball-bearing to keep the balls properly spaced and out of contact with each other.

Cage Valve.　*Ball valve.*

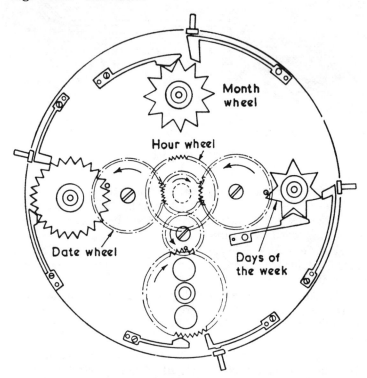

FIG. 19. CALENDAR MECHANISM.

Calculating Machine.　A machine which performs one or more of the five operations, addition, subtraction, multiplication, division and calculation of the square root.　The necessary figures are set, the operation is by hand or power (usually electrical) and the answer is registered.

A *planimeter* is sometimes reckoned to be a calculating machine.

Calendar Mechanism. The mechanism of a calendar watch or clock which indicates the date. In the ordinary mechanism adjustments have to be made by hand if the month is less than 31 days, but in a ' perpetual calendar ' the mechanism makes all corrections including that for leap years. (See Fig. 19.)

Calender. (*a*) A machine in which material is passed through rollers under pressure to impart the desired finish or to ensure uniform thickness, such as in a steel mill.

(*b*) A machine for *calendering* soft materials, usually consisting of a number of rollers.

Calender Rollers. (*a*) Heavy grooved rollers for feeding timber into sawing or planing machines.

(*b*) The rollers in a *calender*.

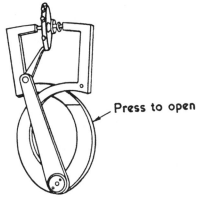

FIG. 20. CROSS-OVER CALLIPERS ADAPTED FOR POISING.

Press to open

Calendering. The series of operations (varying according to the goods) of straightening, damping, pressing, etc., woven goods to give them the desired finish.

Calibre. (*a*) The bore or internal diameter of a cylinder or pipe, a firearm or a piece of ordnance.

(*b*) The arrangement of the various components of a watch or clock.

Caliper. The size of a watch movement.

Calliper Gauge. A horseshoe type of *limit gauge* with two pairs of jaws marked ' Go ' and ' No go ' corresponding to the tolerance allowed on the dimensions for the work. See also **Vernier.**

Callipers, Poising. A form of cross-over callipers with jaws between which a *balance* for a watch can be mounted and rotated to test for truth and poise, that is, static and dynamic *balance*. See Fig. 20 and **Poising.**

Cam

Cam. A shaped component of a mechanism, such as a heart-shaped disc on a shaft, which determines the motion of a *follower*. The motion of the cam may be linear or rotary, usually the latter. A ' two-dimensional cam ' will have either linear or rotary motion, but not both, and may be reversible. A ' three-dimensional cam ' has two input motions which may be either linear or rotary or both (Fig. 21). See also **Wiper** (*c*). Fig. 21 shows (*a*) a wedge type,

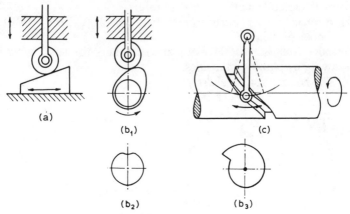

FIG. 21. CAMS: (*a*) wedge type, (*b*) radial or disc type, (*c*) cylindrical.

(*b*) a radial or disc type, and (*c*) a cylindrical type of cam.

Bi-polar Cam. A pair of shaped plates with teeth which engage so that the input and output shafts of the plates have a varying velocity ratio.

Pinwheel Cam. Pinwheel gear.

Cam-ball Valve. A valve actuated by a cam on the axis of a ball-lever, the ball rising with the level of water in a cistern, tank or boiler, so that the cam causes the valve to shut off the supply.

Cam Chuck. A profiling device fitted to the saddle of a lathe for turning irregular forms such as cams. The cutter turns in the *fixed headstock* and the work is manipulated on the *rest*.

Cam Follower. That part of an engine or mechanism which rides on the contour surface of a *cam* and to which motion is imparted by the cam.

Cam Governor. A stepped or differential cam, giving three or four grades of lift, has its action controlled by governor balls which slide a roller on to one or another of the cams according to the centrifugal action on the balls; it is found in *Otto-cycle* gas engines.

Cam Lobe. The raised portions on the contour of a cam which operate the *cam followers*. (See Fig. 200.)

Cam Profile. The cam outline as determined by the form of the flanks and follower's tip.

Cam-type Steering-gear. A steering-gear in motor-vehicles in which the steering-column carries a pair of opposed volute cams which engage with a peg or roller carried by the drop arm. See **Drag Link, Drop Arm.**

Cam Wheel. See **Hammer Wheel.**

Camber Angle. The angle of inclination of an automobile wheel out of the vertical due to inclining the spindle downward out of the horizontal.

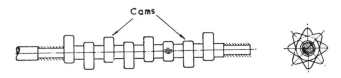

FIG. 22. CAMSHAFT.

Camshaft. A shaft operating the valves of piston engines by means of the cams formed integrally with the shaft or keyed on to it (Fig. 22). Cf. **Eccentric.** See **Overhead Camshaft.**

Cancelling Machine. A rotary cylinder with a ratchet used in post offices for cancelling the stamps on letters.

Cannon. (*a*) A hollow spindle or shaft with a motion independent of an internal spindle or shaft.

(*b*) A wheel in the mechanism of a clock or watch.

(*c*) A watch-key barrel.

(*d*) The wheel or pinion with an extended pipe to which the minute hand is attached.

Cannon Pinion. The pinion with an extended pipe to which the minute hand of a watch is usually attached. (See also **Minute Hand.**) Swiss watches have usually a snap-on cannon pinion on the *centre arbor*.

Cannon Wheel. The wheel on the *centre arbor* which carries the minute hand on a square section of its pipe.

Cannon Wheel Spring. A small plate bent upwards from the front plate and acting as a spring washer for the cannon wheel.

Cantilever Spring. A laminated spring anchored to the axle of a motor-vehicle at its mid-point, and to shackles on the frame at its two ends. Also called semi-elliptical spring.

Canting. Tilting over of the moving part of a mechanism from its proper angle when in motion or tilting machinery at an angle for a special purpose such as for cutting a *bevel* or for cutting at a definite angle.

Cap. The upper half of a *journal bearing;* the lower half is the pillar.

Cap Jewel. A jewel with *endstone* in horology.

Cap Spinning. The spinning of fine yarns in which the spun yarn is led on to a bobbin rotating at high speed via a cap on the top of the spindle.

Capping. The shrouding of gear-wheels. See **Shroud** (*a*).

Caprotti Valve-gear. Two pairs of vertical double-beat *poppet valves* which are operated by cams with an adjustable cut-off obtained by varying the angular position of the inlet cams, as found on some locomotives.

Capstan. A vertical cone-shaped drum or spindle, on which a rope or chain is wound, that is rotated by man, steam, hydraulic or electric power; for example, for warping a ship or hoisting an anchor.

Capstan Engine. A capstan driven by steam power, the general arrangement resembling a *winch* with worm gearing operating the vertical drums.

Capstan-headed Screw. A screw with a cylindrical head pierced by radial holes so that it can be tightened by a tommy bar inserted in the holes.

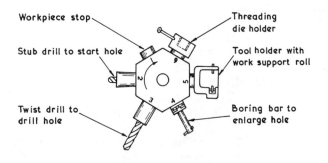

Fig. 23. Capstan Tool Head.

Capstan Lathe. A lathe in which the tools are mounted in a *capstan tool head.* See also **Turret Lathe.**

Capstan Tool Head. The support of the hexagonal tool-post of a *capstan lathe* (Fig. 23). It is mounted on a short slide which in turn is part of a carriage sliding on the *lathe bed.* The capstan rest is actuated by the *star wheel.* This construction gives a short working stroke allowing rapid manipulation of the hexagonal tool-post.

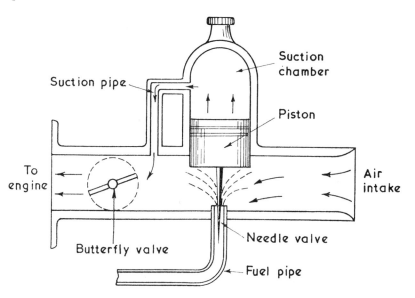

FIG. 24. CARBURETTER WITH BUTTERFLY AND NEEDLE VALVES.

Car. (*a*) A vehicle running on three or more wheels designed primarily for non fare-paying passengers. Cf. **Omnibus.**

(*b*) The cage of a lift or hoist in which passengers or goods are carried.

Carbon Gland. A type of gland in the form of segmented (carbon) rings to prevent leakage along a shaft such as in steam-turbines and on shafts driven from outside chambers full of air at high pressure. See also **Garter Spring.**

Carbonado (Black Diamond). A variety of crystalline carbon found in Brazil used as an abrasive for turning down and truing emery wheels. Unlike a diamond, it has no regular crystal form.

Carburetter (Carburettor). A device in which a fuel is atomized and mixed with air. See also **Float, Float Chamber.**

The pressure difference across a piston caused by the air velocity through the nozzle inlet (choke) regulates the fuel flow by means of a needle valve attached to the piston. The variable demand to an engine is obtained by varying the flow int othe engine through a butterfly valve. A simplified illustration is given in Fig. 24.

Cardan Joint. *Universal joint.*

Cardan Shaft. A shaft transmitting power as in a motor-vehicle or the propeller shaft in a ship. See also **Propeller Shaft.**

Cardan Suspension. A method of suspending the needle of a mariner's compass upon *gymbals.*

Carding Engine. A machine for combing and cleaning cotton fibres for conversion into a continuous strand of fibres (*sliver*). The revolving flat-card type treats fine materials and the roller- and clearer-card type the coarser materials.

Cards. Strips of cardboard that function in a *Jacquard machine* by controlling the cords connecting with the harness mails, which lift or depress the threads to form the desired pattern of the fabric.

Carnot Cycle. The working cycle of an ideal heat-engine of maximum thermal efficiency, consisting of isothermal expansion, adiabatic expansion, isothermal compression and adiabatic compression to the initial state.

Carnot's Law. No engine is more efficient than a reversible engine working between the same temperatures. The engine's efficiency is independent of the nature of the working substance, being dependent on the temperature only. See also **Carnot Cycle.**

Carriage. (*a*) That part of a lathe which slides on the *lathe bed* and carries the cutting tool.

(*b*) The horizontal table of a printing machine which travels to and fro under the cylinder or roller.

(*c*) The moving part of a typewriter which holds the paper.

(*d*) The part of a lace machine carrying the bobbin thread and swinging in an arc on the combs which contain the warp threads.

Carriage Clock. A small portable clock with a platform *escapement*, usually having a brass case and glass panels. Sometimes called ' travelling clock '.

Carriage Spring. Any elastic device, often curved steel strips of varying length, interposed between the bed of a vehicle and its running gear. Cf. **Laminated Spring.** See also **Semi-elliptic Spring.**

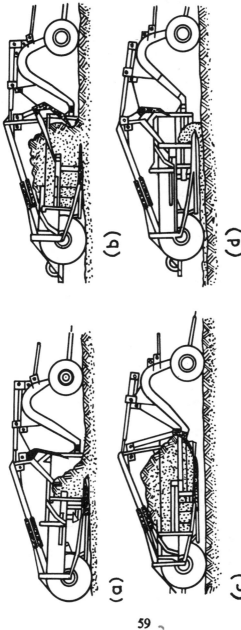

(a)

(b)

(c)

(d)

Fig. 25. CARRYALL: (*a*) and (*b*) loading, (*c*) carrying, (*d*) unloading.

Carrier. (*a*) A receptacle used in connection with a conveyor system.

(*b*) A device screwed to the work and driven by a pin projecting from the *face plate* of a lathe, so that the work is rotated *between centres*.

(*c*) The first roller in a *carding engine* which unwinds the lap and distributes it to the machine.

(*d*) An intermediate roller on a coarse wool-carder (scribbling machine) between the feed rolls and the toothed drum. See also **Carryall.**

Carrier Wheel. *Idle wheel.*

Carryall. A self-loading, self-discharging transport vehicle, either towed or with a built-in power unit (Fig. 25).

Caster (Castor). A small wheel on a swivel.

Caster Angle (*Wheel Action*). The inclination of the *king-pin* in a motor-vehicle in such a way that its axis intersects the ground at a point in advance of the point of contact of the wheel with the ground, thus giving a self-centring tendency to the steerable front wheels after angular deflection by road shocks.

Castle Nut. A hexagonal nut with six radial slots, any two of which can line up with a hole drilled in the bolt or screw for the insertion of a split pin to prevent loosening.

Cataract. A kind of hydraulic brake consisting essentially of a plunger and valves for regulating the action of a pumping engine and other machines.

Cataract Rod. A vertical rod attached to the lever of a cataract which directly operates the above action.

Catch Bar. A long bar of steel forming part of a lace machine and faced with brass on the part, called a ' driving blade ', which engages with the carriages.

Catch Pawl. *Dog.*

Catch Plate. The end flange of a lathe-head speed cone, or of an internal plate driven by the cone through a hole in which a removable peg takes the drive to the lathe *mandrel.*

Catch Points. Trailing points on an up-gradient in a railway to derail rolling stock unintentionally descending the gradient. Cf. **Derailer** and see **Points.**

Caterpillar (Track). Road wheels replaced by endless articulated steel bands of flat plates, or by chains, passing round two or more wheels to enable a vehicle to cope with rough and uneven ground. Removable projecting pieces are sometimes provided to increase the grip on soft ground; the caterpillar also spreads the load.

Cathead (Spider). A turned sleeve, having four or more radial screws in each end, used on a lathe for clamping on to rough work of small diameter and running in the *steady* while centring.

Centre Arbor. The arbor in the centre of the plates of a watch or clock which makes one complete turn per hour.

Centre Bit. A wood-boring tool with a central point and two side wings, one of which scribes the boundary of the hole to be cut and the other cuts away the material.

Centre Distance. The length of a common perpendicular to the axes of two gears in mesh. (See Fig. 92.)

Centre Drill. A small drill with straight flutes, used for drilling holes in the end of a bar, which is mounted between centres in a machine tool; the drill is so shaped that it produces a countersunk hole.

Centre Gauge (Screw-cutter's Gauge). A thin metal gauge with vee-shaped notches, having an included angle of 55°, around its perimeter, used as a template for turning the cone points of lathes, grinding the angles of screw-cutting tools and setting these tools in the tool-post.

Centre Lathe (Engine Lathe). A machine for carrying out turning, boring or screw-cutting operations on work held *between centres* or in a *chuck*, but not for repetition work.

Centre Line. A line parallel to the general direction of a profile and located such that the sums of the areas contained between it and the parts of the profile lying on either side of it are equal. A centre line may be straight or curved and dimensions are normally taken from it and seldom from edges or outer faces. Cf. **Mean Line.**

Centre Line Average Height. The average departure of a profile above and below the *centre line* irrespective of sign.

Centre of Gravity. That point in a body at which its weight may be taken to act and about which it will be balanced though placed in any position.

Centre of Gyration. That point in a revolving body in which its angular momentum is concentrated.

Centre of Mass. That point in relation to a body, any plane passing through which divides the body into two parts of equal weight, and through which acts the resultant force due to the body's inertia when it is accelerated.

Centre of Oscillation. (*a*) That point in the axis of a vibrating

61

Centre of Oscillation

body in which, if the whole matter were concentrated, the body would continue to vibrate in the same time.

(*b*) A point in a ' compound pendulum ' vertically below the point of suspension, when the pendulum is at rest, at a distance equal to the length of a simple ' pendulum ' with the same period. See **Pendulum.**

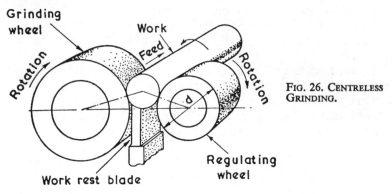

Fig. 26. CENTRELESS GRINDING.

Centre of Percussion. *Centre of oscillation.*

Centre Pinion. The first pinion in a watch or clock train, driven by the *great wheel.*

Centre Punch. A steel punch with a conical point used to mark, by indentation, the centres of holes to be drilled, etc.

An ' automatic centre punch ' contains a spring which releases its energy in a single blow after the operator has applied a certain pressure.

Centre Wheel (Horology). The wheel mounted on the arbor of the *centre pinion*; it usually makes one turn per hour.

Centre-weighted Governor. A high-speed small *governor* with a heavy weight sliding on a central spindle, whose gravity is balanced by the centrifugal force of the balls.

Centreless Grinding. Grinding by means of a ' centreless grinder ' (Fig. 26) where the work is supported and fed to the grinding wheel on a knife-edge support, substituting the two fixed centres of conventional cylindrical grinding machines. Feeding is by an abrasive regulating wheel whose axis is tilted at a small angle (0–10 degrees) to that of the work; the wheel also rotates the work. The knife-edge and regulating wheel can be adjusted to take different diameters of work. The diameter of the regulating wheel ' d ' is always smaller than that of the grinding wheel.

Centrifugal Brake. An automatic brake actuated by revolving brake shoes forced out by centrifugal force into contact with a fixed brake drum.

Centrifugal Clutch. A clutch in which friction surfaces engage at a definite speed with the driving member and remain engaged due to the centrifugal force exerted by weighted levers.

Centrifugal Compressor. An air compressor in which the pressure rise is obtained by the centrifugal forces set up by a rotating impeller. In a ' single-entry compressor ' the air is admitted to one side only of the *impeller* or at one end of the rotating member. In a ' double entry compressor ' the air is admitted to both impellers (back to back) or to both ends of the rotating member (see Fig. 27). Cf. **Axial Compressor.**

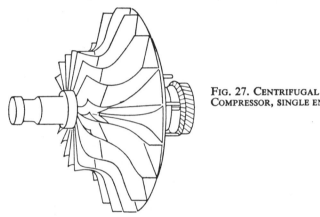

FIG. 27. CENTRIFUGAL COMPRESSOR, SINGLE ENTRY.

Centrifugal Fan (Paddle-wheel Fan). A *fan* of paddle-wheel form in which the air enters axially at the centre and is discharged radially by centrifugal force.

Centrifugal Force; Centripetal Force. The centrifugal force on a body constrained to move along a curved path is away from the centre of curvature of that path. The equal and opposite force directed towards the centre of curvature is called the ' centripetal force '.

Centrifugal Pump. A pump consisting of one or more *impellers* equipped with vanes, mounted on a rotating shaft and enclosed by a casing. The liquid is drawn into the centre(s) of the rotating impeller(s) and flows out radially under centrifugal force; the resulting kinetic energy is converted into pressure energy in the casing or *diffuser*.

Centrifugal Speedometer. See **Speedometer, Centrifugal.**

Centrifuge. An apparatus which rotates at very high speeds.

(*a*) To separate solids from liquids or to separate two liquids of different densities or to drive a liquid to the end of a tube as with a clinical thermometer.

(*b*) To test assemblies under high '*g*' loadings.

Centring (Centering). (*a*) Marking off the centres upon work before turning in a lathe.

(*b*) Adjusting work in a lathe so that the axis of the work coincides with the lathe axis.

Centring Chuck. *Self-centring chuck.*

Centring Machine. A machine for *facing* forgings, etc., and marking the centre for subsequent turning.

Centripetal Force. See **Centrifugal Force.**

Centrode. The path of the instantaneous centre of rotation of a body.

Ceramic Tool. A cutting tool suitable for machining almost all metals and abrasive materials. It can only be dressed by diamond wheels or, under light pressure, by green grit wheels.

Chain. Metal links of oblong or circular shape interconnected to form a flexible cable to be used for hoisting or for power transmission. See **Link.**

Block Chain. A chain consisting of blocks connected together by links and pins and used for comparatively low speeds.

Close Link Chain. An open link chain, with the length of links not greater than five times the diameter and with a width of three and a half diameters; not to be confused with 'circular link chain'.

Coil Chain. A chain consisting of oblong links of circular section usually of welded wrought-iron or steel. The links may be plain or have a stud (or bridge) across the centre; the studs tend to prevent stretching and kinking.

Detachable Link Chain with easily detachable and replaceable links, being used for low-speed and light-load power transmission, and for conveyors and elevators of moderate capacity and length.

Driving Chain. An endless chain of steel links engaging with toothed wheels to transmit power from one shaft to another.

Pintle Chain. A *sprocket chain.*

Plate Link Chain. A chain with flat links united by pins passing through holes near the ends of the links.

Roller Chain. (*a*) Several rollers held in links and used to form a linear bearing. (See Fig. 28.)

(*b*) A chain consisting of alternate links held by connecting-pins which are fastened by *cotters*, the pins serving to carry the rollers which bear on the *sprocket* teeth. Roller chains can transmit more power than block chains and can operate at chain velocities up to 7 m/s. Double, triple and quadruple-strand roller chains are used for larger power requirements.

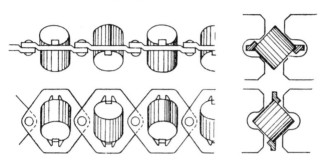

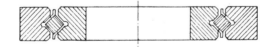

FIG. 28. CROSSED ROLLER CHAINS.

Silent Chain. A chain consisting of alternate flat steel links connected by pins. The links have faces in contact with the *sprocket* and rotate slightly on the pins as the chain bends around the sprocket. Silent chains are used for heavy loads at speeds in excess of 8 m/s. (See Fig. 130.)

Sprocket Chain. A chain suitable for use on a toothed wheel.

Studded Chain. A chain with stud links. See **Link**.

Chain Barrel. A cylindrical barrel, sometimes grooved, on which surplus chain is wound.

Chain Block. *Differential pulley block.*

Chain Conveyor. A *conveyor* with endless chains supporting slats, buckets, etc., as distinct from the use of a single band. See

Chain Conveyor

Apron Conveyor, Bucket Conveyor.   Cf. **Band Conveyor.**

Chain Cutter.   A cutter composed of the links of an endless chain, made of special tool steel and ground with a hook of 25° on the outsides of the links.   The pitch is the length of two links.   (See Fig. 29.)

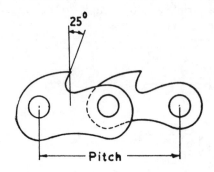

FIG. 29. MORTISE CHAIN CUTTER.

Chain Feed.   (*a*) An endless chain led around sheave wheels for feeding balks of timber.

(*b*) A chain used on machine tools instead of a belt.   See also **Feed, Pitch Chain.**

Chain Gearing.   A gearing using a chain and wheels in which projections on the wheel fit into cavities in the chain or vice versa.

Chain Hook.   The hook at either end of a *fusee chain* for hooking it to the fusee and to the *barrel*.

Chain Pump.   An endless belt provided with buckets which dip into a liquid, lift a quantity and discharge it through a spout while turning over at the top pulley.

Chain Saw.   A power-driven endless chain with saw-like teeth on its links.   See **Chain Cutter.**

Chains, Crossed Roller.   Chains with rollers in each link, alternate rollers being at an angle to each other.   (See Fig. 28.)

Chair.   A cast-iron seating for metal rails spiked to the sleeper and securing a bull-head rail in position.

Joint Chair.   A chair at the joint between two successive lengths of rail.

Chamfer.   A *bevel* produced on edges or corners which are otherwise rectangular.

Chamfered.   (*a*) Rounded or given a radius.

(*b*) Given a *chamfer.*

66

Chamfering Machine. A machine for forming the bevels of nuts and rounding the ends of bolts.

Change Wheels. The gear-wheels through which the *lead screw* of a screw-cutting lathe is driven from the *mandrel;* the wheels are changed to vary the reduction ratio.

Charpy Test. A *notched-bar test* in which a specimen, notched at the middle and fixed at both ends, is struck behind the notch by a striker carried on a pendulum. The absorbed energy is measured by the decrease in height of the swing of the pendulum after fracture. Cf. **Izod Test.**

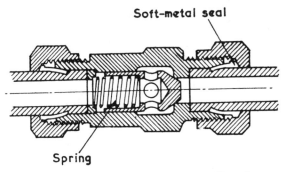

Soft-metal seal

Spring

FIG. 30. CHECK VALVE (NON-RETURN VALVE).

Chaser (Comb Tool). A lathe tool for cutting and finishing internal or external screw threads, usually the latter. The edge of the tool is the counterpart of the screw section.

Chasing. The cutting or finishing of screw threads with a *chaser.*

Chasing Attachment. A special feed motion built into *capstan lathes* and *turret lathes*, the special *lead screw* being driven by a shaft from the *feed box* thus permitting large-diameter threads to be formed with a *chaser*. Small-diameter threaded work is usually formed with a *diehead.*

Chassis. The base-frame of a vehicle.

Chatter. The vibration of a blunt, or badly set or insufficiently rigid cutting tool giving an irregular surface finish on the work-piece.

Check. Any piece or device intended to control or restrain motion.

Check Gauge. A gauge used for checking the accuracy of other gauges, normally for the verification of individual dimensions.

Check Nut

Cf. **Reference Gauge.**

Check Nut. *Lock-nut.*

Check Rail (Guard Rail, Safety Rail, Side Rail). A third rail laid on a curve alongside and near the inner rail to keep the wheel flanges against the main rail.

Check Valve (Non-return Valve). A valve which prevents reversal of flow by means of a check mechanism. The valve is opened by the flow of fluid and closed either by the weight of the mechanism when the flow ceases, or by a spring (Fig. 30) or by back pressure. There are three patterns (*a*) ' horizontal ' with the body ends in line with each other, (*b*) ' vertical ' in a vertical position with the body ends in line with each other, and (*c*) ' angle ' with the body ends at right angles to each other. Cf. **Back-pressure Valve.**

Foot Check Valve. (*a*) A valve fitted to the bottom of a suction pipe, usually with a strainer; the lowest valve in a pump.

(*b*) A non-return valve at the inlet end of a suction pipe.

Lift Type Check Valve. A mechanism incorporating a disc, piston or ball which lifts along an axis in line with the axis of the body seat.

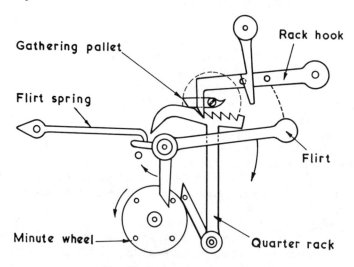

FIG. 31. CHIME—FLIRT ACTION.

Screw-down Stop and *Check Valve.* A check valve with a mechanism to hold the disc in the closed position independently of the flow or to restrict the lift of the disc.

Swing Type Check Valve. A check mechanism incorporating a disc which swings on a hinge.

Checking in (Checking). The fitting together of corresponding parts.

Chilled Rolls. *Rolls* which have had their surfaces hardened by chilling. **Cf. Grain Rolls.**

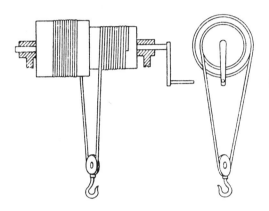

Fig. 32. Chinese Windlass.

Chiming Clock. A clock which strikes the hours and chimes the quarters. **Cf. Clock Striking Mechanism.**

Chiming Mechanism. See **Cannon Wheel, Centre Wheel, Escape Wheel, Gathering Pallet, Great Wheel, Hammer, Hammer Pallet, Hammer Wheel, Lifting Piece, Lifting Pin, Locking Wheel, Minute Wheel, Pallets, Fly, Rack, Rack Hook, Rack Tail, Warning Lever, Warning Wheel.** The mechanism controlling the chiming of a clock is shown in Fig. 31. A self-correcting mechanism is shown in Fig. 161.

Chinese Windlass. Two drums or cylinders of slightly different diameter on the same axis with a single coil of rope wound in opposite directions on each. The rope winds off one drum as it is wound on to the other, giving a slow motion with a considerable mechanical gain and enabling a large weight to be raised very slowly by a small expenditure of power (Fig. 32). **Cf. Differential Pulley Block.**

Chip Breaker. (*a*) A wooden insertion fitted in a bracket at the end of a weighted rod and low enough to contact the surface of the stock when a chain cutter commences to make its exit.

Chip Breaker

(*b*) A groove ground in the cutting tool which prevents the accumulation of the cut material in the form of a long chip which might otherwise damage the finished surface of the work.

Choke (Restrictor). (*a*) A restriction in a pipe to reduce fluid flow.

(*b*) A valve, usually a butterfly valve, in a carburetter-intake to reduce the air supply and thus give a rich mixture for starting purposes or while the engine is still cold. Also called ' strangler '.

Bleed Choke. A choke for releasing fluid, usually to atmospheric pressure.

Chopper. (*a*) Any device which interrupts regularly some quantity, such as light into a photo-cell.

(*b*) A term used in the U.S.A. for a helicopter.

Chordal Height. The shortest distance from the mid-point of the chord, which is to be measured, to the tooth crest of a gear: it is measured on the *back cone* of a bevel gear.

Normal Chordal Height. The shortest distance from the tooth crest to the *reference cylinder* of a gear.

Chordal Thickness. The thickness of the gear-tooth measured at the *pitch circle*. (See Fig. 172.)

Chronograph. A watch with a centre seconds hand which can be started, stopped and will fly back to zero by using a press-button or push-piece, and without affecting the main mechanism of the watch. Cf. **Stop-watch.**

Split-seconds Chronograph. A chronograph with two centre

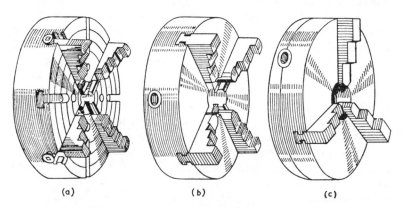

(a)   (b)   (c)

Fig. 33. Chucks: (*a*) four-jaw independent, (*b*) four-jaw self-centring, (*c*) three-jaw self-centring.

seconds hands, which start together but are stopped separately, while a third pressing of the button sends both hands back to zero.

Chronometer. A very accurate timekeeper fitted with (in U.K. and U.S.A.) a spring-detent escapement, but (in Europe) may be fitted with a lever escapement.

Marine Chronometer. A specially mounted chronometer for use on board ship in the determination of longitude.

Chronometer Escapement, Spring-detent Escapement. See **Escapement.**

Chuck. A device attached to the spindle of a machine tool which grips the rotating drill, cutting tool or work, such as work secured to the mandrel of a lathe's headstock (see Fig. 33). See also **Bell Chuck, Combination Chuck, Cup Chuck, Die Chuck, Driver Chuck, Driving Chuck, Independent Chuck, Magnetic Chuck, Oval Chuck, Point Chuck, Scroll Chuck, Self-centring Chuck, Universal Chuck.**

Chucking. Attaching lathe-work to a *chuck*.

Chucking Machine. A machine tool in which the work is held and driven by a *chuck*, but not supported on centres.

Chucking Reamer (Straight Shank Reamer). A *reamer* with a cylindrical shank for use in a *self-centring chuck*.

Ciné Camera. A ' motion picture ' camera that uses film smaller in width than ' standard ' stock (35 mm). The sizes below 35 mm are known as ' substandard ' or narrow-gauge film, namely 8 mm, 9·5 mm and 16 mm. The smallest sizes are popular with amateurs, while the 16 mm-size has achieved almost professional status in recent years, being used for scientific purposes and for sponsored films. The camera takes a series of images on a strip of photographic film, the important mechanical feature being the intermittent motion of the film, controlled in the smaller sizes by a claw movement (see Fig. 34). A similar intermittent motion is found in all ciné cameras and projectors, but the professional projector uses a more robust device known as the Maltese Cross whose toothed sprocket is moved intermittently by a modification of the Geneva escapement found in clocks and watches.

Fig. 34 illustrates the action of a ciné camera. The film on the supply reel A is drawn off by the sprocket wheel B, enters a claw mechanism C which pulls the film past the gate G. The sprocket wheel feeds the film after leaving the gate G to the take-up reel D. E represents a revolving shutter and F the lens. S's are guides for the film. The end L of the claw arm revolves in a circle, the point M

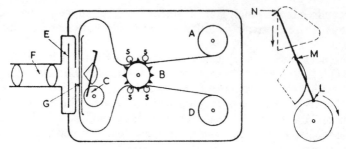

FIG. 34. CINÉ CAMERA.

moves round a cam profile and the claw *N* traces a curve as shown dotted. See also **Maltese Cross Mechanism** and Fig. 109.

Cinematograph. An apparatus for projecting film and reproducing synchronized sound.

Circle, Reference. The circle of intersection of the *reference cylinder* of a helical or spur gear by a transverse plane.

Circular-form Tool. A typical circular-form tool mounted in its holder is shown in Fig. 35.

Circular Motion. A feed mechanism, using an endless screw actuating a quadrant rack, in *shaping* and *slotting machines* whereby the work is rotated so that it can be shaped to circular arcs, as with the end of levers, etc.

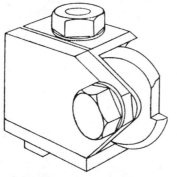

FIG 35. CIRCULAR-FORM TOOL IN HOLDER.

Circular Pitch (Circumferential Pitch). The pitch of wheel teeth measured along the circumference of the pitch circle (rolling circle), upon which one wheel comes into contact with its mate. (See Fig. 173(*a*).)

Circular Saw (Buzz Saw). A steel disc with teeth on its circumference, varying from a few centimetres up to a few metres diameter, used for cutting wood, metal and other materials. See also **Cold-saw.**

Circular Table. A circular cast-iron plate, which sustains the work in drilling and slotting machines.

Circular Vibration. See **Vibration.**

Circulating Pump. A pump for circulating cooling water through the condenser of a steam plant.

Circumferential Pitch. *Circular pitch.*

Clack (of a Pump). That portion of a pump-valve or bucket lifted by the action of the water or air, especially of a *flap-valve.*

Clack (Clack Valve). A ball type of *check valve* admitting water from a feed-pump to the boiler of a locomotive, which makes a characteristic noise when the ball strikes its seat.

Clack Box. (*a*) The chamber in which the *clack* of a pump works.

(*b*) That portion of the chamber which contains the valves that open and close to suction and delivery.

Clapper Box. A tool-head carried on the saddle of a planing or shaping machine. Cf. **Fixed Head.** (See Fig. 163.)

Claw Clutch (Claw Coupling). A shaft coupling for instant connection or disconnection in which flanges carried by each shaft engage through teeth in corresponding recesses in their opposing faces, one flange being slidable axially for disengagement.

Claws. The points or claws that operate on the sprocket holes of a ciné film and thus intermittently feed the film forward through the picture gate for projection.

Cleaning of Surfaces. Removing dust, etc., by blasting, by electrolytic and ultrasonic methods. See **Dry-blast Cleaning, Sand Blasting, Vapour Blast Cleaning.**

Clearance. (*a*) The distance between two objects or between two mating parts when assembled together.

(*b*) The distance between a moving and a stationary part of a machine or between two moving parts.

(*c*) The angular backing-off given to a cutting tool in order that the heel will clear the work.

Major (Minor) Clearance. The distance between the design forms at the root (crest) of an internal screw thread and the crest (root) of the external screw thread.

Clearance Angle. *Angle of relief.*

Clearance Fit. See **Fit.**

73

Clearance Volume. The volume enclosed by the piston and the adjacent end of a reciprocating engine cylinder or of a compressor when the crank is on the *inner dead-centre*.

Clearing Hole (Clearance Hole). A hole full to the specified size so that an object, such as a bolt or stud, of the same nominal size will pass through it.

Click. A *pawl* used in horology in connection with a ratchet wheel to permit rotation in one direction only.

Click Spring. The spring which holds the click in the teeth of a *ratchet wheel*.

Click Wheel. A small *ratchet wheel*.

Climb Milling. The milling process when the work is fed in the same direction as the path of the teeth on the cutter. This is the opposite of conventional *milling*.

Clip Pulley (Clip Drum). A rope pulley with a rim of vee-section constructed of movable clips, about ten centimetres long, hinged on pins whose axes are along the direction of the pulley's periphery. The effect of the rope biting is to pull the clips towards each other and thus increase the bite.

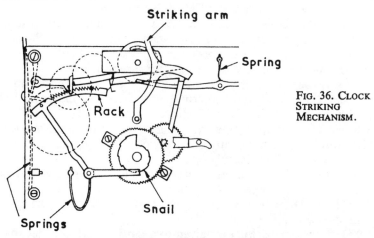

Fig. 36. Clock Striking Mechanism.

Clock. (*a*) A timekeeper, other than a chronometer or a watch. (*b*) A common term for a dial test indicator (*D.T.I.*).

Clock and Watch Mechanism. See **Chiming Mechanism, Pendulum Clock Mechanism, Striking Mechanism, Watch and Clock Mechanism.**

Clock Escapement. See **Escapement,** and **Clock.** Fig. 62.

Clock, Rieffler. A clock in which the pendulum is suspended in such a way that it only contacts the escapement over the central portion of its swing and is virtually disconnected during the rest of its period.

Clock Striking Mechanism. The details of this mechanism are given under *striking mechanism* and illustrated in full in Fig. 36 and in detail in other Figures.

Clock, Transistorised. A clock with a transistorised circuit triggered off by the *balance wheel* passing through a magnetic field and thus causing a transistor to switch the battery current to a driving coil. The return motion is by means of the customary hair spring.

Close Link Chain. See **Chain**.

Club-tooth Escapement. A widely used escapement for watches and platform escapements. See **Escapements—Swiss Lever**.

Cluster Mill. A mill consisting of two small working rolls each supported and driven by a pair of two large rolls. (Fig. 150.)

Clutch. The coupling of two working parts, for example two shafts, in such a way as to permit connection or disconnection at will without the necessity of bringing both parts to rest, and when connected to transmit the required amount of power without slip.

' Friction clutches ' operate by surface friction when two surfaces are pressed together. ' Magnetic clutches ' make use of the attraction of a magnet for its armature. ' Jaw clutches ' give a positive drive by the use of projecting lugs. See also **Fluid Flywheel**.

Band Clutch. A fabric-lined steel band which is contracted on to the periphery of the driving member by means of engaging gear.

Block Clutch. Friction shoes forced inwards into the grooved rim of the driving member or expanded into contact with the internal surface of a drum.

Centrifugal Clutch. A friction clutch which engages automatically at a definite speed of the driving member and is maintained in contact by the centrifugal force exerted by weighted levers.

Coil Clutch. A friction clutch using a coil of steel around a drum.

Cone Clutch. A friction clutch in which the internally coned member can be moved axially in or out of the externally coned member for engaging or disengaging the drive. The cone clutch is either metal to metal running in an oil bath or a fabric to metal running dry. (Fig. 37.)

Clutch

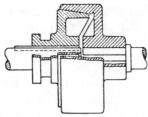

Fig. 37. Cone Clutch.

Disc (Plate) Clutch. A friction clutch in which both the driving and the driven members have flat circular or annular surfaces that are brought into contact and consist of one or more discs running either dry or lubricated, being called respectively *single-plate clutch* or 'multiple-disc clutch'. The former is usually fabric-faced and the latter usually run in oil; both are loaded by springs.

Dog Clutch. A jaw clutch consisting of opposed flanges, or male and female members, provided with projections and slots with one member slidable for engaging and disengaging the drive. (See Fig. 79.)

Split Ring Clutch. A friction clutch consisting of a split-ring which is expanded into a sleeve by a cam or lever mechanism. A clutch commonly used in machine tools.

Coach. A *motor-vehicle* for carrying a number of passengers.

Coal Mill. *Blacking mill.*

Coal-cutting Machinery. Machines operated by electricity or compressed air and using either a bar, chain, pick or rotating disc to cut coal from the face.

Coaxial Propellers. Two propellers mounted in an aeroplane on concentric shafts with independent drives and rotating in opposite directions. Cf. **Contra-rotating Propellers, Counter-rotating Propellers.**

Cock. (*a*) A carrier or bracket for a pivot in horology.

(*b*) A tap or cylindrical valve for controlling the flow of a liquid or gas. (Fig. 176.) See also **Ball-cock.**

(*c*) A lever in a firearm, raised ready to be released by a trigger.

Cock Valve. *Ball valve.*

Cock Wheel. *Idle wheel.*

Cocking Lever. A lever for raising the cock or hammer of a gun in readiness for firing.

Coefficient of Friction. See **Friction.**

Coefficient of Restitution. See **Impact.**

Cogged Bloom. A *bloom* which has been passed through a

blooming mill in readiness for rolling into rails or steel with other sections.

Cogging. (*a*) The operation of rolling or forging an ingot to reduce it to a *bloom* or *billet*.

(*b*) The fitting in, and working of, the cogs of *mortise wheels*.

Cogging Mill. *Blooming mill.*

Cogs (Mortise Teeth). Separate wooden teeth, formerly used in gear-wheels for slow-moving mills. The teeth fit into mortises or slots cast in the rims and are secured by pins or wedges.

Cog-wheel. A wheel with teeth, usually metal. Cf. **Mortise Wheel.**

Coil Clutch. See **Clutch.**

Coiler. The mechanism for delivering the *sliver* in coils into the *coiler cans* in a *carding engine*.

Coiler Can. A slowly rotating upright cylinder for receiving the *sliver* in coils.

Coke Mill. *Blacking mill.*

Cold-saw (Cold Iron Saw). A metal-cutting slow-running *circular saw* cutting steel bars to length. The short teeth may be either integral with the disc or inserted and the saw is thick in proportion to its diameter.

Collar. A rectangular section ring secured to, or integral with, a shaft to provide axial location for a bearing or to prevent axial movement of a shaft. See also **Collet** (*c*).

Collar Bearing. A bearing provided with several *collars* to take the thrust of a shaft, or to provide adequate surfaces for lubrication of a vertical shaft.

Collar-headed Screw. A screw in which the head has an integral collar to stop any fluid leakage past the threads.

Collaring. The wrapping of a rolled bar around the bottom roll of a *rolling mill*. See **Stripping Plate.**

Collective Pitch Control. See **Pitch (Helicopter).**

Collet. (*a*) A slotted sleeve, externally coned, which fits into the

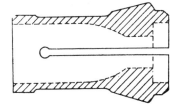

Fig. 38. Collet.

77

internally coned nose of a lathe mandrel and is used to grip small circular work or tools; also called ' collet chuck '. (See Fig. 38.)

(*b*) A disc or ring for holding dies or nuts in a screwing machine.

(*c*) A circular flange or collar.

(*d*) The friction-tight collar on a balance staff to pin the inner end of a balance spring in watches and some clocks.

Comb(s). (*a*) A toothed strip of metal with teeth of different lengths which are struck by the radial pins on the cylinder of a musical box or of a musical clock or watch.

(*b*) The arcs of brass fingers on which the carriages move in a lace machine, the back and front combs forming a well for the warp.

Comb Bars. Steel bars which extend across a lace machine and support the comb leads.

Comb Tool. *Chaser.*

Comber Board. Perforated slips of wood arranged in a frame through which the cords of the *Jacquard machine* pass to the mails.

Combination Chuck. A lathe chuck that can be operated either as an *independent chuck* or as a *self-centring chuck*.

Combination Planer. A machine in which one cutter-block can be used for surface planing and for thicknessing.

Combination Turbine. *Disc-and-drum turbine.*

Combination-turret Lathe. A lathe capable of automatic turning, facing, drilling, boring and threading operations by incorporation of an automatic sliding and surfacing *saddle* and *turret.* See also **Capstan Lathe, Turret Lathe.**

Combine. A *harvester-thresher.*

Combine Baler. A machine which gathers hay or straw from the windrows in a field and forms it into bales.

Combined-impulse Turbine. See **Impulse Turbine.**

Combustion Chamber. (*a*) A chamber in which combustion takes place in an internal-combustion engine, a jet engine or a rocket engine. (See Fig. 194.)

(*b*) Any chamber or space in which the combustion of gaseous mixtures or products takes place. Cf. **Mixing Chamber.**

Comparator. An apparatus for the accurate comparison of length standards or for measuring the coefficients of expansion of metal bars or for indicating progress during the manufacture of a part in relation to specified limits (cf. **Setting Gauge).** The difference between two standards can be measured by various means, including the movement of a pointer over a graduated scale or by sighting with microscopes.

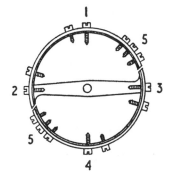

Fig. 39. Compensation Balance with a Bimetallic Rim, Steel (Inside) and Brass (Outside).

1 2 3 and 4 quarter screws

5 compensating screws

Compensated Pendulum. See **Pendulum.**

Compensating Collar. An annular ring fitted on a revolving shaft with adjustment to compensate for wear.

Compensating Screws. Screws near the cut ends of a *compensation balance* for compensation due to changes in temperature. Cf. **Quarter Screws.**

Compensation Balance. A *balance* constructed so as to compensate for dimensional changes and the elastic properties of the *balance spring* and balance wheel caused by temperature variations. (See Fig. 39.)

Compensation Pendulum. See **Pendulum.**

Complementary Gears and Racks. Two gears or racks are complementary if they can be fitted together face to face with completely coincident pitch and tooth faces.

Self-Complementary Racks. Racks whose teeth profiles are identical in shape and pitch and therefore complementary.

Compliance. The displacement in centimetres resulting from the application of the force of one dyne. The reciprocal of stiffness.

Composing Machine ('Linotype' Machine). A 'Monotype' machine composes type-matter in separate letters; a 'Linotype' composes in solid lines or slugs. Both machines have a *keyboard* resembling that of a typewriter.

Composite Engine. A combination of two engines of basically different design, such as a piston-turbine combination. Cf. **Compound Engine.**

79

Compound Engine. (*a*) A gas-turbine engine in which the compression is performed in stages in a number of mechanically subdivided compressors, each driven by a separate turbine. Cf. **Composite Engine, Spool.**

(*b*) A piston-engine with an exhaust-driven *supercharger* whereby any surplus turbine power is fed to the propeller through a fluid, or slip-clutch drive.

(*c*) A steam-engine in which the expansion of the steam from boiler pressure to exhaust pressure is carried out by high- and low-pressure cylinders in series. See **Multiple-expansion Engine.**

Compound Lever. A series of levers for obtaining a large mechanical advantage as in large weighing and testing machines.

Compound Screw. *Differential screw.*

Compound Slide Rest. See **Slide Rest.**

Compounding. Expanding steam in two or more stages, either in reciprocating engines or in steam-turbines.

Compressing Cylinder. (*a*) The cylinder of an air-compressor within which the air is compressed.

(*b*) A cylinder used in some gas-engines for compressing the air in the charge. See also **Working Cylinder.**

Compression Engine. A *gas-engine* in which the mixed charge is compressed previous to ignition.

Compression-ignition Engine. An *internal-combustion engine* in which the heat of compression ignites the mixture of air and injected fuel in the cylinder.

Compression Ratio. (*a*) The ratio of the volume of the mixture in the cylinder of a piston-engine before compression to the volume when compressed. Cf. **Pressure Ratio.**

(*b*) The ratio of the air (gas) pressures across a compressor of a jet-engine. See also **Spool.**

Compressor. See **Air Compressor, Axial Compressor, Blower, Centrifugal Compressor, Reverse-flow Compressor.**

Multi-stage Compressor. An axial-flow compressor with more than one row (stage) of blades or a centrifugal compressor with more than one impeller (stage). The former has usually many stages.

Computer. A calculating and data-handling machine, which may be largely mechanical as in the desk type which is often electrically powered, or wholly electronic, to accept and supply information derived by logical processes. The analogue computer simulates the conditions for some operation and the digital computer calculates from accepted numerical data the solution of problems using a scale of two instead of ten in electronic devices and circuits.

Con Rod (colloquial). *Connecting-rod.*

Concentric (Jaw) Chuck. *Self-centring chuck.*

Concrete Mixer. An appliance in which the constituents of concrete are mixed mechanically by continuous rotation in a large vessel which is also tilted to throw the materials from one side to another. Cf. **Mortar Mill.**

Condenser. A chamber into which exhaust steam from a steam-engine or turbine is condensed; a high degree of vacuum is maintained by an air-pump. See **Jet Condenser, Surface Condenser.**

Cone. (*a*) A solid generated by a straight line, one end of which remains fixed while the other end moves round a closed curve.

Right Circular Cone. A cone formed when the closed curve is a circle. All transverse sections are circular.

The fixed point is the ' apex ', the centre line is the ' axis ', and the constant angle between the moving line and the centre line is the ' cone angle '.

(*b*) The stepped driving pulley used in belting on a machine tool for the governing of different speeds, sometimes called the ' speed cone '.

(*c*) The conical race for the balls in certain types of ball-bearings. See also **Back Cone, Pitch Cone, Root Cone, Tip Cone.**

Cone Bearing. A conical journal running in a correspondingly tapered bush, thus acting as a combined journal and thrust-bearing and used for some lathe spindles. See also **Taper Roller-bearing.**

Cone Clutch. See **Clutch.**

Cone Distance. The distance from the apex of a bevel gear to the *pitch circle*; it is measured along the surface of the *pitch cone.*

Cone Drive. *Cone gear.*

Cone Drums. The driving drums in cotton spinning which vary the speed of delivery rollers or the speed of bobbin spindles.

Cone Gear (Cone Drive). A belt drive between two similar coned pulleys which by lateral movement provide a variable speed ratio.

Cone Plate (Boring Collar). A small bearing bolted to a lathe bed and carrying a circular plate or disc perforated with a series of tapered holes of different diameters. The boring tool is held in the slide rest and fed forward into the spindle shaft, etc., when boring.

Cone Pulley. A belt pulley stepped with two or more diameters to give different speed ratios with a similar pulley.

Conical Pivot. (*a*) A *pivot* shaped as a cone which runs in a screw with a tapered hole, the taper angle being greater than the cone angle. The device is used for the *balance staff* in alarm-clocks and in watches with pin-pallet *escapement*.

(*b*) An English watch type of cylindrical pivot with a conical shoulder.

Conical Spring. A special type of *helical spring*, being based on a cone and not on a cylinder.

Conical Valve. A *lift valve* with cone-shaped sides, no wings and fitting on an annular seating.

Coning Angle. The upward angle that the blades of a helicopter rotor adopt in flight.

Connecting-rod. The rod connecting the piston or cross-head to the crank in a reciprocating engine or pump.

The ' forked connecting-rod ' has the cross-head enclosed in the forked end. (See Fig. 107.) See also **Articulated Connecting-rod, Bow Connecting-rod, Box End Connecting-rod, Marine Pattern Connecting-rod, Master Connecting-rod, Strap End.**

Connecting-rod Bolts (Big-end Bolts). Bolts securing the outer half of a split big-end bearing of a *connecting-rod* to the rod itself.

Constant Mesh Gear-box. A gear-box in which the pairs of wheels providing the various speed ratios are always in mesh. The ratio is determined by the particular wheel which is coupled to the mainshaft by sliding dogs working on splines.

Constant Speed Propeller. A propeller the pitch of which is controlled to vary automatically so as to maintain a constant rotational speed, which is desirable for the engine.

Constant Travel. The travel of a *slide valve* which cannot be varied for purposes of variable cut-off. Cf. **Varying Travel.**

Contact Breaker. A device for breaking repeatedly and remaking an electrical circuit.

Contact Ratio. The ratio of the angle of rotation of a gear between the beginning and ending of contact of a tooth to the angle given by the fraction $360°/$(the number of the teeth in the gear).

Continuous Brake. A brake system in which operation at one point applies the brakes throughout a passenger train.

Continuous Mill. A *rolling mill* in which the stock passes through a series of pairs of rolls, undergoing successive reductions.

Contour Milling Machine. Fig. 113 illustrates the working of a contour *milling machine* by an operator who controls the movement of the milling cutter with the aid of an optical projection from a drawing on to a ground-glass screen.

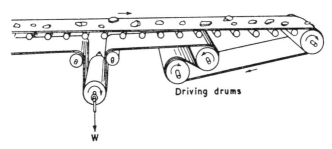

FIG. 40. CONVEYOR BELT MECHANISM.

Contra-rotating. A term used when two shafts rotate in opposite directions. See **Contra-rotating Propellers** and **Ljungström Turbine.**

Contra-rotating Propellers. Two propellers mounted on concentric shafts with a common drive and rotating in opposite directions. Cf. **Coaxial Propellers.**

Contrate Gear (Face Gear, Straight Bevel Rack). A gear with the tooth crests in one transverse plane and the roots in another transverse plane. See **Contrate Wheel.**

Contrate Wheel. (*a*) A toothed wheel with the teeth formed at right angles to the plane of the wheel for transmitting motion between two arbors at right angles. See **Contrate Gear, Crown Wheel.**

(*b*) The fourth wheel in a watch with the *verge escapement*. See **Escapement.**

Control Advance. See **Pitch (Helicopter).**

Control Jets. Jets of gas from pressure cylinders or some power source to control an attitude of a spacecraft in orbit.

Controllable-pitch Propeller. A variable-pitch propeller in which the blades can be set to predetermined pitch angles while rotating.

Convergent-divergent Nozzle. See **Propelling Nozzle.**

Conveyor. A device for moving parts or materials from one place to another, usually over a short distance and/or from one level to another, often on a continuous band called a ' conveyor belt ' (Fig. 40). See **Apron Conveyor, Band Conveyor, Bucket Conveyor, Chain Conveyor, Drag Conveyor, Gravity Conveyor, Pneumatic Conveyor, Pneumatic Tube Conveyor, Roller Conveyor, Vibrating Conveyor.**

Conveyor Screws. Different types of screws which operate conveyors are shown in Fig. 41.

Coolant. (*a*) A liquid (or gas) used as the cooling medium for

Coolant

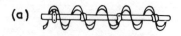

(a)

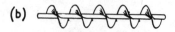

(b)

Fig. 41. Conveyor Screws.

(c)

(d)

an engine; for example, in the jackets of liquid-cooled piston-engines or as a film on the inner wall surfaces of combustion chambers and the nozzles of rocket-engines, or a molten metal, such as sodium in hollow exhaust valves and certain atomic reactors.

(*b*) Fluid used during machining to cool and lubricate the cutting or other tool and the work.

Cooling System. The system by which an engine or mechanism is cooled by air or by a *coolant*.

'Ducted cooling' is obtained by constraining air to flow in ducts.

'Evaporative cooling' uses the latent heat of evaporation by allowing the coolant to boil, condensing the vapour and returning the coolant to the cylinder jackets. See also **Film Cooling** and **Sweat Cooling.**

Cop. A conical ball of thread or yarn wound upon a spindle, varying in size with the type and the count of yarn produced by the *mule*.

Copping Rail (Shaper Rail). A specially shaped rail for controlling the movement of the yarn guide while yarn is being wound on the *cop*.

Copying Carriage. That part of a *copying machine* which travels along the model that is being copied.

Copying Machine. A class of machine for producing similar objects from a master pattern or template by using an engraving tool, end cutter or lathe tool, guided automatically by the pattern or template.

Corliss Valve. A steam-engine valve with an oscillating rotary motion over a port for admission of steam and its exhaust, the motion being controlled by an eccentric-driven wrist plate.

Cornish Engine. A massive type of *beam-engine* for pumping,

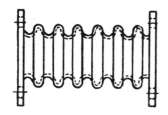

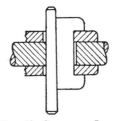

Fig. 42. Corrugated
Expansion Joint (Bellows Type).

Fig. 43. Cottered Joint.

originally single acting, but later double acting and worked expansively. See also **Rising Rod.**

Cornish Valve. *Double-beat valve.*

Corrugated Expansion Joint. An expansion joint of the corrugated type is shown in section in Fig. 42. Corrugated diaphragms are used in pressure transducers. (See Fig. 188.)

Cotter. A tapered wedge, rod or pin passing through a slotted hole in one member and bearing against the end of a second encircling member whose axial position is to be fixed or adjustable. (See Fig. 43.) Cotters for valve springs are illustrated in Fig. 202.

Cotter-pin (Split-pin). A split-pin inserted in a hole, as in a cotter, to prevent loosening under vibration.

Cotter Way. The slot cut in a rod to receive a *cotter* as illustrated by the outer shaft of Fig. 43.

Council of Engineering Institutions. See **Engineering Institutions, Council.**

Counter Balancing. *Balancing (static).*

Counterboring. Boring the end of a hole to a larger diameter.

Counterpart Racks. Racks, the teeth of which, when engaged, each exactly fill the spaces between the teeth of the other.

Counterpoise Bridge. A bridge in which the raising of a platform, roadway, etc., is assisted by *counterpoise weights.* See **Balance Weights.**

Counterpoise Weight. A weight used to give static balance (see **Balancing (Static)**), usually to ease the effort of moving the whole or part of a structure. See **Counterpoise Bridge.**

Counterpoising. *Balancing (static).*

Counter (Revolution Counter). An instrument for recording the number of operations performed by a machine or the number of rotations of a shaft.

Counter-rotating Propellers. A pair of propellers in a ship which rotate in opposite directions but not on the same shaft. Cf. **Coaxial Propellers.**

Countershaft. An intermediate shaft in a line of shafting placed between a driving and a driven shaft to obtain a greater speed-ratio or where direct connection is difficult.

Countersinking. (*a*) The recessing of the orifice of a hole by a conical enlargement.

(*b*) The driving of the head of a screw or nail below the surface so that it can be covered by a plug.

Countersunk Head. A head comprised of a truncated cone whose apex joins the main body of the screw, bolt, etc. The base of the cone is usually slotted to take a screwdriver blade whilst the conical surface is in contact with a cone cut in the mating material. Thus a head inserted in a hole lies flush with the material surface.

Counterweight. *Counterpoise weight.*

Couple. A pair of forces, equal and parallel, but acting in opposite directions, which tend to rotate the body on which they act. The moment of a couple is equal to the magnitude of one of the forces multiplied by the perpendicular distance between their lines of action.

Coupled Wheels. The wheels of a locomotive which are connected by coupling rods to distribute the driving effort.

Coupling. (*a*) A device for connecting two vehicles.

(*b*) A device for connecting railway coaches or trucks, consisting of two links united by a right-handed and left-handed screw (see Fig. 107). See **Claw Coupling, Flange Coupling, Flexible Coupling, Muff Coupling.**

Coupling, Double-slider. See **Oldham Coupling.**

Coupling Rod. A connecting-rod joining two cranks so that they work together as one. (See Fig. 107.)

Cover (of a Gate Valve). *Bonnet.*

Cowburn Valve. *Dead-weight Safety-valve.*

Crab (Crab Winch). (*a*) A jib-less hoisting crane with a snatch block or running pulley pendant from the barrel.

(*b*) The travelling lifting-gear of a *gantry crane* (see **Crane**), mounted on a bogie and running on rails carried by the *gantry*.

(*c*) A *claw coupling.*

Crampon. A pair of grappling irons, working like a pair of scissors, used for gripping loads that have to be hoisted.

Crane. A machine for hoisting and lowering heavy weights using gear-wheels, chain barrel and chain.

Balance Crane.   A two-armed crane, one to take the load and the other counterpoise arrangements, which can be arranged as self-acting.

Cantilever Crane.   A crane in which the jib hangs out from the supporting member and is counterbalanced, such as is used for the transport of excavated materials from the bottom of a cutting to a spoil-bank, often by skips.

Floating Crane.   A large crane carried on a pontoon as used in docks, etc.

Gantry Crane.   A *travelling crane* equipped with legs which support it on rails at ground level.

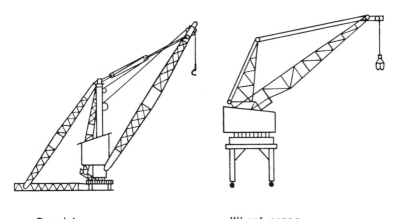

Derrick   Wharf crane

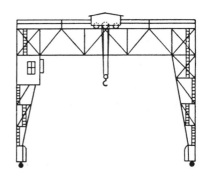

FIG. 44. CRANES—DERRICK,
WHARF CRANE,
GOLIATH CRANE.

Goliath crane

87

Goliath Crane (Fig. 44). A giant *travelling crane.*

Hercules Crane. A steam *travelling crane* with a horizontal swivelling jib used in harbour works for the setting of concrete blocks.

Horizontal Crane. A portable steam *balance crane* with horizontal cylinders.

Hydraulic Crane. A crane operated by hydraulic power.

Jib Crane. A crane with an inclined arm (a jib) attached to the foot of a rotatable vertical post, supported by a tie rod connecting the two upper ends, the chain running from a winch on the post and over a pulley at the end of the arm.

Derricking Jib Crane. A crane with variable radius of action obtained by changing the lengths of the tie ropes between post and jib (Fig. 44).

Travelling Jib Crane. A jib crane with a travelling trolley on a horizontal jib which is mounted on the vertical post.

Level-luffing Crane. A jib crane in which during *derricking* or luffing the load can be moved radially in a horizontal path with consequent power saving.

Overhead Travelling Crane. A workshop crane consisting of a girder mounted on wheels running on rails fixed along the length of the shop near the roof, and a traversing trolley or wheeled cab on the girder. Travelling, traversing and lifting are usually done by power or may be done by hand. This crane is also called a ' shop traveller '.

Platform Crane. A *whip crane* independent of any support at the top of the post.

Portal Jib Crane. A jib crane mounted on a fixed or movable structure with an opening directly under the crane to permit the passage of wagons, etc.

Post Crane. A jib crane with the post supported on fixed pivots at the top and bottom. The radius of the circle in which the hook travels is fixed when a tie-rod connects the post and boom; when ropes are used the radius can be varied.

Rope Crane. A *travelling crane* driven by an endless rope which travels at a very high speed so that minimum of power is needed to lift a heavy load. Any slack in the rope is taken up by tightening pulleys with bearings in sliding frames counterbalanced by suspended balance weights.

Steam Crane. A *crane* operated by a steam-engine.

Titan Crane. A very large steam crane similar to a *Hercules crane* but not provided with motions for slewing.

Tower Crane. A rotatable cantilever crane pivoted at the top of a steelwork tower, either fixed or carried on rails. The lifting machinery and dead weights on the opposite side of the pivot balance the load.

Well Crane. A fixed post crane with one-half of the post in a well with the lower end on a step and the fulcrum at ground level.

Vertical Crane. A steam crane with tall side frames. Cf. **Horizontal Crane.**

Wharf Crane. A travelling or fixed crane on a quay which has a fixed radius of action (Fig. 44). Cf. **Derricking Jib Crane.**

Whip Crane (Dutch Wheel Crane). A light *derrick* with tackle for hoisting but no gearing.

Crane Hook. The hook at the end of the lifting cable of a crane to which the load or sling chain is attached. See also **Ram's Horn.**

Crane Post. The vertical pillar of a jib crane to the top of which the jib is connected by a tie-rod.

Crank. An arm attached to a shaft carrying at its outer end a pin parallel to the shaft. See **Crankshaft.**

Crank Circle (Crank Path). The circle described by the *crank pin.*

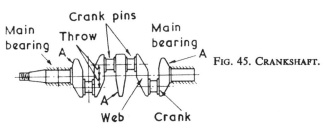

FIG. 45. CRANKSHAFT.

Crank Pin. The pin which unites the web or arm of a crank with the connecting-rod of an engine or pump. (See Fig. 45.)

Crank Throw. (*a*) The radial distance from the mainshaft to the *crank pin* and equal to half the stroke.

(*b*) The web(s) and pin of a *crank.*

Crank Web. The arm of a crank, usually of flat rectangular section.

Crankcase. The housing which encloses the crankshaft and connecting-rods.

Cranking. Hollowing of a tool immediately behind the cutting edge.

Crankshaft. The main shaft of an engine, or other machine, carrying a *crank* or cranks for the attachment of *connecting-rods* by the *crank pin(s)*. (See Fig. 45.)

Crescent (Passing Hollow). A circular notch in the periphery of the safety *roller* to allow the passing of the *safety finger*.

Crest. The surface of a screw thread connecting adjacent flanks at the top of the ridge.

Major (Minor) Crest Truncation. The distance between the generators of the major (minor) cylinders or cones for the basic and design forms of the external (internal) screw thread. The minor crest truncation is additional to the *basic truncation*.

Crimping Machine. A machine for compressing a thin metal ring or cap into corrugations to reduce its diameter such as crimped caps used on bottle tops.

Critical Speed. That rotational speed of a shaft which coincides with the fundamental (or some higher mode) natural frequency of the shaft and any attached masses.

Cropping Machine. *Shearing machine.*

Cross-axle. A driving axle having cranks mutually at right angles.

Cross-cut Saw. A saw designed for cutting timber across the grain with teeth shaped like an equilateral triangle. (See Fig. 46.) Cf. **Rip-saw.**

Crossed Belt. A driving belt which passes from the upper side

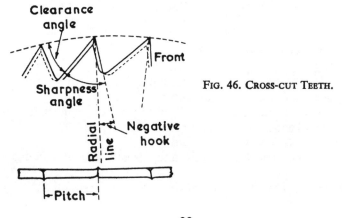

Fig. 46. Cross-cut Teeth.

of one pulley to the lower side of another pulley, the pulleys revolving in opposite directions.

Crossed Helical Gear. See **Helical Gear.**

Crossed Rods. Eccentric rods of *reversing engines* which cross each other to join the ends of the slot-link (see **Link Motion**), the centres of the sheaves being between the axle and the link. When they do not cross, they are called ' open rods '.

Crossed Roller Chains. See **Chains (Crossed Roller)** and Fig. 28.

Crossed-arm Governor (Parabolic Governor). A *governor* in which the balls and the points of suspension of the rods are on opposite sides of the central axis. The path of the balls is arranged to be approximately that of a parabola.

Crosshead. (*a*) A reciprocating block, forming the junction piece between the *piston rod* and *connecting-rod* of an engine, and usually sliding between guides.

(*b*) The upper or transverse beam of a hydraulic press transmitting the pressure to the weight to be lifted.

(*c*) A plate, attached to the plunger top of a hydraulic accumulator, to which the weight case is fastened.

Cross-over Callipers. See **Callipers, Poising.**

Cross-sills. *Sleepers.*

Cross-slide. The slide or bridge which carries the tool on a planing machine or lathe which can be traversed at right angles to the bed of the machine and also raised or lowered.

Cross-ties. *Sleepers.*

Crown Gear. A *bevel gear* with a pitch angle of 90°. See also **Crown Wheel.**

Crown Wheel. (*a*) A *bevel wheel* with its teeth at right angles to its axis. See also **Bevel Gear, Contrate Wheel.**

(*b*) The larger wheel of a bevel reduction gear.

(*c*) A wheel with gear-teeth in the ordinary *spur gear* position and ratchet teeth at right angles to the body of the wheel, as used in Swiss *keyless mechanisms.*

Crown-wheel Escapement. See **Escapement.**

Crowned Pulley. A pulley whose circumferential surface is convex to prevent lateral movement of the belt which drives it.

Crowning. The progressive reduction of tooth thickness towards the ends of a gear-tooth.

Crusher. A machine for the mechanical subdivision of solids; including a ' jaw crusher ' which consists of a set of vertical jaws, one fixed and the other moved back and forth; a ' gyratory crusher ' consisting of inner and outer vertical crushing cones, the inner with vertex upward and the outer with vertex downward; ' crushing rolls ' consisting of two horizontal cylinders close together and rotating in opposite directions; *hammer mills* and *ball mills.* Cf. **Grinder.**

Crutch (Clock). A brass arm on the *pallet arbor* with an upright steel post, at right angles to the plate or horizontally standing out. The post engages in the slot cut in the pendulum rod and passes on the impulse from the *escapement* to make up the losses due to friction in the suspension and from air resistance and thereby to keep the pendulum swinging.

Crystal Puller. A mechanism for growing single crystals of metals. It enables the crystal to be rotated at high speed (up to 700 rad/s) whilst it is slowly extracted from the liquid at a rate as low as 35 μm/s.

Cumulative Pitch. The distance between corresponding points on any two thread forms of a screw thread measured parallel to the axis of the thread, whether in the same axial plane or not.

Cup Chuck (Bell Chuck). A hollow cylindrical chuck screwed to the nose of a *mandrel* in which small articles are held by screws in the walls of the chuck.

Cup Drum. A *sheave-wheel* whose rim is recessed for individual chain links.

Cup Wheel. A grinder in the form of a cylinder, the grinding being done on the revolving edge.

Cupping (Tool). A punch used to form a hollow pressing from a blank.

Curb (Curb Ring). An internal ring of teeth on which a pinion, revolving in a fixed bearing on the upper portion of a crane, travels via suitable gearing and thus slews the crane.

Curb Pins (Index Pins). The pins fixed to the *index* to control the balance spring, making it vibrate faster or slower to regulate the watch. Compensating curb pins have one pin as a fixture and the other fitted with a bimetallic arm which at high temperatures moves closer to the fixed pin to make the watch gain to compensate for losses due to the expansion of the balance and the spring. In some Swiss watches a rod, called a ' turnboot ', covers the pins and may replace one pin.

Curling Tool. A tool designed and made to curl the edge of a metal article; it curls, coils, rolls, laps or bends the edge and the lap may be formed either internally or externally.

Cut-off. The point at which a valve closes the port opening of an engine cylinder to the admission of steam, generally expressed as a percentage or fraction of the stroke, but sometimes as so many inches. See also **Fixed Expansion.**

Cut-off Valve. (*a*) A separate *slide valve* fitted to control the admission of steam in a steam-engine when it is necessary or desirable to cut off the supply at a period earlier than half stroke.

(*b*) A valve actuated by the governor of certain gas engines to cut off the supply of mixture before the completion of the charging stroke, thus regulating the quantity of mixture.

Cutter. A single-point tool used in a lathe, planer or shaper, or a multi-toothed tool for milling and broaching, including drills and reamers. The multi-toothed cutter includes (*a*) the ' profile type ' which is limited to straight-line cutting edges of pointed teeth, and (*b*) the ' form-type ' in which the face of each curved tooth is sharpened. See **Angles of Cutting Tools.**

Rotary Cutter. A cutter rapidly rotated on its spindle. See also **Router Cutter.**

Cutter Bar. *Tool-holder.*

Cutter Head. *Boring head.*

Cutting Angles. *Angles of cutting tools.*

Cutting Fluid. Any fluid used for lubricating a cutting tool and washing away chips and swarf.

Cutting Tools. Steel tools for the machining of metals. The action of two cutting tools is shown in Fig. 47. See **Angles of Cutting Tools, Balance Turning Tool, Boring Bars, Broach, Cutter, Lathe Tool, Milling Cutters, Planer Tools, Reamer, Screwing Die, Slotting Tools, Tap,** Figs. 3, 5, 14.

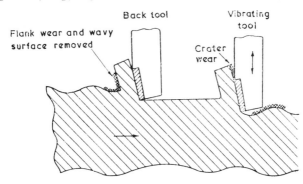

FIG. 47. DIAGRAMMATIC ACTION OF TWO CUTTING TOOLS.

93

Cycle. (*a*) The sequence of values of a periodic quantity throughout a complete period.

(*b*) The sequence of operations in an *internal-combustion engine* namely, induction, compression, ignition and exhaust. See **Carnot Cycle, Diesel Cycle, Four-stroke Cycle, Otto Cycle, Steam-engine Cycle, Two-stroke Cycle.**

Cyclic Pitch Control. The control of a helicopter rotor by which the blade angle is varied *sinusoidally* with the blade azimuth position; also called 'azimuth control'.

Cycling. See **Hunting.**

Cycloid. The curve traced on a plane by the motion of a point fixed on the circumference of a circle which rolls along a straight line. See also **Epicycloid** and **Hypocycloid.**

Cycloidal Curves. The shapes of gear-teeth with profiles which are *cycloids, epicycloids* or *hypocycloids* or combinations of the last two.

Cycloidal Gears. Gears with *cycloidal teeth.* When in mesh the faces of the teeth in one are epicycloids and the flanks of the teeth of the other are hypocycloids, both curves being generated by the same rolling circle. See also **Cycloid.**

Cycloidal Propeller. See **Propeller.**

Cycloidal Teeth. The teeth of gear-wheels whose flank profiles are cycloidal curves. See **Cycloid, Epicycloid** and **Hypocycloid.**

Cylinder. (*a*) A solid generated by a straight line moving round a closed curve and remaining parallel to a given direction.

(*b*) The cylindrical (tubular) chamber in which the piston of an engine or pump reciprocates. See also **Bore** and **Stroke.**

Cylinder Barrel. The wall of an engine cylinder.

Cylinder Bit (Half-round Bit, D-bit). A boring tool with the section at the cutting face a semicircle. The cutting face is sloped at an angle of about 4°.

Cylinder Bore. The internal diameter of the cylinder of a piston-engine. See also **Bore.**

Cylinder Bore Gauge. A gauge, usually with a centralizing shoe to give accurate location on each side of the gauging point, and by this means, to give a true bore measurement, using a transmission system to record on a circular dial.

Cylinder Cover. The end cover of the cylinder of a compressor or of a reciprocating engine.

Cylinder Escape Valve. A spring-controlled valve fitted to the cylinders of marine engines.

Cylinder Escapement. See **Escapement.**

Cylinder of Generation (of a Gear). The *pitch cylinder* when meshed with its generating cutter.

Cylinder, Reference. In helical and spur gears, the right circular cylinder on which the normal *pressure angle* has a specified standard value.

Cylindrical Grinding. Using a high-speed abrasive wheel to finish accurately cylindrical work by rotating the work in the *headstock* and automatically traversing the wheel along it under a copious flow of *coolant*. See also **Centreless Grinding.**

Cylindrical Slide Valve. See **Valve, Cylindrical Slide.**

D

δ. The symbol for deflection.

db. *Decibel.*

D.B. pull. Draw bar pull.

D.H.N. *Dynamic Hardness Number.*

D.P.N. Diamond Pyramid Hardness Number. See **Vickers Hardness Test.**

D Slide Valve. A *slide valve* of ' D ' section which slides on a flat face in which ports are cut.

D.T.I. Dial Test Indicator. See **Dial Gauge.**

Dalton's Law. The pressure exerted on the interior walls of a vessel containing a mixture of gases is equal to the sum of the pressures which would be exerted if each of the gases occupied the vessel alone.

Damper. A device for dissipating energy in a mechanical system by the suppression of vibrations of unfavourable non-linear characteristics.

Blade Damper. A damper to prevent the *hunting* of a helicopter rotor.

Roll and Yaw Dampers. Dampers for the suppression of rolling and azimuth oscillations of an aeroplane respectively.

Shimmy Damper. A damper for the suppression of *shimmy.*

Friction Damper. A mass frictionally driven from a crankshaft at a point remote from a node to dissipate the energy of vibration in heat.

Vibration Dampers. Dampers fitted to an engine crankshaft to suppress or reduce stresses resulting from torsional vibration at critical running speeds.

Torsional Vibration Damper. A flywheel mounted on a shaft

with the relative motion damped by viscous friction.

Tuned Torsional Vibration Damper. A flywheel coupled to a shaft by a spring to form a resonant system effective at frequencies near its natural frequency. See also **Detuner.**

Damping. The process by which the energy of a vibrating system is dissipated.

Damping Coefficient. The constant coefficient of the velocity term, $\dot{x}$, in a motion defined by the differential equation $m\ddot{x} + c\dot{x} + kx = 0$, where m is the mass and k is the stiffness of the system.

Damping Factor. Decay factor.

Damping Ratio. The ratio of the *damping coefficient* to the *critical damping coefficient.*

Critical Damping Coefficient. The smallest value of the damping coefficient required to prevent vibration.

Coulomb Damping. Damping in which the force opposing a motion has a constant magnitude.

Internal Damping. Damping intrinsic to the materials.

Magnetic Damping. Damping due to eddy currents set up by the movement of a system in a magnetic field.

Structural Damping. Damping due to the total effect of a built-up structure.

Viscous Damping. Damping in which the opposing force is proportional to the velocity.

Dart. *Safety finger.*

Dashpot. (*a*) A damping device consisting of a piston and cylinder whose relative motion is opposed by the fluid friction of a liquid or of air. It provides forces proportional to the rate of movement of the piston when a spring is added to the device. A one-way valve may be incorporated to give a differential damping action. (See **Air Dashpot** and Fig. 1.)

(*b*) A cylinder employed in steam-engines fitted with trip gears for closing the admission valves suddenly as soon as they are released by the trip.

Datum. (*a*) A point from which all measurements are made.

(*b*) A line from which all measurements are made.

(*c*) A horizontal plane from which all vertical measurements are made.

See **Datum Line, Datum Plane, Datum Point.**

Datum Dimension. See **Dimension.**

Datum Feature. See **Feature.**

Datum Level. A base line of a section from which all heights and depths are measured.

Datum Line. A defined line or base from which dimensions are taken or calculations are made. It establishes an exact geometrical reference.

Datum Plane. (*a*) A plane occupying a defined position from which dimensions are taken or calculations are made. It establishes an exact geometrical reference.

(*b*) That plane of a *rack* in which the ratio of tooth thickness to pitch has a specified value, normally 0·5.

Datum Point. (*a*) A point occupying a defined position from which dimensions are taken or calculations are made. It establishes an exact geometrical reference.

(*b*) The fixed starting point of a scale.

D-bit. *Cylinder bit.*

Davits. Curved or F-shaped uprights fitted with tackle for raising, lowering or suspending a boat.

De Laval Turbine. An early single-wheel *impulse turbine.*

Dead Angle. The angle of movement of the crank of a steam-engine during which the engine will not start when the *stop valve* is opened due to the ports being closed by the *slide valve.*

Dead Axle. An axle which does not rotate with the wheels carried by it. Cf. **Driving Axle.**

Dead-beat Escapement. An *escapement* with no recoil of the *escape wheel*, used for *regulators.*

Dead-centre Lathe. (*a*) A lathe in which the work alone revolves between dead-centres; the *mandrel* does not revolve.

(*b*) A small instrument-maker's lathe in which both centres are fixed, the work alone being revolved, e.g. by a small pulley mounted thereon.

Dead-centres (Dead Points). The least and greatest extension of a piston or a crank, where it exerts no effective power, the piston rod being coaxial with the cylinder. See **Inner Dead-centre, Outer Dead-centre.**

Dead End (Poppet). The tail stock of a lathe containing a *dead spindle* and a *back centre*. Cf. **Live Head.**

Dead Eye. A light type of bearing answering the same purpose as a *plummer block*. It may consist only of a hole in a sheet of metal or other material.

Dead Head. The *poppet* of a lathe or the fixed *headstock* of a machine tool.

Dead Hole (Blind Hole). A hole bored in metal for a certain distance but not right through the material.

Dead Points. *Dead-centres.*

Dead Spindle. (*a*) The arbor of a machine tool that does not revolve.

(*b*) The spindle of the *poppet* of a lathe.

Dead Time. The period (or time) before an instrument can respond to a second impulse. Also called ' Insensitive time '.

Dead-weight Pressure-gauge. A device for measuring fluid pressure in a *Bourdon gauge* by balancing the force on a vertical piston with weights.

Dead-weight Safety-valve (Cowburn Valve). A safety-valve loaded by a heavy metal weight, the valve being usually small and the pressure low.

Dead Wheel. The wheel in an *epicyclic gear* around whose centre the remainder of the gear (or train) revolves.

Decay Factor (Decay Coefficient, Damping Factor). The *logarithmic decrement* divided by the period.

Deceleration. The rate of decrease in the speed of a vehicle or of a moving part; the opposite of *acceleration.*

Decibel. The unit of difference of power (or noise intensity) level, measured logarithmically. If P_1 and P_2 are two amounts of power and N the number of bels, then $N = \log_{10}(P_1/P_2)$. A decibel is the one-tenth power of a bel.

Decimal Gauge. *Wire gauge.*

Decrement. The ratio of the amplitudes of two successive waves in a train of damped vibrations, the amplitudes decreasing with time. *Logarithmic decrement* is normally used.

Dedendum. The whole depth of a wheel tooth less the appropriate *addendum.* (See Figs. 84, 93, 172.)

Dedendum Angle. The difference between the *pitch angle* and the *root angle* of a bevel gear.

Dedendum (Screw Thread). The radial distance between the pitch and minor cylinders (or cones) of an external thread; the radial distance between the major and pitch cylinders (or cones) of an internal thread. See **Pitch Cylinders (Cones), Screw Thread Diameters,** and Fig. 156.

Degrees of Freedom. The minimum number of co-ordinates which are required to specify the possible motion of a mechanical (or other) system.

Delay Period. (*a*) The time or crank angle between the passage of the spark and the pressure rise in a piston engine.

(b) The time between fuel injection and pressure rise in an oil engine.

Delivery Valve. The pump valve discharging a liquid into a delivery pipe.

Demagnetizer. A device for removing residual magnetism from metal parts of a mechanism or from a piece of work held by a magnetic chuck or from some other cause.

Depth. (a) The amount by which the teeth of two mating wheels (or pinions) intersect.

(b) The depth of cutting teeth. (See Fig. 17.)

Depth Gauge. A gauge or tool for measuring the depths of holes by the use of a narrow rule or cylindrical rod working within, and at right angles to, a crossbar.

Derailer. An arrangement of rails for turning off, deliberately runaway trucks from the track. Cf. **Catch Points.**

Derrick. A type of *crane* in which the radius of the *jib* can be altered by means of ropes or chains passing over the top of the post or mast. Also called a ' Derricking jib crane '.

Derrick Barrel. The winding drum for the ropes used in a *derrick*.

Derrick Chains (Ropes). The chains (' Ropes ') by which the jib can be raised or lowered.

Derricking. The raising or lowering of the jib of a *derricking jib crane*. See also **Crane** and **Derrick.**

Desaxe Engine. A reciprocating engine mechanism in which the line of stroke does not pass through the axis of the crankshaft. The obliquity of the connecting-rod is diminished during the forward stroke, the pressure on the guide bars is reduced and the turning moment is slightly more uniform; the return stroke gives less advantage; on the whole there is a slight gain.

Design Form. The form of a surface, profile or screw thread which defines their design requirements, including the limits of *tolerance*. Cf. **Basic Form.**

The design forms of external and internal screw threads are shown in Fig. 156.

Detent. A catch or checking device, the removal of which allows machinery to work such as the detent which regulates the striking of a clock (Fig. 36, 108). In the *escapement* of a chronometer the detent carries a jewel for locking the *escape wheel*.

Detent Cams. Small roller cams fitted to the follower of a larger cam having a *detent notch*.

Detent Escapement. An *escapement* using a *detent* as in a chronometer.

Detent Notch. An axial groove in the peripheral face of a *cam*.

Detent Spring. The flat spring in a chronometer *escapement* attaching the detent blade to the detent foot; the detent blade carries the locking pallet.

Detonation. The spontaneous combustion of part of the charge in the cylinder of a petrol engine after the passage of the spark, accompanied by *knocking*. Cf. **Pre-ignition.**

Detruding. Thrusting or forcing down or pushing down forcibly, as when a hole is punched in a plate, leaving the plate strained. Cf. **Extrusion.**

Detuner (Dynamic Damper). An auxiliary vibrating or rotating mass driven through springs to modify the vibration characteristics of the main system to which it is attached, thus eliminating a critical vibration or critical speed. An auxiliary mass controlled by a non-linear spring is one type of detuner.

Deviations. The amounts, plus or minus, by which the limits of size are greater or less than the *basic size*.

Diagonal Winch. A steam *winch* the cylinders of which are placed diagonally on the side frames.

Diagram Factor. The ratio of the mean effective pressure developed in a steam-engine cylinder to its theoretical value; in practice varying between $0 \cdot 5$ and $0 \cdot 9$.

Dial. (*a*) The observable part of an indicating instrument with a moving pointer indicating a reading on a graduated scale.

(*b*) The graduated plate immediately behind the hands of a clock or watch, from which the time, date, etc., is read.

Dial Foot. Circular pins on the back of a *dial* for its correct location in corresponding holes in the pillar plate (see **Pillar**); the feet are sometimes pinned to this plate.

Dial Gauge (Dial Test Indicator). A sensitive measuring instrument, indicating small displacements of a plunger in thousandths of an inch by a pointer moving over a circular scale.

Dial Plate. *Bottom plate.*

Dial Test Indicator (see **Dial Gauge**). Also called colloquially ' clock '.

Diameters of Screw Threads. See **Screw-thread Diameters.**

Diametral Pitch (Manchester Principle). The number π divided by the spacing of adjacent teeth in inches.

Diametral Normal Pitch. The number π divided by the *normal pitch* in inches.

Transverse Diametral Pitch. The number π divided by the transverse pitch of a gear in inches; also the number of the teeth in the gear divided by the diameter of the *reference circle* in inches. This is sometimes called the ' diametral pitch '. See also **Transverse Pitch.**

Diamond. A crystalline form of carbon; with boron nitride the two hardest substances known (10 in *Mohs' Scale*). Diamond is used as a tip for tools and diamond dust as an abrasive.

Diamond-tip Turning Tool. A typical turning tool with a cutting diamond tip is shown in Fig. 48.

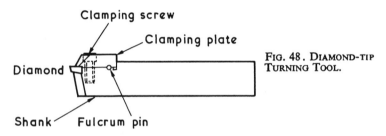

FIG. 48. DIAMOND-TIP TURNING TOOL.

Diamond Tool (Diamond Point). A name given to a tool when the surface of the cutting plane is formed like a lozenge or diamond.

Diamond Wheel. An abrasive wheel with diamond particles bonded in it.

Diaphragm Pump. A pump in which a flexible diaphragm has a reciprocating rod of short stroke attached at its centre.

Diaphragm Valve. A valve which relies upon the deflection of a flexible diaphragm by fluid pressure applied to shut off fluid flow.

Die. (*a*) An internally threaded steel block with cutting edges for producing screw threads. The larger dies are made in halves and set in a die stock. The block is called a ' screwing die ' or ' chaser ' (Fig. 49).

(*b*) A metal block used in stamping operations by pressing down on to a blank of sheet metal.

(*c*) A metal block with a correct internal cross-section through which hot metal is rammed when making *extrusions*. Cold metal and warm thermoplastics are also extruded. (Fig. 116.)

(*d*) When piercing is followed by blanking the die is called a ' progressive die '.

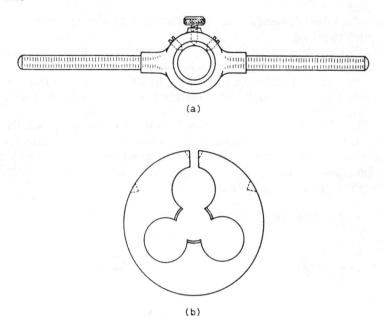

FIG. 49. (*a*) DIE HOLDER, (*b*) DIE.

(*e*) A male member of the correct shape for *stretch forming*. (Fig. 72.)

(*f*) Shapes for moulding dies are shown in Figs. 115, 116.

Die Chuck. A small two- or three-jaw *independent chuck*.

Die Nut. A square nut of hardened steel with grooves cut in it, used for cleaning the threads of *studs*.

Diehead. Dieheads (see Fig. 50) are commonly used on the turret of a capstan lathe where they automatically form an external thread on the workpiece up to a pre-set length when the diehead opens and enables withdrawal of the tool from the work. Cf. **Machine Tapper.**

Diesel Cycle. The cycle of a *compression-ignition engine* in which air is compressed, heat added at constant pressure by injecting fuel into the compressed charge, expanded to do work on the piston and the products exhausted, either in a *four-stroke* or a *two-stroke cycle*. This involves reversible heating and reversible cooling at constant pressure. Cf. **Otto Cycle.**

Diesel Engine. A *compression-ignition engine* in which the oil

102

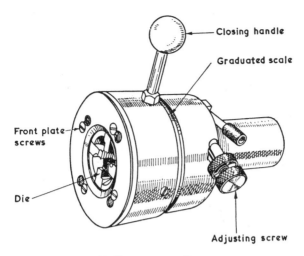

Closing handle

Graduated scale

Front plate screws

Die

Adjusting screw

FIG. 50. SELF-OPENING DIEHEAD.

fuel is injected into the heated compressed-air charge, originally by a blast of air. A two-stroke or four-stroke engine in which the fuel is ignited by the heat of compression.

In the ' hot-bulb type ', ignition is by means of a very hot bulb-shaped surface, and the operations of compression and fuel injection are performed separately.

Diesel-electric Locomotive. A locomotive in which a *diesel engine* drives an electric generator and the latter supplies current to electric motors which are connected to the driving axles.

Die-stock (Screw Stock). The frame holding a die for screw cutting which is held and turned by a pair of handles.

Difference Gauge. A cylindrical gauge fitted with two plug gauges. The difference between the two plug gauges corresponds to the easy movement or tight fittings of a mandrel in a cylinder.

Differential. Creating a difference. **See Differential Motion** and **Differential Gear.**

Differential Cam. An arrangement of cams with different outlines in the valve gearing of gas-engines, the cams sliding upon a shaft under the control of a governor to come into contact, in turn, with the roller and thus to vary the admission to suit the load.

Differential Gear (Differential). An assembly of *bevel* or *spur gear* wheels with two co-axial shafts and a third co-axial member

103

Differential

with a rotation proportional to the sum or difference of the amounts of rotation of the other two.

Differential Lever (Floating Lever). A rigid link carrying three pivots, the third of which has a displacement dependent on the input displacements of the other two.

Rack-type Differential. A linear displacement mechanism consisting of two racks engaging with opposite sides of a pinion wheel so that the linear movement of the wheel axis is half the algebraic sum of the two linear displacements of the racks.

Synchro-control Differential Transmitter. A synchro with mechanically positioned rotor for transmitting information corresponding to the sum or difference of the synchro and electrical angles. **See Synchro.**

Synchro-torque Differential Receiver. A synchro with freely turning rotor, which develops a torque dependent on the difference between the two electrical angles received from its connected torque transmitters.

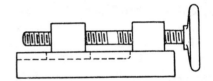

FIG. 52. DIFFERENTIAL SCREW MECHANISM

FIG. 51. DIFFERENTIAL PULLEY.

104

Synchro-torque Transmitter. A synchro with a mechanically positioned rotor for transmitting electrical information corresponding to the angular position of the rotor.

Differential Motion. The motion of one part to and from another in a mechanical movement, the speed (velocity) of the driven part being equal to the difference of the speeds of the two parts connected to it.

Differential Pulley Block (Chain Block). An endless chain runs over two *sheaves* and a movable load pulley below. The sheaves are fixed together on one axis and the mechanical gain is proportional to the difference in diameter of the two sheaves. Rotation of the sheave axis by a hanging loop shortens a second loop supporting the load pulley to give a large mechanical advantage. Cf. **Chinese Windlass** and **Differential Windlass.** (See Fig. 51.)

Differential Screw (Compound Screw). A screwed spindle working within a nut which is threaded internally for the reception of another screw of the same hand but of slightly finer pitch, the nut being attached to the diehead of the press. This device gives a high pressure through the prolonged action of a small power. (See Fig. 52.)

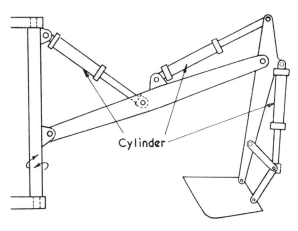

Cylinder

FIG. 53. MECHANICAL DIGGER.

Differential Windlass. A *windlass* in which the power exerted is due to the difference between the velocity of the rope upon two drums of unequal diameters. Cf. **Differential Pulley Block.**

105

Diffuser. A chamber in which kinetic energy is converted into pressure energy by a gradual increase in the cross-sectional area of the flow of a gas.

Digger. A mechanical excavator as shown in Fig. 53 with a digging shovel.

Dimension. An element whose size is specified in a design such as a length or an angle, and the word ' element ' may also refer to weights, capacities, areas, etc.

Auxiliary or Reference Dimension. A *dimension* given solely for information or convenience of reference.

Constructional Dimension. A *dimension* which defines a positional or angular relationship between two or more features, or the form of a surface including a profile.

Datum Dimension. A dimension which fixes the position of a *datum* plane, line or point.

Dimension Saws. Two circular saws mounted on a common spindle, one for cross-cutting and the other for ripping, and moved upwards or downwards by a *worm* and worm wheel.

Dip. The amount by which the upper edge of a *paddle-wheel* blade is immersed when the blade is vertical.

Dipper (or **Dipper-bucket**) **Dredger.** A dredger with a single large bucket at the end of a long arm which can be swung in a vertical plane by gearing.

Direct-acting Engine. A steam-engine in which the action of the piston is transmitted directly to the crankshaft.

Direct-acting Pump. A steam-driven reciprocating pump with the steam and water pistons carried on opposite ends of a common rod. See also **Scotch Crank.**

Direct-acting Slide Valve. A *slide valve* whose length of travel is equal to the throw of the eccentric.

Direct-injection Pump. A pump which meters the fuel and injects it direct under high pressure into piston-engine cylinders.

Disc (Disk) (in a Valve). The closing component on which the *disc face* is formed or to which the *disc facing ring* is secured. The disc may be integral with the stem of a *needle valve.*

Disc Face. A machined face making contact with the body seat when the valve is closed.

Disc Facing Ring. A ring of different material to the disc, permanently secured to the disc, on which the disc face is machined.

Disc Guide. That part of a valve in which the disc or disc holder is guided.

Disc Guide Pin. The pin which engages with the disc guide.

Disc Guide Wings. That part of the disc assembly in the form of wings which guides the disc to the body seat.

Disc Area. The area of the maximum circle described by the tips of the blades of a propeller or a helicopter rotor. The former is also called ' screw area '.

Disc Brake. A brake in which the friction is obtained from pads acting upon a disc on a vehicle's wheel or on the landing wheel of an aircraft's undercarriage or, similarly, for braking machinery. Cf. **Brake Shoes.** See **Hydraulically-operated Disc Brake.**

Disc Clutch. See **Clutch.**

Disc Crank (Crank Plate, Balanced Crank). A crank of circular outline on which the metal is sometimes so disposed as to balance the varying motion of the connecting-rod.

Disc Feed. A feed on some machine tools using two discs, the large disc driven at a constant speed and the other at right angles driven therefrom at speeds depending on the distances from the centre of the large disc. Cf. **Planimeter.**

Disc Loading. (*a*) The thrust of a propeller divided by the *disc area*.

(*b*) The lift of a helicopter-rotor divided by the *disc area*.

Disc Separator. A sorting machine for grain or seeds of various sizes which pass between revolving iron discs on a horizontal shaft. The faces of the discs have numerous small indents which remove the unwanted seed by discharging it at a higher level.

Disc Valve. A light steel or fabric disc resting on a ported flat seating used as a suction and delivery valve in pumps and compressors. Steel discs are usually spring-loaded.

Disc Wheel. A wheel with a solid disc of metal connecting the *hub* and rim.

Disc-and-drum Turbine (Combination Turbine, Impulse Reaction Turbine). A type of steam-turbine with a high-pressure impulse wheel, followed by intermediate and low-pressure reaction blading, mounted on a drum-shaped rotor. See also **Turbine.**

Discharge Valve. (*a*) A self-acting valve for controlling the rate of discharge of a fluid from a pipe or centrifugal pump.

(*b*) Any *delivery valve*.

Disengaging Clutch (Coupling). A clutch for throwing out of gear a line of shafting or a train of wheels.

Disengaging Gear. The mechanism for operating a *disengaging clutch*.

Disengaging Nut. The clasp nut of a lathe, being disengaged when not required for use or for screw cutting.

Dished Plate. A plate forged or pressed into a dish-like shape to increase its stiffness under pressure on the convex side. Cf. **Swage.**

Disintegrator (Disintegrating Mill). A mill consisting of fixed and rotating bars in close proximity for reducing lump material to granular product.

Disk. See **Disc.**

Displacement. (*a*) The mass of fluid displaced by a body.

(*b*) The volume of fluid displaced by a pump plunger per stroke or per unit time.

(*c*) The swept volume of a working cylinder of an engine.

Displacer Piston. An auxiliary piston in some gas-engines for expelling the residual products of combustion from the cylinder.

Distributor Rollers. The rollers which distribute ink on the inking table in a cylinder printing press or those carrying ink in a duplicating machine.

Dither Mechanism. A mechanism to remove *stiction* by providing an oscillatory motion of small amplitude between two relatively moving parts.

Divergence. A disturbance from a stable position, or uniform motion, which increases in magnitude without oscillation about that position or condition.

Divergent Nozzle. A nozzle with a cross-section increasing continuously from entry to exit, as used in compound impulse turbines. Cf. **Convergent-divergent Nozzle.**

Diverting Pulley. *Idle pulley.*

Divided Bearing. *Split bearing.*

Divided (Axial) Pitch. The axial distance between corresponding points (such as centres of successive helices) on successive threads of a multiple-threaded screw.

Dividing Engine. An instrument for marking or engraving accurate subdivisions on scales; it may be linear, circular or cylindrical. The linear machine consists of a carriage, adjusted by a micrometer screw and holding a marking tool, which moves on a very accurate lead screw. It is sometimes called a ' Dividing machine '.

Dividing Head (Indexing Head). An attachment used on a milling-machine table for accurately dividing the circumferences of components for grooving or fluting, gear-cutting, the cutting of splines, etc. Other machine tools have similar attachments.

Dividing Machine. *Dividing engine.*

Division Peg (Index Peg). The peg for fitting in the holes of a *division plate* to hold it steady.

Division Plate (Dividing Plate). A plate used for positioning the plunger of a *dividing head* which is provided with concentric rings of holes accurately dividing the circumference into various equal subdivisions.

Dobbie (Dobby). A shedding mechanism on a loom for the production of fabrics of complicated structure.

Doble Vane. A double cup with central cutwater which splits the jet impinging on the vanes of a *Pelton wheel* and turns the water through an angle of about 160° to clear the wheel at exit.

Doffer. A wire-covered cylinder of a *carding engine* which removes fibres from the wire-covered surface of the main cylinder.

Doffing. Removing full *cops* or bobbins from a textile machine.

Doffing Comb. A steel blade extending across and oscillating over, the *doffer* of a *carding engine* to strip the carded material in the form of a sheet or web.

Dog. (*a*) A jaw of a *chuck*.

(*b*) The *carrier* of a lathe.

(*c*) A *pawl* (*d*).

(*d*) An adjustable stop in a machine tool.

(*e*) A spike for securing rails to sleepers.

(*f*) In general, a gripping implement. See also **Clutch**.

Dog-chuck (Jaw Chuck). A lathe *chuck* usually with four independent jaws or dogs. Also called a ' jaw chuck '.

Dog Wheel. *Ratchet wheel.*

Dolly Bolt. A bolt passing through the laminated springs on a vehicle and continued into the axle to prevent the U-bolts from slipping.

Dome. A domed cylinder often attached to a locomotive boiler to act as a steam space and to house the regulator valve. (See Fig. 107.)

Donkey Boiler. A small vertical boiler for supplying steam to drive small engines and machinery, especially on board ship.

Donkey Engine. An auxiliary engine for pumping and light work, especially on board ship.

Donkey Pump. A small steam reciprocating pump used on board ship.

Dorr Mill. A *tube mill* designed for operation as a closed-circuit wet-grinding unit.

Double-acting Engine

Double-acting Engine. A reciprocating engine in which the working fluid acts alternately on each side of the piston. All steam-engines were originally single-acting, but all are now double-acting. Very few internal-combustion engines are double-acting.

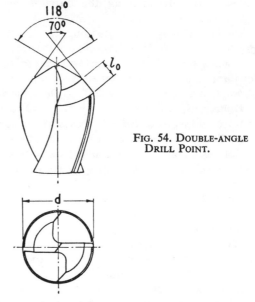

FIG. 54. DOUBLE-ANGLE DRILL POINT.

Double-acting Piston. A piston acted upon on both sides alternately by gaseous or fluid pressure. See **Double-acting Engine, Double-acting Pump.**

Double-acting Pump (Double-action Pump). A reciprocating pump in which both sides of the piston act alternately to give two delivery strokes per cycle.

Double-acting Steam Hammer. A *steam hammer* which admits steam above and below the piston.

Double-angle Drill Point. A drill point with the clearance face ground in the form of two cones (see Fig. 54). The double-cone decreases the side clearance angle along the outer cutting edge, l_0, and increases the value of the tip angle compared with conventional single-angle drill point. A longer tool life, quietness, smaller axial force for given diameter, d, and better machining accuracy, for a given power input is obtained.

110

Double-beat Valve (Cornish Valve). A hollow cylindrical *lift valve* with two seating faces at the two ends of only slightly different areas. When exposed to pressure the valve is nearly balanced and easily operated. It is used for controlling high-pressure fluids. Cf. **Parallel Slide Valve.**

Double Cards. *Indicator cards* taken from both ends of the cylinder of a steam-engine.

Double-cutting Drill. A drill which is ground to cut with equal facility in a right- or left-hand direction. The cutting edge lies in the longitudinal axis and the angles are symmetrical on the two sides.

Double-cylinder Engine. A steam-engine with two cylinders, one piston being at half and the other at full stroke when the cranks are at right angles to each other.

Double Disc (Gate) Valve. A *gate valve* in which the gate consists of two discs forced apart by a spreading mechanism at the point of closure against both parallel body seats to ensure effective sealing of the valve without the aid of the fluid pressure. Cf. **Wedge Gate Valve.** See also **Spreader.**

   Discs. The components, attached to the stem or the spreading mechanism, on which the disc faces are machined.

Double Driver Chuck. *Double driver plate.*

Double Driver Plate. A *driver plate* with two pins which engage and drive a carrier attached to the revolving work. It is used for work requiring special accuracy.

Double-ended Bolt. A bolt screwed at each end for the reception of a nut having a central portion with flats or a tommy-bar hole.

Double-ended Machine. A punching and shearing machine, the two sets of operations being carried on at the same time.

Double-entry Compressor. (*a*) A *centrifugal compressor* with two-vaned *impellers* mounted back-to-back so that air is admitted from both sides and ejected in the central plane.

   (*b*) A *centrifugal compressor* with air admitted to both ends of the rotating member.

Double-flow Turbine. A turbine in which the working fluid enters at the middle of the casing and flows axially towards each end.

Double-helical (Herring-bone) Gear. A gear-wheel having teeth of helical pattern in opposite directions like a ' herring-bone ', thus eliminating axial thrust.

Double-ported Slide Valve. A slide valve in which steam is admitted through two steam ports at each end of the cylinder face, thus reducing the travel of the valve.

111

Double Purchase. A lifting arrangement consisting of two pinions and two wheels. Cf. **Single Purchase.**

Double-roller Safety Action. A lever escapement with two rollers, one carrying the impulse pin and the other being used for the *guard finger*. See **Impulse.**

Double-row Radial Engine. A *radial engine* with two rows of cylinders arranged radially one behind the other and operating on two crank pins 180° apart.

Double Shaper. A shaping machine for a double set of operations with two rams and two tool boxes; e.g. one operation being for straight and the other for circular cutting.

Double Sleeve Valve. A *sleeve valve* arrangement as shown in Fig. 55, which illustrates an open inlet port.

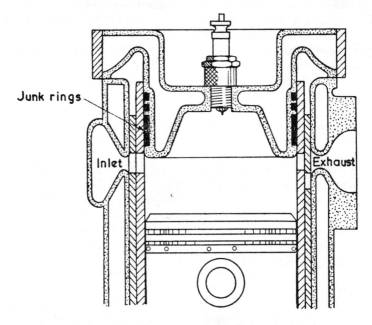

FIG. 55. DOUBLE SLEEVE VALVE.

Double Slider Coupling. *Oldham coupling.*

Double Strap Butt Joint. A butt joint with a double strap, one on each side, riveted, bolted or welded together.

Double-threaded Screw (Two-start Thread). A screw with two threads, half the true pitch apart. It gives an increase in the rate of travel.

Double Thrust Bearing. A *thrust bearing* for taking axial thrust in either direction.

Doubling Frame (Twisting Frame). A spinning machine of either the flyer cap or ring type in which yarns are folded or twisted together.

Dovetail Cutter. A specially shaped rotary cutting tool for shaping dovetail grooves.

Dovetail Key. See Key.

Dowel. A pin used for the accurate reassembly of two parts. Cf. Steady Pin (*a*).

Draft. (*a*) The order in which the warp threads are drawn through the eyes of the *healds* in weaving.

(*b*) The thinning-out (attenuation) of a textile material on passing out from the feeding end to the delivery end, as of a *carding engine*.

Draft Tube. A pipe discharging from a water-turbine to the *tail race*, which, by running full, decreases the pressure at outlet and increases the turbine efficiency.

Drag-bar. *Draw-bar*.

Drag Conveyor. A *conveyor* for feeding loose material along a trough by means of an endless chain with wide links carrying projections or wings.

Drag Link. (*a*) The suspension rods or links of a valve gear controlling the forward and backward gear of an engine.

(*b*) A link for connecting and disconnecting the cranks of coupled engines.

(*c*) A rod by which the link motion of a steam-engine is moved for varying the cut-off.

(*d*) A link from the *drop arm* of the steering-gear of a motor-vehicle which acts on the steering-arm through ball joints at its ends. See **Drop Arm**.

Drag Plate. The casting which forms the footplate or platform of a locomotive.

Drag Roller. A sprocket roller in a *cinematograph* or sound-on-film mechanism, which puts tension in the film by an internal friction arrangement.

Drag Rope. A *drum head* that is driven through gearing from a pulley on the spindle of a circular saw to pull logs up to the saw.

Drag Surface. The forward face of a screw propeller used for ship propulsion.

Drain Boss (of a Valve). See **Boss.**

Draught Bar. *Draw-bar.*

Draw. (*a*) The outward and inward run of a *mule* carriage.

(*b*) The action whereby one part of a watch or clock is drawn into another part, such as in a lever *escapement*.

Draw-bar (Drag-bar). (*a*) The bar connecting a locomotive to the rolling stock and by which the tractive force is transmitted.

(*b*) The bar connecting a *tractor* to a trailer in road vehicles; it is usually part of the trailer.

Draw-bar Cradle. A closed frame or link for coupling the ends of *draw-bars.*

Draw-bar Plate. A heavy transverse plate through which the *draw-bar* is attached to a locomotive frame.

Draw-bar Pull. (*a*) The tractive effort of a locomotive drawing a train in specified conditions.

(*b*) The pull exerted by a *tractor* on its trailer.

Draw-bar Spring. A shock-absorbing spring fitted between the *draw-bar* and the frame of a railway carriage, or the springs fitted within the draw-bar of a trailer to absorb sudden changes in *draw-bar pull.*

Draw Box. Two or more pairs of fluted rollers between the *doffer* and the *coiler* of a *carding engine.* The web of fibres coming from the doffer passes through the draw box to become *sliver* for delivery to the coiler.

Draw Hook (Drag Hook). The central hook of a railway wagon attached to the *draw-bar* and flanked by side hooks.

Draw Plate. The plate on which dies are supported in *drawing* (*b*) operations.

Drawing. (*a*) The representation by lines on paper of machines, etc., to be manufactured.

Assembly (Sub-assembly) Drawing. A drawing limited to an individual assembly.

Detail Drawing. A drawing giving manufacturing requirements for a specific detail or details.

General Arrangement Drawing. A drawing of a complete finished product showing (*a*) the components which make up the final assembly, and (*b*) the means of identification with other drawings.

Operation Drawing. A document giving the procedure to be adopted by the operator in carrying out the operations to make a specified part.

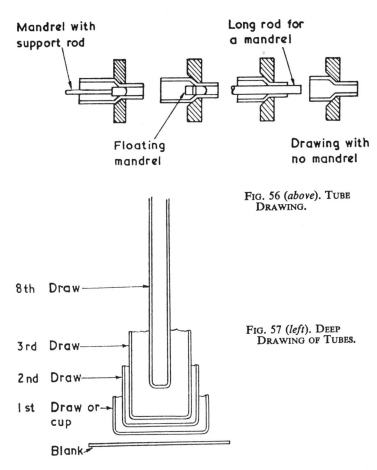

Mandrel with support rod

Long rod for a mandrel

Floating mandrel

Drawing with no mandrel

Fig. 56 (*above*). Tube Drawing.

8th Draw

3rd Draw

2nd Draw

1st Draw or cup

Blank

Fig. 57 (*left*). Deep Drawing of Tubes.

(*b*) The manufacture of wire, tubing, etc., by pulling or drawing the material through dies of progressively decreasing size. (See Figs. 56, 57.)

(*c*) The running together and constricting in diameter of a number of *slivers*, usually six, preparatory to spinning.

(*d*) The reduction of worsted *tops* to a roving suitable for spinning.

Drawing-down. The operation of reducing a dimension, such as the diameter of a bar, by forging.

Drawing-in. (*a*) The cutting of a tool deeper than the set dimensions.

(*b*) The placing of the warp threads in the eyes of the *healds* in weaving, in accordance with the *draft*.

Dredger. A chain of buckets, scoops or suction pumps for removing sand, alluvial deposits, etc., from under water and loading the material on to the barge or raft carrying the necessary machinery or on to another vessel. See also **Hopper Dredger, Suction Dredger.**

Dredger Excavator. An *excavator* similar to the *bucket dredger* but designed to work on land.

Dresser. A tool for facing and grooving millstones or for truing grinding wheels. (See Fig. 58.)

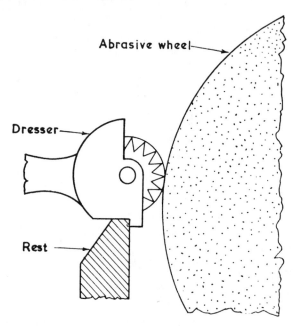

FIG. 58. DRESSER.

Drill. (*a*) A tool for boring cylindrical holes with cutting edges at one end and having flats or *flutes* for the release of chips. See also **Centre Drill, Double-angle Drill Point, Lip Drill, Twist Drill.** Cf. **Boring Tool.**

(*b*) A compressed-air-operated rock drill.

(*c*) A straight steel bar with a shank at one end and a cutting edge at the other. See **Twist Drill.**

(*d*) The machine used to drive a drilling tool. *Morse taper* drill nomenclature is given in Fig. 114.

Drill Bush. A bush of hardened steel used to guide drills and reamers, enabling accurate positioning of holes in the workpiece when the bushes are held in a jig or fixture. (See Fig. 96.)

Drill Chuck. A self-centring chuck, usually with three jaws, for holding small drills by contraction of the chuck with an internally coned sleeve encasing the jaws.

Drill Feed. A mechanism for feeding a drill into the work.

Drill Plate (Boring Flange). A circular plate fitted over the nose of a lathe *mandrel* to take the pressure of a workpiece which is being drilled.

Drill Socket. A tapered sleeve bored out to a standard size to receive the shank of a twist drill. See also **Morse Taper.**

Drill Spindle. The vertical revolving shaft in a drilling machine for holding the drill and through whose vertical movement the feed is operated.

Drill Template (Drilling Job). A metal plate attached to the work and providing a *drill bush* where each hole is to be drilled.

Drilling Machine. A machine tool for drilling holes, usually composed of a vertical standard with a table for supporting the work and an arm provided with bearings for the drilling spindle. Several drilling spindles are frequently operated together on the same machine.

Pillar Drill. A drilling machine with brackets on a pillar to carry the spindle and table, the latter sliding on the pillar. (Fig. 126.)

Radial Drill. A large machine with the drilling head movable along a rigid horizontal arm carried by a pillar.

Sensitive Drill. A small machine with the drill fed into the work by a hand lever directly attached to the drilling spindle, thereby maintaining a sensitive control of the rate of drilling.

Drilling Pillar. A vertical pillar with a sliding adjustable arm from which pressure is exerted on the revolving drill.

Drilling Spindle. The revolving spindle holding the drill in a *drilling machine.*

Drilling Table. The table on which work is supported and clamped in a *drilling machine.*

Drip-feed Lubricator. An arrangement for supplying lubricating oil in drops to a bearing from a small reservoir. See also **Sight-feed Lubricator.**

Driver Chuck. A point *chuck* in a *mandrel* for carrying work which is being turned between centres. Cf. **Driving Chuck.**

Driver Plate. A disc screwed to the mandrel nose of a lathe and carrying a pin which engages with and drives a *lathe carrier* attached to the work. See **Double Driver Plate.** Cf. **Driving Plate.**

Driving Axle (Live Axle). An axle rigidly attached to its wheels and through which the driving effort is transmitted to the wheels.

Driving Band. *Belt.*

Driving Blade. (*a*) A brass strip on the front end of a *catch bar*. (*b*) A brass strip on the rollers of locker bars of a lace machine.

Driving Chain. See **Chain.**

Driving Chuck. A *driving plate* of a lathe provided with slots by which *dogs* are attached for gripping the work. Cf. **Driver Chuck.**

Driving Drum. A power-driven drum which transmits its energy to other parts of a mechanism, such as a *conveyor* belt or a *bobbin*.

Driving Fit. A term applied to two mating pieces when the inner is slightly too large and has to be driven by a hammer or a press into the outer piece.

Driving Gear. Any system of gears, belts, pulleys, shafting, etc., through which power is transmitted to a machine.

Driving Plate. A flange screwed to a lathe spindle to carry a projecting pin for engaging and driving the carrier attached to the work. Cf. **Driver Plate.**

Driving Pulley. That pulley on a driving shaft which imparts motion to a countershaft of a machine or a second line of shafting.

Driving Side. That side of a belt which drives a pulley; tension side of the driving belt.

Driving Springs. The springs carrying the axle boxes of a locomotive's driving axles.

Driving Wheel. (*a*) The first member of a train of gears. (*b*) A wheel on a *driving axle* of a locomotive or road vehicle.

Drop Arm. An arm actuated by the steering-gear which converts rotary motion of the steering shaft to turning of the wheels. See also **Drag Link.**

Drop Forging (Pressure Forging). The process by which a steam or mechanical hammer shapes metal parts between two *dies*, one fixed to the hammer and the other to the anvil. The process is used for the mass production of parts.

Drop Hammer (Drop Stamp). A heavy hammer used for forging which slides between vertical guides to provide the force required to form a shape in metal against a *die*. The hammer is raised by hand or power and falls under gravity.

Drop Stamp. *Drop hammer.*

Drop Test. (*a*) An environmental test in which the retardation is varied to give the desired *g* force on the equipment under test in a drop test-rig. Retardation can be controlled pneumatically, hydraulically or by use of a suitable diameter of spike which enters a pad of lead.

(*b*) The dropping of a steel tyre on a steel rail from a specified height, varying with the tyre diameter, as a strength test.

Drop Valve. A conical-seated valve with rapid operation by a trip-gear and return spring. Cf. **Trip Gear.**

Drop Worm. A *worm* which can be dropped out of engagement with its wheel.

Dropping Valve. A valve reducing the supply pressure by a constant amount.

Drum. (*a*) The rotor of a *reaction turbine*. See also **Brake Drum, Disc-and-drum Turbine, Turbine, Winding Drum.**

(*b*) A cylinder on which lace threads are wound to a definite length, for transfer to brass bobbins.

(*c*) A textile term for revolving parts which transmit motion to other parts by surface contact.

(*d*) A *stepped pulley.*

Drum Head. The upper portion of a *capstan.*

Drum Movement. A clock movement housed in a cylindrical metal case.

Drum Pump. *Rotary pump.*

Drum Scanner. A rotating drum used in mechanical scanning systems in television and carrying the required picture-scanning elements.

Drum Winding. Using a *driving drum* to wind yarn on to a flanged *bobbin* in contact with the drum.

Drunken Saw. A circular saw tightened on its spindle at an angle to the axis, by means of bevelled collars. It cuts a wide groove, the width being dependent on the angle at which the saw is set. See also **Kerf.**

Dry-blast Cleaning. Blasting the surface of a workpiece with abrasive material travelling at a high velocity, including *sand-blasting.* See also **Vapour Blast Cleaning.**

Ducted Fan. A fan which functions inside a duct. In a turbojet engine, the blades of the turbine are lengthened so that they extend into an annular space outside the engine, the fan being driven by the low-pressure turbine, independent of the high-pressure turbine of the main engine. (See Fig. 59.)

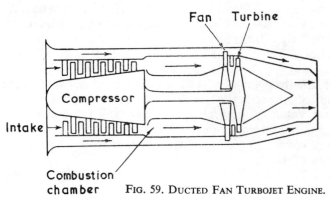

FIG. 59. DUCTED FAN TURBOJET ENGINE.

Ducted Propulsor. A propeller for propelling underwater vehicles.

Dukey. (*a*) A train of tubs travelling on an inclined haulage road underground.

(*b*) A platform on wheels for tube or trams to be lowered on steep self-acting inclines.

Dumb Pintle. The pin of a hinge giving radial but not axial location of its members.

Dummy Piston (Balance Piston). A disc placed on the shaft of a *reaction turbine* so that steam pressure can be applied on one side to balance the end thrust. See also **Balance Cylinder.**

Duplex Carburetter. A *carburetter* in which two mixing chambers are fed from a single float chamber.

Duplex Escapement. See **Escapement.**

Duplex Lathe. A lathe using two cutting tools simultaneously, one on each side of the work, either to increase the rate of working or to avoid springing of the work. Cf. **Multiple-tool Lathe.**

Duplex Planing Machine. A planing machine with two beds and two tables.

Duplex Pump. A pump with two working cylinders side by side.

Duplex Slide Rest. A *slide rest* fitted with two tool-posts for taking two cuts off the same piece of work.

Dutch Wheel Crane. *Whip crane.* See **Crane.**

D-valve. *D slide-valve.*

Dwell. The angular period of a *cam* during which the *cam follower* is allowed to remain at a given lift.

Dynamic Balance. The condition wherein centrifugal forces due to rotation produce neither couple nor resultant force in the shaft of the rotating part, such as a flywheel, rotor or propeller.

Dynamic Damper. *Detuner.*

Dynamic Hardness Number (Rebound Hardness). The number given by a *Herbert pendulum* or a *Shore scleroscope.*

Dynamic Load (Live Load). A rolling or a moving load. Cf. **Dead Load.**

Dynamic Model. (*a*) A model having the correct proportions for the simulation of full-scale dynamic conditions.

(*b*) A free-flight aircraft model, a wind-tunnel model or a model of a seaplane hull (or float) with dimensions, masses and moments of inertia correctly proportioned, so that the model will simulate correctly the full-scale behaviour.

Dynamic Similarity. A term used to define the conditions of comparative experiments in which there must be a similarity in mass distribution and a scale reproduction of the acting forces.

Dynamic Stiffness. The *stiffness* of a massless constraint which, if substituted for a part of a system at a given frequency, will give the same vibratory force or torque for the same vibratory amplitude as the original part. Cf. **Mechanical Impedance.**

Dynamics. See **Mechanics.**

Dynamo. A machine converting mechanical into electrical energy.

Dynamometer. (*a*) An instrument or machine for measuring the brake horse-power of any prime mover. (See Fig. 88.)

(*b*) An apparatus for measuring power or force, either mechanical or muscular. See also **Absorption Dynamometer, Hydraulic Dynamometer, Transmission Dynamometer.**

Dyne. The C.G.S. unit of force. A force of one dyne acting on a mass of one gram imparts to it an acceleration of one centimetre per second per second. (1 dyne $= 10^{-5}$ newtons.)

E

E. (*Young's modulus.* Also the symbol for energy (see *T* and *V*).

η. *a*) The symbol for coefficient of viscosity, $\eta = \mu/\rho$.

(*b*) The symbol for efficiency.

eff. Efficiency.

e.h.p. Effective horse-power.

eng. Engine.

Ear (Lug). A permanent projection on an object for its support or for the attachment of anotherpart to it by a *pivot*.

Early Cut-off. (*a*) Any cut-off shorter than one half the stroke in a steam-engine cylinder.

(*b*) The occasion when the motors of a rocket or a launcher vehicle switch themselves off before the programmed time.

Eccentric. (*a*) Not concentric.

(*b*) A mechanism for converting the rotary motion of a crankshaft into a reciprocating rectilinear motion, used chiefly for short *throws*. Cf. **Cam.**

(*c*) A crank, in which the pin diameter exceeds the stroke resulting in a disc eccentric to the shaft, such as is used for operating steam-engine valves, pump plungers, etc. Cf. **Crank.**

Eccentric Crank. A crank in a locomotive valve gear, which is substituted for the *eccentric*.

Eccentric Hoop. *Eccentric strap.*

Eccentric Key. A key recessed into the shaft which drives the *eccentric sheave*, the latter being slotted in its larger half for the reception of the key.

Eccentric Lug. The projecting portion of an *eccentric strap* to which the *eccentric rod* is attached.

Eccentric Rod. The rod attached to the *eccentric strap* which transmits the motion of the strap to the valve or pump.

Eccentric Sheave. An *eccentric* disc, commonly made in two halves so that it can be placed upon the crankshaft or axle and the portions joined by bolts and cotters, or formed integral with the shaft or directly keyed to the shaft.

Eccentric Strap. A narrow split bearing, or a metal hoop, which encircles the *eccentric sheave* or cam and transmits its motion to the valve rods and thus to the valve gearing.

Eccentric Throw-out. A device by which the back gear shaft of a lathe runs in eccentric-bored bearings to bring the gears in and out of mesh with those on the mandrel. Cf. **Back Gear.**

Eccentricity. (*a*) Half the ' radial run-out ' in a gear.

(*b*) The distance from the centre of a figure or revolving body to the axis about which it turns. In an *eccentric* this distance is half the *throw*.

(*c*) The deviation from a centre.

(*d*) A conic is the locus of a point which moves so that the ratios of its distances from a fixed point, called the focus, and a fixed line, called the directrix, is constant. This constant is called the eccentricity of the conic.

(*e*) The perpendicular distance from the centre of application of a load to the centroid of the section of the structural member supporting the load non-axially.

Edge-planing Machine. *Plate-edge planing machine.*

Edge Runner. A grinding mill in which the cylindrical stones or rollers are mounted to run on their edges in a circular pan to grind loam, mortar, putty, pigments, etc.

Effective Inertia. The moment of inertia of a mass which, if substituted for a part of a system at a given frequency, would have the same vibratory acceleration under the same vibratory torque at the point of separation as the original part. It is measured by the value of the torque. Cf. **Effective Mass.**

Effective Mass. That mass, which, if substituted for a part of a system at a given frequency, will have the same vibratory acceleration under the same vibratory force at the point of separation as the original part. It is measured by the value of the force. Cf. **Effective Inertia.**

Efficiency. (*a*) The performance of a machine as a percentage of its theoretical maximum performance.

(*b*) The ratio of the energy output to the energy input of a machine. See also **Efficiency Ratio.**

(*c*) The ratio of the *mechanical advantage* to the *velocity ratio*.

See also **Ideal Efficiency, Mechanical Efficiency, Rankine Efficiency, Relative Efficiency, Thermal Efficiency.**

Efficiency Ratio. The ratio of the actual *thermal efficiency* of a heat-engine to the *ideal efficiency* corresponding with the cycle on which the engine is operating.

Ejector. (*a*) A device for extracting cartridge-shells from a fire-arm.

(*b*) A *jet pump.*

(*c*) A pneumatic device for creating a partial vacuum.

Elastic Axis. (*a*) *of rotation.* An axis about which a rigid body on resilient supports will rotate under the action of a couple in a plane perpendicular to the axis; the axis will not undergo any lateral displacement.

(*b*) *of translation.* An axis along which a rigid body on resilient supports is displaced by a force along the axis without lateral displacement or rotation.

Elastic Limit. The limiting value of the force deforming a body beyond which it does not return to its original shape and dimensions after the force has been removed, that is, no permanent deformation. For steels it is usually assumed to coincide with the *limit of proportionality*. See also **Permanent Set.**

Elastic Strain. (*a*) A strain in a material which disappears with the removal of the straining force.

(*b*) The amount of such a strain. See also **Elastic Strength.**

Elastic Strength. The greatest stress which a bar or structure is capable of sustaining within the *elastic limit*.

Elasticity. The property of a body which returns, or tends to return, the body to its original size or shape after deformation by external forces.

Elasticity, Modulus of. See **Young's Modulus.**

Electric Locomotive. A locomotive driven solely by electric power.

Electrogyro. A device which uses the kinetic energy of a large flywheel to power short-haul vehicles, the flywheel being accelerated at charging points along the route.

Electrolytic Polishing. The production of a smooth lustrous metal surface by making it the anode in an electrolytic solution and dissolving the protuberances.

Electronic Control. The application of electronic techniques to the control of machines and machine tools, power, data and material processing, moving vehicles, aircraft, missiles, satellites, etc.

Elevator. (*a*) A *lift*.

(*b*) A *conveyor* for raising and lowering material temporarily carried in buckets, etc., attached to an endless chain. See also **Barrel Elevator.**

Ellicott Pendulum. See **Pendulum.**

Elliptic Chuck. A special chuck used in a lathe to enable pieces of material to be turned with elliptical cross-sections.

Elliptic Trammel. An instrument for drawing ellipses consisting of two straight grooves at right angles to each other and a bar with a pencil at one end and two adjustable studs that slide in the grooves.

Elliptical Gears. Toothed wheels which are elliptical in form, each rotating about a shaft located at a focus of the ellipse. The rotary motion varies from point to point during the rotation of the wheels. They are used occasionally for producing a quick return motion on machine tools so as to give a slow cutting and a quick return stroke.

Elongation. The total extension produced in a test specimen during a tensile test, usually expressed as a percentage of the original length of the specimen.

Emery Surfacer. A broad *emery wheel*, revolving at high speed, under which the work passes on a sliding table.

Emery Wheel. A *grinding wheel* made of powdered emery cemented by a bonding material.

End Gauge. A metal block or cylinder with the ends made parallel within very narrow limits: the distance between the ends defines a specified dimension.

End Links. The links at both ends of a chain, made slightly stronger than the other links to withstand the extra wear in use.

End Measuring Instruments. Instruments which measure lengths by contacting both ends of an object, such as micrometers.

End Mill. A *milling cutter* with radially disposed cutting teeth on its circular end face for *facing* operations. Cf. **Face Cutter.**

End Play. The longitudinal movement (or play) possible at the end of a revolving part in a mechanism.

End Shake. The movement between the shoulders of an *arbor* and its plates or between the ends of a *staff* and its *endstones.* See also **Shake.**

Endless. A term applied to a belt, chain or rope to imply that the ends are joined together so that it can be used for the transmission of power.

Endless Rope Haulage. The haulage of trucks underground, using a long loop of rope guided by pulleys along the roads and actuated by a power-driven *winding drum.*

Endless Saw. *Band saw.*

Endless Screw. *Worm.*

Endstone. A flat jewel or stone limiting the pinion movement along its axis and acting as a bearing surface for the end of the pinion.

Endurance. See **Limiting Range of Stress.**

Energy. The capacity of a body for doing work.

Potential Energy. The energy of a body by virtue of its position or of its state of tension or compression as in a spring. A body of mass m at a height h above the ground has a potential energy of mgh.

Kinetic Energy. The energy of a body by virtue of its motion. A body of mass m moving with velocity v has a kinetic energy of $\frac{1}{2}mv^2$, or rotating with angular velocity w and moment of inertia I, has a rotational kinetic energy of $\frac{1}{2}Iw^2$.

Energy

Intrinsic Energy. The store of energy possessed by a material system. Changes in this energy depend solely upon the initial and final conditions. See also **Mechanical Equivalent of Heat.**

Engagement of Screw Thread. (*a*) *Depth.* The radial distance by which the thread forms of two mating threads overlap each other.

(*b*) *Length.* The axial distance over which two mating threads are designed to make contact.

Engine. A prime mover; a machine in which power is applied to do work, often in the form of converting heat energy into mechanical work. Cf. **Heat-engine.** See **Air Engine, Atmospheric Engine, By-pass Engine, Compression-ignition Engine, Diesel Engine, Gas-engine, High-speed Steam-engine, Hot-air Engine, Jet Engine, Marine Engine, Piston-engine, Radial Engine, Reciprocating Engine, Rotary Engine; Engine (Servomotor Types), Steam-engines, Turbojet Engine, Turboprop Engine, Wankel Engine.**

Engine (Servomotor Types). A mechanism for transferring energy from hydraulic fluid to a rotating shaft.

Abutment Engine. An engine in which an oscillating piston or a vane (usually double-acting) rotates a shaft through an angle less than 360°.

Swashplate Engine. An engine constructed similar to a *swash-plate pump* with a fluid flow input and a rotating shaft output.

Vane Engine. An engine constructed similar to a *vane pump* with fluid under pressure for the input and a rotating shaft output.

Variable Capacity Engine. An engine in which the capacity per revolution can be changed by varying the porting, rotor eccentricity or stroke.

Variable Stroke Engine. A *radial engine* or *swash-plate engine* with a variable crank throw or swash angle respectively for altering the fluid consumption per revolution. When the angle is varied automatically it is called an ' auto-variable stroke engine '. See also **Ram.**

Engine Beam. The beam of a *beam engine*.

Engine Bed. *Bed plate.*

Engine Indicator. An instrument for recording the pressure in an engine against either a time base or the crankshaft position. Fig. 60 shows how the pressure, fed to below a piston (shown stippled) in a cylinder, moves the piston upwards against a calibrated spring and is recorded on a chart by a pen to which it is linked. The drum on

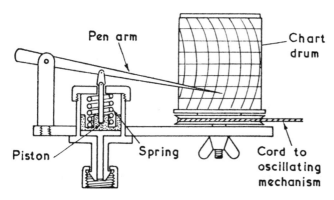

FIG. 60. ENGINE INDICATOR.

which the chart is wrapped oscillates synchronously with the engine piston by a cord linked with the piston motion and kept taut by a spiral spring inside the drum. The pen makes a cyclic diagram on the chart showing cylinder pressure against crankshaft position.

Engine Lathe. *Centre lathe.*

Engine Pit (Pit). (*a*) A hole in the floor for the inspection of a locomotive or a motor-vehicle.

(*b*) The place reserved on a motor-racing track for the inspection, maintenance and refuelling of the racing vehicles.

(*c*) The box-like lower part of the crankcase of an engine; also called ' crank-pit '.

(*d*) A large pit for giving clearance to the flywheel of a large gas-engine or winding-engine.

Engine Power. See **Brake Horse-power, Indicated Horse-power, Thrust.**

Engine Shaft. *Crankshaft.*

Engine Speed. The speed in revolutions per minute of the driving shaft of an engine, or of the main rotor assembly in a turbine engine.

Engineer. One qualified to design, construct or supervise the execution of mechanical, electrical, hydraulic, pneumatic, electronic and other devices, public and military works, etc. See **Engineering Institutions Council, Institution of Mechanical Engineers, Licensed Aircraft Engineer.**

Engineering Institutions, Council. A joint council of fifteen British institutions and societies concerned with different branches of engineering; namely, Royal Aeronautical Society,

Engineering Institutions, Council

Institution of Chemical Engineers, Institution of Civil Engineers, Institution of Electrical Engineers, Institution of Gas Engineers, Institution of Marine Engineers, Institution of Mechanical Engineers, Institution of Mining Engineers, Institution of Mining and Metallurgy, Institution of Municipal Engineers, Institution of Production Engineers, Institution of Electronic and Radio Engineers, Institution of Structural Engineers, Royal Institution of Naval Architects, Institute of Fuel. Full members are known as Chartered Engineers (C. Eng.).

English Lever. See **Escapement.**

Enlarging Drill. A combination of *drill* and *reamer*; a double-fluted twist drill presenting three cutting edges.

Entering Tap. *Taper tap.*

Epicyclic Gearing. A system of gears with one or more wheels travelling round the outside or inside of another wheel whose axis is fixed. See also **Planetary Gear, Solar Gear** and **Star Gear.**

The darker lines in Fig. 61 show the relative movement of the sun and planet wheels respectively.

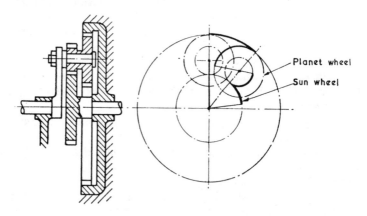

FIG. 61. EPICYCLIC GEARING.

Epicyclic Train. A system of epicyclic gears in which one or more wheel axes revolve about other fixed axes.

Epicycloid. The curve traced out by a point on the circumference of a circle as it rolls round the outer circumference of another circle. Cf. **Hypocycloid.**

128

Equilibration. The production of balance as when providing balance weights for a lift. Cf. **Balancing (Static).**

Equilibrium. The state of a body when the resultant of the forces acting on it is zero, when the body is at rest and when the body is moving with uniform velocity.

Neutral Equilibrium. That state of a body when a slight displacement does not alter its *potential energy.*

Stable Equilibrium. That state when a body will return to its original position after a slight displacement; in other words, a slight displacement will increase its *potential energy.*

Unstable Equilibrium. That state when a body will move further from its original position after a slight disturbance; in other words, a slight displacement will decrease its potential energy.

Equilibrium Ring (Valve Ring). A ring between the back of the *slide valve* and the cover in the steam chest of large steam-engines to lessen the amount of friction of the valve by connecting the enclosed space to the exhaust.

Equilibrium Slide Valve. A large *slide valve* balanced by the use of an *equilibrium ring.*

Equilibrium Valve. A valve in which the pressures on the two sides are made as nearly equal as possible, such as in an *equilibrium slide valve* and a *double-beat valve.*

Equipoise. Static balance. See **Balancing (Static).**

Erecting Shop. (*a*) That part of an engineering works where finished parts are assembled together.

(*b*) A fitting shop for relatively large machines. See also **Fitting Shop.**

Erosion. A wearing away by frictional forces including the action of fast-flowing hot gases. Cf. **Corrosion.**

Error. (*a*) The difference between the correct value and unavoidable defects in the measuring instrument or in inaccuracy of observation; ' accidental errors ' may be of either sign and ' systematic errors ' are always positive or always negative.

(*b*) The difference between the design and the actual size of a dimension.

(*c*) The difference between the actual value of a quantity arising in a *servo*, or other control, system, and the adjusted value in the controller.

Escalator. A moving staircase for carrying passengers up or down. A continuous series of steps are carried on an endless chain and the steps are guided to flatten out to a horizontal platform at the top and bottom of their run. See also **Travelator**.

Escape Motion. *Box of tricks.*

Escape Pinion. The *pinion* on the arbor of the *escape wheel.*

Escape Valve. *Safety-valve* or *relief valve.*

Escape (Escapement) Wheel. The wheel in a clock or watch, the teeth of which act on the *pallets.*

Escapement. The mechanism of a watch or clock connecting the motive power and the *regulator.* See also **Balance Arm**.

Escapement, Clock. The device which releases the energy stored in a driving wheel in a pendulum clock, or in the spring in a household clock, with an even and regular movement and in correct proportional relation to the passage of time. The teeth of the escape wheel as shown in Fig. 62 engage in turn with the *pallets*, usually two in number, which are so pivoted that as one advances and engages with a tooth on the escape wheel, the other recedes a corresponding distance.

The types of clock escapements given below are (*a*) *recoil*, (*b*) *dead-beat* or (*c*) *half dead-beat* and various modifications of these three main types. See **Escapements**.

Escapement, Watch. The types of watch escapements given are either ' frictional-rest ' like the *cylinder escapement* or detached like the *lever escapement.*

Escapements. Crown-wheel Escapement (Verge Escapement). A kind of ' recoil escapement ' which is never detached and there is no locking. The teeth are set at right angles like a *contrate wheel* and the *verge* (*pallet arbor*) is set centrally across the wheel. When the one pallet is receiving impulse the other pallet is falling into position to receive the next impulse, both pallets being set at right angles to the escape wheel axis. This is the oldest form of escapement. **Recoil Escapement** Fig. 62.

Cylinder Escapement. The balance of a watch, mounted on a hollow cylinder, with a little more than half cut away, with a tooth of the escapement wheel giving an impulse by pressing against the lips of the cylinder alternately. As the wheel tooth finishes its impulse, the tooth drops. The succeeding tooth drops on to the solid part of the cylinder, resting there until the cylinder returns and unlocks that tooth so that it can give its impulse. Also known as ' horizontal escapement '.

Dead-beat Escapement. An escapement in which there is no recoil to the escape wheel, obtained by making the locking faces of

130

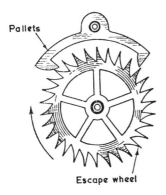

Pallets

Escape wheel

FIG. 62. PENDULUM CLOCK
RECOIL ESCAPEMENT.

the pallets as arcs of circles with the pallet staff (arbor) as centre. See **Pallet** and **Regulator**.

Half Dead-beat Escapement. An escapement in which the locking faces are struck with a large radius from the pallet arbor to introduce recoil. Cf. **Recoil Escapement**.

Duplex Escapement. An escapement in which the escapement wheel has two sets of teeth, one horizontal and the other perpendicular, the former locking and the latter giving impulse to the impulse roller on the balance staff. The horizontal teeth are pointed with one side radial and the others are triangular in shape. The ruby or locking roller fitted on the *balance staff* has a slot cut to permit the passing of the wheel to give impulse every alternative vibration.

Gravity Escapement. An escapement in which a weight is lifted and falls through a constant distance to give a constant impulse to the pendulum. The clock gear train lifts two gravity arms, one on each side of the pendulum rod and the weight in question is the weight of these arms falling alternately. This escapement is used in the clock of Big Ben.

Lever Escapement. An escapement in which the impulse from the *escape wheel* is transmitted to the *balance* by a pivoted lever carrying the *pallets* and the roller carrying the *impulse pin*.

English Lever. A lever escapement in which the escape wheel with ratchet teeth is planted at right angles to the line joining the pallet staff and balance staff centres, and the impulse is due entirely to the pallets and given to the escape wheel teeth.

Swiss Lever. The escape wheel, the pallet staff and the balance staff centres are in a straight line and have a club tooth lever escapement. The impulse is divided between the pallets and the escape wheel teeth.

Pin-pallet Escapement. A *dead-beat escapement* and a type of *lever escapement* in which the *pallets* are semi-circular vertical steel pins or jewels and the impulse is derived entirely from the teeth of the escape wheel. The pallets usually embrace ten teeth and with an escapement wheel of thirty teeth have about 4° of impulse. It is found in cheap watches, alarm clocks and small drum movements.

Pinwheel Escapement. A *dead-beat escapement* in which D-shaped pins standing at right angles to the plane of the wheel give impulse to the pallets.

Platform Escapement. A large form of watch escapement mounted on an independent plate with the escapement pinion extended and a long drop *cock* to enable it to engage with a wheel on a different plane; it is also found in *carriage clocks, chronometers* and *chronometer clock* types. Many platform escapements have *spring detents.* See **Spring-detent Escapement** (with platform).

Recoil Escapement (*Anchor Escapement*) Fig. 62. An escapement in which, during part of the action between the pallets and the wheel, the pendulum pushes the escape wheel back against the power in the train of wheels; this makes the escapement partly self-correcting. Cf. **Dead-beat** and **Half Dead-beat Escapements** above.

Remontoire Escapement. One type has two escapement wheels and two *detents* and in another type the second wheel is replaced by a pivoted steel arm or lever, although the action is almost identical in both types. The first wheel is mounted on the escape pinion and the second has a hairspring mounted on a collet which is fitted on an arbor. The hairspring provides the motive power.

Resilient Escapement. (*a*) An escapement in which the *banking pins* yield to excess pressure caused by over-banking so that the impulse pin can pass the lever, which has no horns.

(*b*) An escapement in which the escape wheel teeth are so formed as to provide a recoil.

Revolving Escapements. See **Karrusel** and **Tourbillon Movements.**

Spring-detent Escapement. A chronometer escapement in which locking is performed by a pallet on a spring *detent* and unlocking by a discharging pallet carried on a roller on the *balance staff.* The gold spring is screwed lengthwise to the body

132

of the detent and extended slightly beyond its horn. The spring is acted upon by the discharging roller which lifts it and releases the escapement wheel tooth locked there. The roller then drops or releases the spring and travels on until on its return it again lifts the gold spring; as it is only resting on one side of the detent it is free to move. The wheel is unlocked only when the discharging roller is travelling in one direction which is why the gold spring is referred to as the ' passing spring '.

Spring-detent Escapement (with platform). A chronometer escapement mounted on a platform, with the *detent* mounted on top of this platform under the *escapement wheel* controlled by a spring. **See Detent Spring.**

Vertical Escapement. An escapement in which the axis of the *balance* is at right angles to that of the escape wheel.

Virgule Escapement (*Hook Escapement*). A *cylinder escapement* with all the impulse on the cylinder and none on the wheel. The wheel teeth are removed leaving only stalks. The impulse pallet on the cylinder is in the form of a hook—hence the term ' hook escapement '. This type of escapement is now obsolete.

Evacuator. A vacuum pump. **See Air-pump.**

Evaporative Cooling. The cooling of a system, or part thereof, by utilizing the latent heat of evaporation, usually of some liquid, such as a fuel; also the sublimation of solids, such as carbon dioxide (dry ice). **See Sweat Cooling.**

Even Pitch. The pitch of a screw thread cut in a lathe which is equal or a multiple of the pitch of the lathe-head screw (or *lead screw*). Cf. **Fractional Pitch.**

Evolute. The evolute of a given curve is the locus of its centre of curvature. A curve has only one evolute but an infinite number of *involutes*.

Excavator. A power-driven machine for digging out and removing earth. See **Dredger Excavator, Grab, Grabbing Crane, Power Drag-line, Power Shovel, Steam Shovel.**

Exhaust. (*a*) The working fluid discharged from an engine cylinder after expansion or from the exhaust pipe of a jet-engine or from a steam-engine to the condenser.

(*b*) That period of an engine cycle occupied by the discharge of the used fluid.

Exhaust Cone. An assembly which leads the exhaust gas from the annular turbine discharge to the *jet pipe* in a *turbojet engine*. (Fig. 194.)

Exhaust-driven Supercharger. See **Supercharger, Exhaust Turbine.**

Exhaust Fan. (*a*) A *fan* usually placed at the entrance to, or in, a pipe for creating an artificial draught out of a building.

(*b*) A fan in the smoke uptake of a boiler to exhaust the flue gases.

Exhaust Gas. The gaseous products of a *piston-engine, rocket engine, turbojet engine* or other form of gas-driven turbine.

Exhaust Injector. An *injector* operated by the exhaust steam of a steam-engine.

Exhaust Lap (Inside Lap). The distance moved by a *side valve* from the mid-position on the port face before uncovering the steam port to exhaust. It promotes cushioning by closing the exhaust early.

Exhaust Line. The lower line of the area drawn on an *indicator diagram,* indicating the back pressure on the piston during the exhaust stroke of a steam or other engine.

Exhaust Manifold. See **Manifold.**

Exhaust Nozzle. *Propelling nozzle.*

Exhaust Pipes. The pipes through which the exhaust products of an engine are discharged.

Branch Pipe. A short pipe which conveys exhaust gases from a piston-engine cylinder to an *exhaust manifold.*

Ejector Pipe. An exhaust pipe from a piston-engine which gives appreciable forward thrust, as with some aeroplanes.

Stub Pipe. A short pipe which conveys exhaust gases direct from a piston-engine cylinder to the atmosphere.

Tail Pipe. A pipe which leads exhaust gases away from an *exhaust manifold.*

Exhaust Port. The port in a cylinder through which a valve allows the escape of exhaust steam or gases.

Exhaust Relief Valve. A *slide valve* which opens to a greater width for exhaust than for steam inlet, being fully opened for the exhaust.

Exhaust Silencer. An expansion chamber fitted to the exhaust pipe of a piston-engine to reduce the noise level.

Exhaust Stator Blades. See **Stator Blades (Exhaust).**

Exhaust Stroke (Scavenging Stroke). The piston stroke of a reciprocating engine during which the steam or exhaust gases are ejected from the cylinder.

Exhaust System. The duct or ducts through which the exhaust gases from a combustion system are discharged.

Exhaust Turbine. A turbine driven by the exhaust gases of a piston-engine to provide the power for a *supercharger, generator* or aircraft *propeller.*

Exhaust Valve. One or more independent valves fitted to those types of engines which have separate ports for exhausting and admitting. See **Inlet Valve.**

Exhaust Vanes. Gyroscopically controlled vanes fitted in the nozzle of a rocket, for steering while the rocket engine is functioning.

Expanding Bit. A boring bit carrying an adjustable cutter on a radial arm for boring holes of different diameters.

Expanding Clutch. A friction clutch which engages by forcing shoes radially against the inner rim of a disc, cone or drum by the interposition of a *toggle joint.*

Expanding Mandrel. See **Mandrel.**

Expanding Reamer. See **Reamer.**

Expanding Ring Clutch. A form of clutch or coupling, in which the effective agent is the friction of a split-ring against a bored hole or recess. A metallic ring, divided transversely, is forced outwards by a wedge against the sides of a parallel-bored hole in the female portion of the clutch, the wedge being actuated by a lever or a screw.

Expansion. (*a*) The increase in the volume of the working fluid in an engine cylinder.

(*b*) The piston stroke during the above increase.

(*c*) The increase in one or more dimensions of a body caused by a rise of temperature, or by a decrease of pressure or by ageing.

Expansion Coupling. A form of *coupling* in which two thin steel sheets with circumferential corrugations are interposed between two flanged coupling faces which they connect and thus allow a certain amount of end play. See also **Expansion Joint.**

Expansion Curve (Line). The line on an *indicator diagram* showing the pressure during the expansion stroke.

Expansion Engine. An engine which utilizes the expansion of a working fluid.

Expansion Gear. That part of the valve gear of a steam-engine through which the degree of expansion can be varied.

Expansion Joint. A joint to permit linear expansion or contraction with changes in temperature.

Expansion Pulley. A pulley wheel, for use with *vee-belts* which, by axial movements of the two halves of the vee, can create different working diameters. They are used in pairs to give variable-speed drive without the use of a clutch and gear-box.

Expansion Ratio. The ratio between the pressure in the combustion chamber of a rocket or a jet pipe and that at the outlet of the propelling nozzle.

Expansion Rollers. Rollers placed underneath the ends of long steel girders, one end being fixed, or under the end of a bridge to give support and at the same time allow linear expansion or contraction with variation of temperature.

Expansion Valve. (*a*) An auxiliary valve, working on the back of the main slide valve of some steam-engines, to provide an independent control of the point of cut-off.

(*b*) A regulation valve in refrigeration machines to control the escape of the refrigerant from a liquid state under pressure to a gaseous state.

Expansive Working. The use of the expansion of a working fluid in an engine, such as when steam, cut off at a fractional part of the stroke, is caused to do work by its own expansion.

Experimental Mean Pitch. See **Pitch (Propeller).**

Explosive Forming. See **Forming.**

Extensometer. An instrument attached to gauge-points on a test piece for measuring accurately small strains under load.

Extension Lathe. A lathe in which the *headstock* can be slid longitudinally to provide an adjustable gap of a length to suit the work.

External Screw Thread (External Thread). A screw thread cut on the outside of a cylindrical bar. Also called a ' male thread '.

External Screw Tool. A tool adapted for cutting external screw threads.

Extraction Turbine. A steam-turbine from which steam for process work is tapped at a suitable stage in the expansion, the remainder expanding down to condenser pressure.

Extrusion. The process for producing long lengths of metal of constant cross-section, including rods and tubes, by forcing hot metal through a suitable *die* by means of a *ram*. When the metal is forced by the ram direct through the die, the process is called ' direct extrusion '; when the die is contained on the end of the plunger and is literally pushed through the billet it is called ' indirect extrusion '. (Cf. **Detruding.**) Examples of direct extrusion are shown in Figs. 190, 191.

Impact Extrusion. Extrusion caused by the sudden action of a *punch*, Fig. 63. (See also Fig. 116 (*c*).)

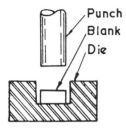
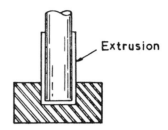

FIG. 63. IMPACT EXTRUSION.

Eye. (*a*) A loop formed at the end of a steel wire or a cable attachment, rod, lever or bolt to facilitate removal.

Eye Bolt. A bolt with an eye at the end instead of a head.

(*b*) The central hole or bore of a wheel or pulley through which an axle or shaft passes.

(*c*) The central inlet passage of an *impeller* in a centrifugal pump or compressor.

F

F. The symbol for force.

f. A symbol signifying stress; also δ.

f. A symbol for frequency or acceleration.

ft-lb. Foot-pound.

ft/m, f.p.m. Feet per minute.

ft/s, f.p.s. Feet per second.

f.p.s. Foot pound second.

f/s. *Factor of safety.*

Face. (*a*) The flat surface of greatest area on a piece of work.

(*b*) The working surface of any part, implement or tool.

(*c*) The curved outline flank of a tooth in a gear-wheel which lies beyond the *pitch circle.* Cf. **Flanks.**

(*d*) The surface of a slide valve and that upon which a slide valve moves.

(*e*) The surface of a valve which comes into contact with its seat.

(*f*) The dial of a clock or registering instrument.

Face Chuck (Face Plate). A large disc, for screwing to the *mandrel* of a lathe, which is provided with slots and holes for securing work of flat or irregular shape.

Face Cutter. A milling cutter with the teeth radially disposed upon the surface of a disc, either solid with teeth or inserted in grooves. Cf. **End Mills.**

Face Gear. *Contrate gear.*

Face Lathe. A lathe chiefly used for surfacing work of large diameter and short length such as large wheels and discs.

Face Plate. See **Face Chuck** and **Surface Plate.**

Face-plate Coupling. *Flange coupling.*

Face-width. (*a*) The width over the teeth measured parallel to the axis of a helical, spur or worm gear. (See Fig. 92.)

(*b*) The width over the teeth measured along the *pitch cone* generator of a bevel gear.

Face-width, Minimum. The face-width necessary to ensure continuity of tooth action in the crossed-axis type of *helical gear* and in *worm* gears.

Minimum Desirable Face-width. The minimum face-width necessary to ensure overlap of tooth action in helical, bevel and hypoid gears.

Facing. (*a*) Turning a flat face on a piece of work in a lathe.

(*b*) A raised machined surface to which another part is to be attached.

Facing Machine. A *centring machine* with suitable cutters for *facing* work.

Facing Points. See **Points.**

Factor of Safety. The ratio of the breaking load on a member, structure or mechanism and the safe permissible load on it. This ratio is allowed when designing the member, bearing in mind the normal conditions of service operation and providing for the possibility of uncertainties of various kinds including variation of strength resulting from deterioration in service.

Proof Factor of Safety. The factor of safety based on the *proof load.*

Ultimate Factor of Safety. The factor of safety based on the *ultimate load.*

Falling Weight Test. *Drop test.*

False Key. A circular key driven into a hole parallel with a shaft axis, half drilled in the shaft and half in the hub which is keyed to it.

Fan. (*a*) A rotating paddle wheel or a specially designed propeller for delivering or exhausting large volumes of gas or air with but a low pressure increase. See **Centrifugal Fan, Propeller Fan.**

(*b*) A small vane to keep the wheel of a *wind pump* at right angles to the wind.

(*c*) A wheel in a clock mechanism whose velocity is regulated by air resistance.

Fan Cooling. The use of an engine-driven fan in motor-vehicles to produce a greater flow of air through the radiator at low speeds than would result from the forward movement of the vehicle.

Fang. The spike of a tool held in the *stock*.

Fast Coupling. A permanent coupling of two shafts consisting of flanges formed integral with the shafts. Cf. **Loose Coupling.**

Fast Feed. See **Feed.**

Fast Head. *Fixed headstock.*

Fast Pulley. A pulley fixed to a shaft by a *key* or *set screw*. Cf. **Loose Pulley.**

Fatigue. The process leading to the failure of metals (or other materials) under the repeated action of a cycle of stress. The failure depends on the mean stress, the range of stress and the number of cycles. With a decreased amount of stress a material can withstand a greatly increased number of repetitions before failure or failure may not occur after millions of stress cycles (see **Fatigue Life**). With a large amount of stress, failure may occur after a relatively small number of reversals.

Fatigue Life (see **Life Factor**). The life of a test-piece, or of a part of a structure or mechanism, expressed as the number of applications of a load before failure.

Safe Fatigue Life. The period of time during which the continued applications of a load are extremely unlikely to result in failure.

Fatigue Limit. The upper limit of the range of stress that a metal can withstand indefinitely. See also **Limiting Range of Stress.**

Fatigue Testing Machine. A machine for applying rapidly alternating or fluctuating stresses to a test-piece to determine its *fatigue limit*.

Fearnought. A machine which opens and mixes woollen material, preparatory to carding. See **Carding Engine.**

Feather. A parallel key partially sunk into a recess in its shaft to permit a wheel to slide axially but not to rotate. It enables machine parts to be thrown into and out of gear. (See Fig. 98.) Cf. **Key: Sunk Key; Splines.**

Feathering Board. *Feathering (float) paddles.*

Feathering (Float) Paddles. Paddle wheels which are controlled so that the floats enter and leave the water at right angles to the surface, thus economizing power.

Feathering Propeller. A propeller the *pitch* of which is altered so that it gives no thrust.

Feature.   An individual characteristic.

Concentric Feature.   A feature required to conform to a specified concentric relationship with other features in the same group.

Datum Feature.   A *positional feature* which serves as a reference for the location of other features.

Grade of a Feature.   The relationship between the tolerance and the allowance, if any, on the feature and its basic size, accuracy and fit.

Positional Feature.   A feature required to conform to a specified positional relationship with other features in the same group.

Feature Tolerance.   See **Tolerance.**

Feed.   (*a*) The rate at which a cutting tool of a machine is advanced.

(*b*) The advance of material operated upon in a machine or the provision of materials or requirements necessary for a process or operation.

Different rates and types of feed are called continuous, fast, intermittent, slow, fine, etc.   See also **Chain Feed, Disc Feed, Rack Feed. Cf. Speeds.**

Feed Box.   On preoptive and automatic lathes the feed box is mounted on the bed below the headstock to provide a range of feeds to both the *saddle* and *slide*.   The feeds are independent and reversible.

Feed Gear.   The mechanism in a machine tool whereby a cutter is fed to its work or vice versa.   A self-acting mechanism.

Feed Mechanism.   The mechanism which controls the movements of the carbons of an arc lamp at such a speed as to keep the arc length constant while the carbons burn away.

Feed Pump.   (*a*) A *force pump* for supplying water for steam boilers.

(*b*) A *force pump* for supplying air to gas-engines.

Feed Screw.   A screw used for controlling the motion of the feed mechanism of a machine tool.   See **Lathe Traverse.**

Feedback (Back-coupling).   A transfer of energy from an output back into an input.   Feedback is said to be positive when it tends to increase, and negative when it tends to decrease, an amplification. See **Pressure Feedback Unit.**

Feeder.   A machine which passes paper, sheet by sheet, into a printing machine.

Feeler(s).   (*a*) Thin strips of hardened steel of known thickness, mounted like blades in a pocket knife for gauging small distances between surfaces.

(*b*) A device which determines when automatic weft replenishing is necessary in a weaving machine.

Fell. The edge of a cloth in a *loom* to which the weft yarn is placed during weaving.

Felloe (Felly). The circumference or a segment of the circumference of a wheel fitted with *spokes*.

Female Gauge. *Ring gauge.*

Female Screw. An internal screw thread. See also **Male and Female.**

Fence. (*a*) A stop to limit motion.

(*b*) A guide for material such as in a planing machine.

(*c*) An adjustable limitation or direction of movement of one piece with respect to another during a grinding or machining operation.

Fenders. See **Bumpers.**

Ferris Wheel. The giant revolving wheel with a horizontal axis supporting passenger cars which hang freely at its periphery.

Ferrule. (*a*) A metal band forming or strengthening a joint.

(*b*) A watchmaker's *bow drill* using a small grooved pulley.

Fiddle Drill. *Bow drill.*

Fielding Panels. Blocks, usually two, fitted on a spindle one above the other with spacing washers interposed when making heavy cuts, the slots in each block being positioned over each other and a cutter of suitable length fitted in both. The blocks, as keys, securely lock the cutters together.

Figure-of-eight Callipers. Callipers with jewelled jaws which are used to hold watch components (Fig. 64). Cf. **Callipers, Poising.**

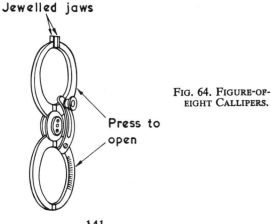

Jewelled jaws

Press to open

FIG. 64. FIGURE-OF-EIGHT CALLIPERS.

141

Filled Rail. A rail which has one or both sides filled up flush to provide extra strength at points, etc.

Fillet. (*a*) A narrow strip of metal raised above the general level of a surface.

(*b*) A radius at the intersection of two surfaces.

(*c*) That portion of the tooth surface of a gear-wheel which joins the tooth flank and the bottom of the tooth space.

Fillet Radius. The smallest radius of curvature of a fillet.

Film Cooling. The injection of a *coolant* directly into a combustion or other chamber, usually through small orifices, so as to provide a cooling film on the inner surface; also called ' transpiration cooling '. See also **Sweat Cooling**.

Fine. An adjective usually implying small dimensions, such as gear-teeth, abrasive particles in emery wheels, etc.

Fine Pitch. See **Pitch (Propeller)**.

Finger. (*a*) A narrow projection or a pin used as a guide or guard in a mechanism.

(*b*) A pointer or index.

Finger Guard. (*a*) Any device which serves to guard the fingers of operators from injury by machinery.

(*b*) Protection for a pointer or index of an instrument.

Finishing. An adjective used to describe the final process in the completion of a piece of work such as cutting, rolling, polishing, etc. Fig. 65 illustrates the terminology of surface finish.

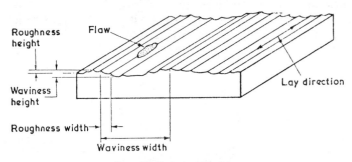

FIG. 65. SURFACE FINISH.

Finishing Teeth. Teeth which complete the formation of a hole when using a *broach*. (See Fig. 17.)

Finishing Tool. A lathe or planer tool for making a final cut, usually cutting on a wide *face*.

142

Fins. (*a*) Thin projecting strips of metal on air-cooled piston-engine cylinders to increase the cooling area.

(*b*) Any thin projecting edges on a metal piece.

Fire Ring. A top piston-ring of a special heat-resisting design found in two-stroke *compression-ignition engines*.

Firing. (*a*) The ignition of a charge in a *gas-engine*.

(*b*) The quickening of the source of heat in a steam boiler by adding fuel.

(*c*) The overheating of a bearing.

(*d*) The ignition of a cartridge starter of a *gas-turbine engine*.

(*e*) The operation of a *rocket engine*.

(*f*) The operation of *control jets* on spacecraft and high-altitude aircraft.

Firing Chamber (Lighting Chamber). The small chamber through which a charge of a *gas-engine* or other engine is ignited.

Firing Order. The sequence in which the cylinders of a multi-cylinder piston-engine are fired; for example, 1, 3, 4, 2 for a four-cylinder engine.

Firing Stroke. The expansion stroke of an internal-combustion engine.

Firing Top-centre. The *top dead-centre* before the *firing stroke*.

Firth Hardometer. A *hardness* tester using a diamond *indenter* with loads of 10, 30, or 120 kg for hardened steels or a steel ball for soft materials.

Fish Bolt. A bolt for fastening *fish-plates* and rails together.

Fished Joint. See **Fish-plates.**

Fish-plates (Splice Pieces). Steel cover-plates or cover-straps fitted on the sides of a fished (butt) joint between successive lengths of rail or beam.

Fit. The relationship between two *mating* parts when *clearance* or *interference* is present on assembly.

Clearance Fit. A fit in which the limits for the mating parts always allow assembly of the parts. See also **Allowance.**

Force Fit. A fit which requires considerable force or pressure in order to mate two parts, such as a shaft into its hole. See **Interference Fit.**

Grade of a Fit. See **Grade.**

Interference Fit. A fit in which the limits for the mating parts are such that *interference* always occurs when a pair of parts are brought together for assembly. See also **Allowance.**

Running Fit. A term used for parts which are assembled so that they are free to rotate.

Types of Fit. Fits are often described by their functioning requirement, such as drive fit, force fit, free fit, push fit, running fit, sliding fit, etc., being dependent on other factors than dimensions.

Fitting. The operations, usually manual, necessary to complete an assembly other than those done in the foundry and machine shop.

Fitting Allowance. The total length of screw thread beyond the *gauge plane* on the pipe end, needed for assembly with the maximum permitted over-size coupling.

Fittings. (*a*) Small auxiliary but essential parts of an engine, machine or mechanism.

(*b*) Accessories, especially for boilers, such as valves, gauges, etc.

Fitting Shop. The department of an engineering workshop where *fitting* is done. Cf. **Erecting Shop.**

Fixed Cutters. Cutters fixed in a machine for the work to be moved over them.

Fixed Eccentric. An *eccentric* permanently keyed to a shaft. Cf. **Loose Eccentric.**

Fixed Expansion. A steam-engine with a constant *expansion ratio* in which the *cut-off* cannot be altered.

Fixed Head. The head of a shaping machine which is attached to the ram and moves only in the direction of its stroke. Cf. **Clapper Box.**

Fixed Headstock (Fast Head). The casting bolted to the left-hand end of a lathe bed. It houses the bearings supporting the mandrel and driving pulleys.

Fixed Pulley. A pulley keyed to its shaft. Cf. **Fast Pulley.**

Fixture. A fixture holds the work without controlling the tools. Cf. **Jig.**

Flame Cutting. The cutting of metal with an oxy-acetylene or oxy-hydrogen flame.

Flame Trap. A gauze or wire grid in the intake of a *carburetter* to prevent the emission of flame from a *back-fire*.

Flame Tube. The perforated inner tubular ' can ' of the *combustion chamber* of a jet-engine in which the actual burning of the fuel takes place.

Flange. (*a*) A projecting rim, as on the rim of a wheel which runs on rails.

(b) A disc-shaped rim formed on the ends of shafts (or pipes) for coupling them together or on to an engine cylinder.

(c) The top and bottom members of an I-beam.

(d) A projecting offset on a piece of work.

Flange Coupling. A *shaft coupling* consisting of two accurately faced flanges which are keyed to their respective shafts and bolted together.

Flanged. Provided with a projecting flat rim, *collar* or rib.

Flanged Chuck. *Face chuck.*

Flanged Nut. A nut with a broad flange integral with its solid face. See also **Collar-headed Screw.**

Flanged Pipes. Pipes provided with *flanges* for connecting them together by means of bolts. See **Flanged.**

Flanged Rail (Flat-bottomed Rail). A rail section of inverted T-shape with a broad flat flange at the bottom and the top enlarged locally to form the *head* of the rail.

Flanged Socket. A very short pipe with a flange at one end and a socket at the other.

Flanged Spigot. A very short pipe with a flange at one end and a *spigot* at the other.

Flanged Wheel. A wheel with a *flange* on one or both sides to keep it on the rails.

Flanging Machine (Press). A machine (press) for bending over the edges of plates to make a *flange.*

Flank Angles. The angles measured in an axial plane between individual flanks and the perpendicular to the axis of the screw thread. (See Fig. 157.)

Flanks. (a) The curved outlines of teeth in a gear which lie within the *pitch circle* or below the *pitch line.* Cf. **Face.**

(b) The parts of the surface on both sides of a screw thread that intersect an axial plane in straight lines. (See Fig. 157.)

(c) The working faces of cams.

Flanks. *Clearing Flank.* The flank of a screw thread that does not take the thrust or load in an assembly.

Following Flank. The flank of a screw thread opposite to the *leading flank.*

Leading Flank. The flank that faces the mating screw thread just before assembly.

Pressure Flank. The flank that takes the thrust or load in a screw-thread assembly.

Flap. Any hinged or pivoted surface which can be adjusted,

Flap

usually either automatically or through controls. The term is used extensively in relation to aircraft with an adjective, largely self-descriptive or after an inventor, that explains its action or position such as dive-recovery, extension, Fowler, leading edge, slotted, etc. See also **Air Brake.**

Flap-valve. *A non-return valve* in the form of a hinged flap or disc used for low pressures as found in lifting pumps. It is sometimes faced with leather or rubber. (See Fig. 66.)

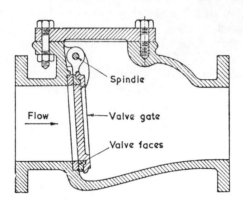

FIG. 66. FLAP-VALVE.

Forms of flap valves are called ' reflux ', ' clapper ' and ' swing check ' valves.

Flapping Angle. The angle between the *tip-path plane* of the *rotor* of a rotorcraft and the plane normal to the *hub* axis.

Flaring. The shaping out of the end of a pipe to increase its diameter towards the end. Cf. **Flange.**

Flash-point. The temperature at which a liquid heated in a special apparatus gives off sufficient vapour to flash momentarily on the application of a small flame.

Flat-bottomed Rail. *Flanged rail.*

Flats. (*a*) Iron or steel bars of rectangular section.

(*b*) Flat portions of a nut or bolt head.

Fleam. The angle of rake between the cutting edge of a saw-tooth and the plane of the blade.

Fletcher's Trolley. A trolley, designed for the investigation of the motion of a body every part of which has the same velocity at a given moment, and mounted on wheels of small mass. An accelerating force is applied to the trolley by a weight over a pulley at the end of the plane on which the trolley runs or its motion is controlled by a spring or springs.

Flexible Coupling. A shaft *coupling* connecting two shafts, not in rigid alignment, with the drive transmitted through a resilient member such as a steel spring, rubber disc (Fig. 67), *bushes* or belt and pins (Fig. 68).

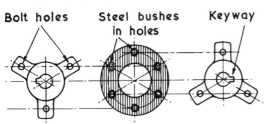

Bolt holes   Steel bushes   Keyway
                in holes

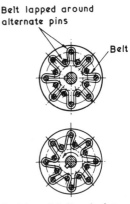

Belt lapped around
alternate pins

Belt

FIG. 67 (*above*). FLEXIBLE DISC COUPLING COMPONENTS.

FIG. 68 (*right*). LOOPED BELT AND PIN-TYPE COUPLING.

Position of belt and pins,
when coupling rotates

Flexible Shafting. A shaft consisting of a number of concentric spiral coils of wire, wound alternately—right and left hands—over each other, thus giving flexibility when revolving like an ordinary shaft.

Flexural Axis. The line joining the *flexural centres*.

Flexural Centre. A point in a plane which, when a force is applied at it in any direction in that plane, will not cause twisting in the plane.

Flexural Rigidity. See **Flexure**.

Flexure. The bending of a thin bar under forces or moments so that its displacement is perpendicular to its length. The ratio of the applied force to the displacement is the ' flexural rigidity '.

Flirt. A device, including a lifting pin and spring, for bringing about the sudden movement of part of the mechanism of a clock or a repeater. See **Chiming Mechanism** and Fig. 31. Cf. **Trigger**.

Float. (*a*) A small buoyant cylinder placed in the float chamber of a *carburetter* for actuating a valve controlling the petrol supply from a main tank.

147

(*b*) A buoy used to indicate the height of the water in tanks or boilers.

Float Board. (*a*) The rectangular boards attached to the arms of a *paddle wheel.*

(*b*) The boards which receive the impulse of the water in an *undershot wheel.*

Float Chamber. The petrol reservoir in a *carburetter* from which the jets are supplied for the engine cylinders, and in which the petrol level is maintained constant by means of a float-controlled valve.

Floating. A term applied to various cases of exact balance, as of a chemical balance, a testing machine, a weighbridge, etc.

Floating Axle. An axle on which the shaft is relieved of all loads or stresses except turning the wheel.

Floating Crane (Floating Derrick). A large *crane* carried on a pontoon for use in docks.

Floating Gudgeon Pin. See **Gudgeon Pin.**

Floating Mandrel. See **Mandrel.**

Floating Mill Wheel. A *water wheel* with its bearings in a boat moored in a rapidly flowing river. The stream turns the wheel to provide power.

Floor Rest. A *rest* carried on a heavy standard resting on the floor and used for work of large diameter in pattern-makers' lathes.

Flow-meter, Registering Element. That part of the meter which enables the rate of flow or total volume past the meter to be recorded. Fig. 69 shows a centrifugal type of recording element which gives a visual indication of the flow together with a permanent record. It also stores the integrated value on a counter geared from the meter spindle. With rotation the *centrifugal force* causes the mercury to flow into the annular chamber; thus the cast-iron float drops, altering the indicator needle and recorder pen positions. A near linear scaling of the flow-meter is possible with careful design of the internal bowl shape.

Fluid Coupling. *Fluid flywheel.*

Fluid Drive. (*a*) A constant-torque drive mechanism built into the flywheel of an automobile, consisting of two rotors with vanes operating in oil. The amount of slip of the oil between the driving and the driven plate varies inversely as the speed, permitting a smooth starting of a car in any gear.

(*b*) A vehicle propulsion system incorporating a *fluid flywheel.*

Fluid Flywheel. A device for transmitting a rotary drive through the medium of a change in momentum of a fluid (usually oil). The

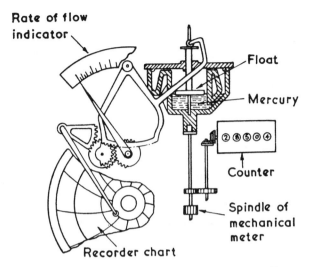

Fig. 69. Flow-meter—Centrifugal Type Registering Element.

coupling is achieved by the action of the fluid on the vanes of the driving and driven tori. Cf. **Froude Brake.** (See Fig. 70.)

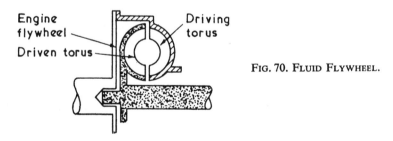

Fig. 70. Fluid Flywheel.

Fluid Lubrication. Lubrication when the bearing surfaces are completely separated by a viscous oil film, induced and sustained by the relative motion of the surfaces.

Fluid Pressure. (*a*) The pressure per unit area exerted equally in all directions by a fluid. The total force on a horizontal area is equal to the product of the area, its depth and the density of the fluid.

(*b*) The pressure transmitted by a fluid.

Fluid Seal. A seal for a bearing in which a fluid (usually oil) envelops the bearing to prevent the escape of a gas from a high-pressure region to a lower one.

Fluidity. The inverse of *viscosity* or loosely, the ability to flow.

Flume. A channel conducting the water to a *water wheel*, ore-washing plant, etc.

Flute. The grooves straight, or spiral, running along the length of a twist drill. (See Fig. 114.)

Fly. (*a*) The loaded lever which actuates the screw of a *fly press*.

(*b*) A tiny fan with two or four blades which acts as an air brake to maintain uniformity in the hammer blows of striking and chiming clocks; hence ' hour fly ' and ' quarter fly '.

Fly Cutter. (*a*) A single-point tool-holder which can be revolved around its axis and by a length adjustment of the horizontal bar in which the tool is mounted a hole can be made of almost any diameter.

(*b*) A narrow milling cutter for cutting slots such as *keyways* in shafts.

(*c*) A cutter with edges set diagonally and sloping towards the centre for shaping the ends of metal rods, etc.

Fly Frames. A series of machines used to attenuate *roving* in the preparation of cotton for the cotton-spinning frame.

Fly Pinion. A *pinion* on the arbor of which a clock's *fly* is mounted.

Fly Press. A press for punching holes and stamping out thin works in metal. The press is actuated by a vertical square-threaded screw and a cross-piece terminating in one or two heavy steel balls to give additional impetus to the descent of the die attached to the bottom end.

Fly Shuttle. The mechanism for propelling a *shuttle* across a *loom*.

Flyback Action. The action in a stop-watch or chronograph which causes the hands to fly back to zero when the button is pressed. See also **Zero (Setting).**

Flyballs. An American term to describe the out-flying weights on a *governor*.

Flyer. See **Flyer Spinning.**

Flyer Spinning. A method of *spinning* in which a flyer guides a coarse yarn on to the *bobbin*. Flyer and bobbin revolve at different speeds.

Flywheel. (*a*) A heavy-rimmed wheel on a revolving shaft to

150

absorb fluctuations in the speed and thus even out the torque output of machinery.

(*b*) A heavy-rimmed rotating wheel which is run up to accumulate power for use during a short interval.

Folding Machine. See **Book-folding Machine.**

Follower. (*a*) A toothed wheel driven by another wheel.

(*b*) A pinion driven by a toothed wheel.

(*c*) That part of a mechanism, such as a lever arm, driven by a *cam* and usually returned by a spring.

Following Steady. A *steady* attached to the back of the side rest of a lathe, which embraces the work behind the tool and follows it along with the rest.

Foot Board. *Treadle.*

Foot Brake. A pedal operating the brake shoes on wheels of a vehicle, hydraulically, pneumatically or through levers and cables. See **Hydraulic Brake.**

Foot Lathe. A light lathe driven from a *treadle* and crank by the foot.

Foot-pound. The unit of work in the old British system of units and equal to the work done in raising a mass of one pound through a vertical distance of one foot against gravity. See **Work, Acceleration due to Gravity.** Cf. **Kilogramme-metre.** (NOTE. The pound was defined by a British Parliament's Act in 1962–3 exactly as 0·453,593,37 kilogrammes.)

Foot-rail. *Flanged rail.*

Foot-ton. The work done in raising a mass of one ton through a vertical distance of one foot against gravity, equal to 2240 *foot-pounds*. See also **Acceleration due to Gravity.**

Foot Valve. (*a*) The suction valve of a pump. Cf. **Head Valve.**

(*b*) A non-return valve at the inlet end of a suction pipe. See **Check Valve.**

Footstep Bearing (Footstep). A *thrust bearing* used to support the lower end of a vertical shaft.

Fopple Card. A geometrical device for measuring small amplitude vibrations. Two lines are drawn on the card at a small angle and symmetrically about a horizontal line. Vertical vibrations make the lines appear to cross over each other at a point which is horizontally displaced from the true intersection point. This horizontal displacement can, however, be clearly read by observation of vertical lines drawn at positions where the vertical displacement has been measured between the inclined lines. With the displacement thus measured and a knowledge of the frequency the acceleration force on a body can be calculated.

Force Feed. (*a*) The lubrication of an engine by forcing oil to the main bearings and through the hollow crankshaft to the big-end bearings. See **Forced Lubrication.**

(*b*) The use of a motor-driven pump in a central heating system. See **Forced-circulation Boiler.**

Force Fit. See **Fit.**

Force Pump (Plunger Pump). (*a*) A pump consisting of a barrel with a solid plunger and a valve chest with suction and delivery valve, which delivers liquid at a pressure greater than its suction pressure. Cf. **Lift Pump.**

(*b*) An air pump used to clean out gas and other service pipes by blowing air through them.

Forced-circulation Boiler. Steam boilers in which water and steam are continuously circulated over the heating surface by pumps (as opposed to natural circulation) in order to increase the steaming capacity. Velox and Löffler boilers are forced-circulation boilers.

Forced Draught. An air supply to a furnace with the aid of fans or steam jets (as opposed to natural draught created by a chimney) to increase the rate of combustion. Cf. **Induced Draught.**

Forced Lubrication. The supply of oil under pressure for the lubrication of engine bearings and of machine tools. See **Force Feed.**

Forced Vibration. See **Vibration.**

Fore Carriage. The *bogie* under the two front wheels of a portable engine.

Fore-gear Eccentric. *Forward eccentric.*

Forge. (*a*) A plant where forging is carried out.

(*b*) A mill where the rolling of puddling bar is carried on.

Forge Dies. Shaped metal *dies* used in *forging*.

Forge Rolls. The train of rolls by which *slabs* and *blooms* are converted into puddled bars.

Forge Train. The series of rolls for rolling out shingled *bloom* after leaving the steam hammer.

Forge Tests. Rough workshop tests, including bending, made to check the malleability and ductility of iron and steel.

Forging. The operation of shaping metal parts when hot by means of hammers or presses. See **Drop Forging, Forge, Forge Dies, Press Forging, Steam Hammer, Upset Forging.**

Forging Machines. Power hammers and presses used for *forging*.

Forging Press. See **Press Forging.**

Fork. (*a*) When *beading* in a lathe the rolling tool is held in a fork carried by the tool slide.

(*b*) The end of the lever which receives the impulse pin in the lever *escapement* of a watch or clock.

(*c*) A double-pronged clip on a tub or wagon for the haulage rope or chain. See also **Belt Fork.**

Fork Chuck. *Prong chuck.*

Fork-lift Truck. A vehicle with power-operated prongs which can be raised or lowered at will, for stacking and loading, for transporting and unloading, packages of goods; the last are called *pallets.*

Forked Connecting-rod. See **Connecting-rod.**

Forked End. The bifurcated end of a lever or rod which receives the end of another rod in its fork, being connected by a *joint pin.*

Forked Strap. The forked end of a pump-rod embracing the end of a wooden rod.

Form. The shape of one complete profile of a screw thread in an axial plane.

Form Diameter. The diameter of the circle from which the involute of a gear is designed.

Form Grinding. *Profile grinding.*

Form Tolerance. See **Tolerance.**

Form Tool (Forming Cutter). A cutter with a profile similar to, but not necessarily identical with, the shape and contour desired for the workpiece.

Former. A templet used for the cutting of gear-teeth, etc., in copying machines.

Forming. When two or more bends are made simultaneously, the bending operation by plastic deformation of a metal sheet is called forming. A forming tool is usually made for each particular job. A combined blanking and forming tool is illustrated in Fig. 185.

Brake Forming utilizes standard dies in a brake press which has a long narrow forming bed: standard presses can handle plates up to thirty feet long.

Electric Discharge Forming. A high-energy spark is discharged between two electrodes immersed in water along with the metal to be formed. The discharge induces a shock wave throughout the water which forms the metal instantaneously.

Explosive Forming. A process in which the metal is shaped by the energy released when an explosive charge is fired and its success lies in the high velocities, 70–5000 m/s, at which the metal forming takes place. It is accomplished by the shock wave, gas pressure or a combination of the two, with water or oil distributing the pressure uniformly. Sometimes called ' high kinetic energy forming '. (See Figs. 71, 133.)

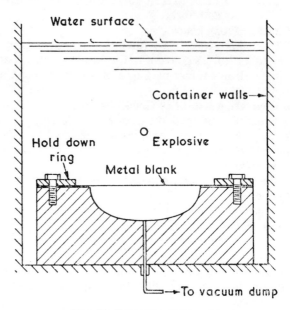

FIG. 71. EXPLOSIVE FORMING.

High-temperature Forming. The utilization of high energy in the form of heat using either dies integrally heated by cartridge-type heaters or hot-fluid forming, in both of which heat and pressure are supplied by a hot liquid metal alloy.

Rubber-pad Forming. Forming using a confined rubber pad instead of the conventional mating die, when the resistance offered by the rubber forces the metal to conform to the punch shape. See **Rubber Press.**

Stretch Forming. A process which entails wrapping sheet metal around a male die, pulling and then trimming the ends. It is a suitable and rapid process for making complicated shapes. (See Fig. 72.)

Forming Cutter. *Form tool.*

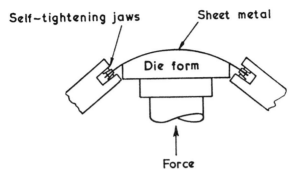

Self-tightening jaws   Sheet metal

Die form

Force

FIG. 72. STRETCH FORMING.

Forward Eccentric. The *eccentric* on a steam-engine with link motion reverse gear, that imparts a forward motion to the valve gear.

Forward Gear. The gear in which an engine moves forward.

Föttinger Coupling (Föttinger Transmitter). An outward-flow water turbine driving an inward-flow turbine, within a common casing, acting as a coupling, gear or clutch for transmitting power, such as from an engine to a ship's propeller. See also **Fluid Drive, Fluid Flywheel.**

Foucault's Pendulum. A heavy metal ball suspended by a very long fine wire which, when left to swing freely, changes its plane of oscillation by 15°, multiplied by the sine of the latitude, per sidereal hour, thus demonstrating rotation of the earth.

Four-cutter Machine. A machine for cutting four faces of a piece of wood at the same time.

Four-cylinder Engines. Compound steam-engines with two high-pressure and two low-pressure cylinders.

Four-high Mill. A rolling mill composed of two small working rolls supported and driven by the larger rolls. (Fig. 150.)

Four-stroke (4-stroke) Cycle. A piston-engine cycle completed in four strokes, involving two crankshaft revolutions, namely, suction or induction, compression, expansion or power stroke and expulsion of exhaust. See **Two-stroke Cycle.** Cf. **Diesel Cycle, Otto Cycle.**

Four-way Canting Work Table. A work table that can be tilted backwards or forwards and to right or left. It is commonly fitted to band-saws.

Fourneyron Turbine. See **Turbine.**

155

Fourth Pinion. The pinion on which the fourth wheel of a clock or watch movement is mounted.

Fourth Wheel (Watch). (*a*) The wheel in a watch which drives the escape pinion.

(*b*) If the fourth wheel makes one turn per minute, the seconds hand is carried on an extension of the fourth wheel arbor.

Fractional Pitch. A pitch of a screw thread cut in a lathe which is not an integral multiple or sub-multiple of the pitch of the lead screw. Cf. **Even Pitch.**

Frame. (*a*) The structure of the chassis of motor-vehicles.

(*b*) Any structure built up of compression and tension members.

Frames per Second. The rate of taking or projecting ciné pictures.

Framing. (*a*) The skeleton structure of a locomotive.

(*b*) The vertical adjustment of the picture gate in a ciné projector or the adjustment of the picture-repetition frequency in a television receiver to keep the picture stationary on the screen.

Francis Water Turbine. A *reaction turbine* type with water flowing radially inwards into guide vanes and on into a runner from which it emerges axially. It is used for heads from about 70 to 500 m

Frazing Machine. A machine for removing the fin from forged nuts and bolts.

Free-piston Engine or Gas Generator. An engine in which the reciprocating portion serves as a gas generator yielding no nett mechanical power but supplying hot compressed gas at about 770 K and 4 bars to the turbine. A symmetrical-opposed engine has the air pistons attached direct to the diesel pistons with a linkage between opposed pistons but no crankshaft. The air is sucked in and compressed in the air cylinders either on the outward or return stroke. (See Figs. 73, 74.)

Free-sprung. A free-sprung watch has no *index* and *curb pins* for the correction of its rate. The balance and spring in chronometers are so proportioned and adjusted as to make these unnecessary.

Free Turbine. A separate turbine which is not mechanically connected to the remainder of the engine, as for example the free turbine in some turboprop engines which drives the propeller.

Free Vibration. See **Vibration.**

Freewheel. (*a*) An automatic overrunning device with similar operation to overrunning the ratchet in a pedal bicycle.

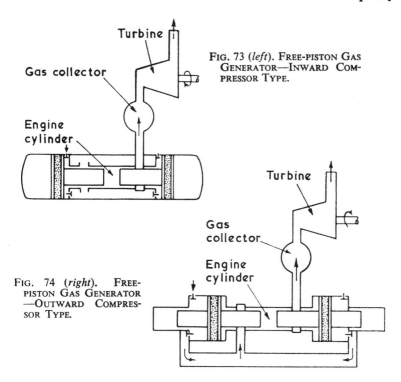

FIG. 73 (*left*). FREE-PISTON GAS GENERATOR—INWARD COMPRESSOR TYPE.

FIG. 74 (*right*). FREE-PISTON GAS GENERATOR —OUTWARD COMPRESSOR TYPE.

(*b*) A mechanical one-way clutch, as in an automobile, depending on the wedging action of rollers in the transmission line to transmit torque only when the engine is driving.

Fremont Test. A *notched-bar impact* test in which a beam specimen notched with a rectangular groove is broken by a falling weight.

Frequency (Periodicity). The number of vibrations, cycles or waves of a periodic phenomenon per second. (Units: Hertz, Hz.)

Natural Frequency. The frequency of a *free vibration.* See **Vibration.**

Fundamental Natural Frequency. The lowest natural frequency.

Undamped Frequency. The frequency of free *vibration* of a system when undamped.

Angular Frequency. The frequency of a *sinusoidal* quantity multiplied by 2π.

157

Frequency

Anti-resonance Frequency. A frequency at which *anti-resonance* occurs.

Resonance Frequency. A frequency at which resonance occurs.

Dominant Frequencies. The frequencies of significant maxima in the *amplitude spectrum.*

Fundamental Frequency. The reciprocal of the *period.*

Order of Frequency. The number of cycles of an engine vibration occurring during one revolution of the engine shaft, such as, half-order, first-order, etc.

Sub-harmonic Frequency. A frequency of which the *fundamental frequency* is an integral multiple, such as half-order.

Fret Saw. (*a*) A very shallow and narrow saw with small teeth held under tension in a frame used for cutting ornamental patterns and small wooden parts.

(*b*) An endless band saw of the same type as (*a*) mounted on a stand and worked by a treadle.

(*c*) A keyhole saw, a saw with a straight, long and narrow tapering blade.

Fretting. Wearing away slowly by friction between two surfaces similar to sharpening a cutting tool on an oil stone. Fretting is an undesirable phenomenon.

Friction. The sliding resistance to the relative motion of two bodies in contact with each other.

Kinetic Friction. The value of the *limiting friction* after slipping has occurred, being slightly less than the *static friction.*

Limiting Friction. The frictional force, which when increased slightly, will cause slipping.

Rolling Friction. The frictional force during rolling as distinct from sliding.

Static Friction. The value of the *limiting friction* just before slipping occurs.

Coefficient of Friction. The ratio of the *limiting friction* to the normal reaction between the sliding surfaces; the ratio is constant for a given pair of surfaces under normal conditions.

Friction Angle. The angle that the resultant force makes to the normal to the surface over which a body is sliding when friction is present.

Friction Axis. In a link chain, the actual line of thrust or pull on the link, which will not coincide with the axis of the link due to the effect of friction.

Friction Back Gear. A lathe device which puts the back gear in and out of engagement by means of a friction clutch, within or next to the cone pulleys, while the lathe is running.

Friction Block. *Brake block.*

Friction Brake. See **Dynamometer.**

Friction Clutch. See **Clutch.**

Friction, Coefficient of. See **Friction.**

Friction Coupling. *Friction clutch.* See **Clutch.**

Friction Disc. A pair of revolving discs with a limited axial movement enclosing a smoothed turned wheel with its axis at right angles. When contact is made between the smooth wheel and one or other of the disc wheels, the foremost rotates on its spindle, and by moving one or other of the disc wheels, which rotate in the same direction, the driven wheel and spindle will rotate in opposite directions. This device is used for reversing the motion of traversing cranes.

Friction Drive. A drive in which one wheel, pressed into contact with a second wheel, causes the second wheel to rotate by the agency of the friction force between them.

Friction Gear. A gear transmitting power from one shaft to another through the tangential friction between a pair of wheels pressed into rolling contact. One contacting surface is usually faced with fabric and the gear is only suitable for small powers. Grooves on the circumferences, which are counterparts of one another, are sometimes provided. The gear uses the same principle as a *disc clutch.* See **Clutch.**

Friction Hoist. A light *hoist* driven by the agency of the friction force between the smooth surfaces of pulleys in contact with each other.

Friction Horse-power (f.h.p.). That part of the *indicated horse-power* developed in an engine cylinder which is absorbed in frictional losses, being the difference between the indicated and *brake horse-power*.

Friction Pulley. A friction wheel with the rim in contact with the rim surface of another pulley.

Friction Ring. A loose metallic ring cut through at one point which is pressed outwards against a female portion by means of a lever. The device is used in some forms of friction clutches.

Friction Rollers. See **Anti-friction Bearing.**

Friction Wheel. Any wheel which drives or is driven by friction, the contact being between smooth or grooved surfaces.

Frictional Damper. See **Damper.**

159

Frictional-rest Escapement.   See Escapement.

Frog. (*a*) The point of intersection of the inner rails, in the form of a vee, where a train crosses from one set of rails to another.

(*b*) A metal stop on a power *loom* to stop its machine whenever the shuttle is trapped in the warp.

Front Rake.   See Rake.

Froude Brake.   A rotor inside a casing, with both free to rotate, and the space between the two filled with water. The torque absorbed by the dissipation of the energy by eddy formation and heat is measured by the torque necessary to prevent rotation of the casing.   Cf. **Fluid Flywheel.**

Frozen Stress Technique.   A technique in *photo-elasticity* where the stresses, which occur under load at room temperature, are retained in the no-load condition by heating the photo-elastic material to about 390 K whilst under load and cooling down over 10–12 hr.   The material can then be examined and sliced up under no-load conditions at room temperature.

Fuel Manifold.   See Manifold.

Fuel Trimmer.   A device for resetting in flight the automatic fuel regulation of a gas-turbine by *barostat* to meet changes in ambient pressures.

Fulcrum Plate.   The metal plate in an ordinary lift pump which receives the stud about which the handle turns.

Full Gate.   The working of a water turbine when the regulator is opened so that the whole width of the vanes receives the impact of the water.

Full Gear.   When the valves of a steam-engine are set at the position of maximum travel and cut-off for full power.

Full Shroud.   See Shroud.

Full Thread.   A parallel *screw thread* cut to the correct depth for size and pitch.   Cf. **Complete Thread** and **Thread.**

Length of Full Thread.   The distance for which the thread is fully formed at the root (see also **Root Diameter).**   It is often measured from the normal plane defining· the end of the thread.

Fullering. (*a*) Producing circumferential grooves on circular forged work using a ' fullering tool ', a tool which consists of a split block internally grooved that is placed round the work and hammered.

(*b*) Caulking a riveted joint to make it pressure tight.

Full-force Feed.   Forced lubrication of an engine, in which the oil goes first to the main and big-end bearings and thence by drilled holes or attached pipes to the gudgeon pins and cylinder walls.

Full-load. The normal maximum load under which an engine or machine is designed to operate continuously.

Full-plate Watch. A watch with circular top plate and the balance mounted above the plate.

Fundamental Circle. *Base circle.*

Fundamental Frequency. See **Frequency.**

Fundamental Triangle. A triangle with two sides representing the form of a screw thread with sharp crest and roots, and pitch and flank angles the same as the *basic form*; and with the base parallel to a generator of the cylinder, or cone, on which the thread has been formed. (See Fig. 157.)

Height (or Depth) of the Fundamental Triangle. The length of the perpendicular from the *apex* to the base.

Funicular Railway. A railway with rack and pinion rails or clutches to grasp the sides of the rails, as found when the gradients are very steep.

Furnace Hoist. A hoist for raising materials to the platform of a melting furnace.

Furrowing. *Grooving.*

Fusee. A spirally grooved pulley of gradually increasing diameter to equalize the pull of the main spring. The *great wheel* is attached to the fusee which is connected to the main-spring fusee barrel by a fine-linked *fusee chain* or gut line.

Fusee Arbor. The *arbor* on which the fusee is mounted.

Fusee Stop-work. A clock mechanism to prevent over-winding consisting of an arm which fits freely into a slot in a brass block screwed to the front plate and is positioned in direct line with the fusee chain (wire or gut line), so that, when the last turn is reached, a poke engages with the hook on the stop-work arm and the fusee cannot then be turned. See **Fusee Poke.**

Hollow Fusee. A fusee with its top pivot sunk into the body in order to reduce the height of the movement.

Fusee Barrel. See **Barrel** (*f*).

Fusee Chain. A fine-linked chain connecting the fusee to the main-spring *barrel*.

Fusee Engine. A special lathe for the cutting of fusees.

Fusee Great Wheel. See **Fusee.**

Fusee Poke. A snail-shaped piece on the smaller end of the *fusee*. See also **Snail.**

161

G

g. The symbol for gravitational acceleration.

Gab Lever. The lever connecting the *slide-valve* spindle and the *eccentric rod* in some marine-engine valves.

Gage Block (U.S.). *Gauge.*

Gaiting (Gait). Preparing a loom for weaving by placing the *warp* in position with *healds* and *reed.*

Gang. (*a*) A train of mining tubs or trucks.

(*b*) The joining together of two prime movers or locomotives.

(*c*) A series of gears or machine tools connected together.

Gang Die. A die for a multiple punching machine.

Gang Milling. Using several milling cutters on one spindle to produce a required profile or to mill the sides and face of a work in one operation. Cf. **Straddle Milling.**

Gang Saws. A number of parallel saws in one frame to cut simultaneously a log into many strips.

Gang Tool. A tool-holder with a number of cutters.

Gantrees. Wooden supports above a loom upon which the *jacquard* is carried.

Gantry. The trussed girders in an *overhead travelling crane.*

Gantry Crane. See **Crane.**

Gap Bed. A lathe bed with a gap near the *headstock* to permit the turning of large flat work of greater radius than the centre height.

Gap Bridge. A casting of the same cross-section as the bed of a *gap-bed* lathe to close the gap when not required for use.

Gap Gauge. A pair of anvils held in a rigid frame, often of C-shape, to check the dimensions of shafts, external threads, etc. (see also **Snap Gauge**). It may be solid for a go or no-go gauge, or adjustable. See also **Micrometer.**

Gap Lathe. A lathe provided with a *gap bed.*

Garter Spring. An endless band formed by connecting the two ends of a long helical spring, and which, when stretched round a circular piece, exerts a uniform radial force. It is found in some *carbon glands.*

Gas-engine. An *internal-combustion engine* working on the *Otto cycle* in which gaseous fuel is mixed with air to provide the combustible mixture in the cylinder, the mixture being fired by spark ignition. In the two-port two-cycle engine, admission and exhaust occur at the same time; in the three-port engine, a carburetter can be used without a check valve, as the port is opened and closed by the piston.

Gas-engine Starter. (*a*) A small engine used for pumping a gas/air mixture into the cylinder of a large gas-engine. Cf. **Gas Generator.**

(*b*) A compressed-air supply for starting a large gas-engine.

Gas Exhauster. A large, low-pressure rotary vane pump or a centrifugal blower used to exhaust gas from gas-works' retorts.

Gas Generator. A gas-producing unit forming a source of power which is used as a starter for a turbojet engine or to drive an auxiliary power unit in an aircraft. A common type of gas generator is a compact compressor/annular combustion/single turbine assembly. Cf. **Free-piston Gas Generators, Gas-engine Starter, Gas Starter.**

Gas Meter. A mechanical device for measuring the amount of gas flowing through a pipe. Fig. 75 shows a mechanical gas meter of the semi-positive type which works on the principle of the *Roots blower*. The two horizontally mounted impeller shafts are geared together at each end. The impeller shape ensures that the two are always in contact with each other and with the walls via a scraper tip at each end. Gas enters from the top, the *pressure differential* then causes the impellers to rotate thus allowing the gas to pass down by the outside case. The pressure drop in the meter is proportional to the gas velocity. The flow range is from approximately 10% of the maximum up to 150% at overload. The accuracy guarantee is about ±1%.

Gas Meter, Wet Type. The gas meter consists of a drum, revolving in a cylinder which is more than half filled with water and divided into compartments by partitions. (See Fig. 76.) One end of each partition must always be below the water surface. Gas enters by the inlet *I* and escapes by the outlet *O*, the water level *WL* being as shown and filling two of the compartments. When the gas is drawn from one compartment, the pressure of incoming gas rotates the drum and brings another compartment into communication with the outlet. As each compartment fills and empties a revolution indicator attached to the drum reads the delivered volume of the gas.

Gas Ports. (*a*) The inlet passages to a gas-engine cylinder.

(*b*) The inlet and outlet passages to the cylinders of internal-combustion engines.

(*c*) A general term used for the tubes or pipes leading into a larger volume and usually closed by an inlet valve.

163

Gas Pump

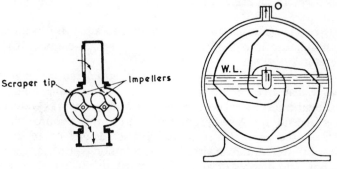

FIG. 75. GAS METER. FIG. 76. WET-TYPE GAS METER.

Gas Pump. A small pump for forcing gas into the combustion chamber of some gas-engines. See also **Humphrey Gas Pump.**

Gas Regulator. (*a*) An automatic valve for maintaining a steady gas pressure in gas-supply mains.

(*b*) The throttle valve of a gas-engine.

(*c*) The manually set thermostat in domestic gas-fired equipment. (See Fig. 77.)

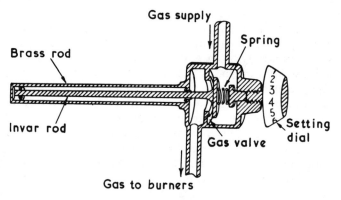

FIG. 77. GAS-OVEN REGULATOR.

Gas Ring. A spring ring for maintaining a gas-tight seal between the piston and the cylinder wall. Cf. **Junk Ring.**

Gas Slide. The slide valve of a gas-engine regulating the gas supply to the combustion chamber.

164

Gas Starter. An aero-engine starter which supplies the normal explosive mixture to the cylinder from an external source and explodes it at the beginning of the power stroke.

Gas Thread. Screw threads of fine pitch, standardized for the wrought-iron tubes used in commercial and domestic gas supplies, which ensure that the joints are gas tight. Also known as *British Standard Pipe Screw Thread (B.S.P.).*

Gas Turbine. An internal-combustion engine in which the burnt gases of combustion are utilized by doing work on a turbine. Cf. **Jet Engine.**

Gaseous Steam. *Superheated steam.*

Gasket. (*a*) A sandwich of an asbestos compound between thin copper sheets for making gas-tight joints between engine cylinders and heads, etc.

(*b*) Packing material such as cotton rope impregnated with graphite grease for packing stuffing-boxes on pumps, etc.

(*c*) Any ring or washer of packing material.

(*d*) A soft thin metal sheet with ridges which partially flatten on assembly.

Gate. (*a*) A valve controlling the supply of water in a conduit. See **Gate Valve.**

(*b*) The annular opening through which the water passes into the vanes of a turbine.

(*c*) In general, a movable barrier to the passage of a fluid or gas, especially sliding and tilting gates.

Rolling Gate. A gate supported on a cylindrical drum which can be rolled up a steep incline such as by chains wrapped round the cylinder ends and connected to hoists.

Taintor Gate. A gate widely used in the United States for crest control. The face of the gate is part of the peripheral surface of a cylinder and the centre of the segment is the pivot point so that the resultant of the water pressure against the gate has to be handled when opening and closing it.

Gate Valve. A valve which provides a straight-through passage for the flow of a fluid. The gate is moved between the body seats by a stem whose axis is at right angles to that of the body ends which are themselves in line. The actuating thread of the *stem* is either contained inside the valve or is exterior to the *bonnet.*

The following are gate valves: *wedge gate valve, sluice valve, double disc gate valve, parallel slide valve.*

Gate Wheel. A toothed wheel controlling the various gates by which the opening and closing of the ports of an inward flow hydraulic turbine are effected. A key or crank turns a small pinion gearing into the gate wheel.

Gathering Pallet. See **Pallet** and Fig. 31.

Gauge. (*a*) An instrument to determine dimensions, capacity, etc.

(*b*) An accurately dimensioned piece of metal for checking dimensions, such as *master gauge* and *workshop gauge.*

(*c*) A measuring tool, such as a *micrometer gauge.*

(*d*) The distance between the inner edges of rails or tramways.

(*e*) The diameter of wires and rods on some specified schedule.

See also **Check Gauge, Go Gauge and No-Go Gauge, General Gauge, Inspection Gauge, Limit Gauge, Plate Gauge, Plug Gauge, Position Gauge, Projection Gauge, Receiver Gauge, Reference Gauge, Ring Gauge, Rail Gauge, Setting Gauge, Slip Gauge, Snap Gauge, Standard Gauge, Wire Gauge** and **Gauges commonly used.**

Gauge Cocks. Small test *cocks* fitted to the outside of a vessel to indicate the liquid level within.

Gauge Diameter. The basic major diameter of a taper screw thread, whether external or internal.

Gauge Length. The distance on an external taper screw, at the pipe end, from the *gauge plane* to the small end of the screw, measured parallel to the axis.

Gauge Plane. In a taper screw thread, the plane perpendicular to the axis at which the *major cone* (see **Screw Thread Diameters**) has the *gauge diameter.*

Gauge Plate. An adjustable plate fitted to shearing and similar machines to ensure the uniform length of short pieces of bar or plate cut off by the machine.

Gauge Rod. *Plug gauge.*

Gauges commonly used (dimensions in inches):

Name	Gauge no.			
	0	1	20	36
S.W.G.	0·324	0·300	0·036	0·0076
B.W.G.	0·340	0·300	0·035	0·004
A.W.G.	0·325	0·289	0·032	0·0050
Steel wire	0·307	0·283	0·035	0·0090
B.G. (plate)	0·396	0·353	0·039	0·0061
U.S. steel plate	—		0·0359	0·0067

No I.S.O. recommendations exist for SI dimensioned gauges for wire and sheet metal, but BS preferences are given in BS 4318 : 1968 and BS 3737 : 1964.

Gear. (*a*) Any mechanical system for transmitting motion.

(*b*) The transmission of rotation by *gear-wheels*.

(*c*) A *gear ratio* as in transport vehicles, such as first gear, etc.

(*d*) The positions of the valve mechanism in a steam-engine such as astern gear, etc.

(*e*) A set of tools for performing a particular task. See also **Gears.**

Anti-backlash Gear. A double gear comprising two gear-wheels next to each other as shown by *A* and *B* in Fig. 78. One is

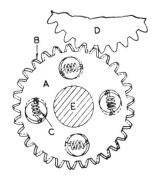

FIG. 78. ANTI-BACKLASH GEAR.

fixed rigidly to shaft *E* whilst the other is spring-loaded against the first through spring *C*. The combination eliminates *backlash* between *D* and *E*. Cf. **Split Gear.**

Gear-box. (*a*) The complete system of gear-wheels for changing the speed from that of an input shaft to that of an output shaft, or changing the direction of rotation, or changing the actual direction of a shaft with or without a speed change. (See Fig. 79.)

(*b*) The box containing the system of gear-wheels.

Gear Cluster. A set of gear-wheels integral with, or permanently attached to, a shaft such as the *lay shaft* in the *gear-box* of an automobile.

Gear-cutters. *Milling cutters*, *hobs*, etc., with the correct tooth form for cutting teeth on gear-wheels.

Gear-grinding Machine. A machine, for grinding a gear to remove slight distortion after heat-treatment, which uses either a formed wheel which fits the space between two gear-teeth or two flat wheels.

Gear Miller. A milling machine for cutting gear-teeth with a milling cutter of the correct shape.

Gear Pump

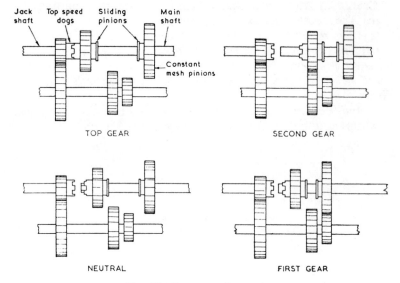

Jack shaft — Top speed dogs — Sliding pinions — Main shaft — Constant mesh pinions

TOP GEAR

SECOND GEAR

NEUTRAL

FIRST GEAR

FIG. 79. GEAR-BOX OPERATION.

Gear Pump. A small pump used for lubricating systems and the like, which delivers fluid through the tooth spaces of a pair of gear-wheels in mesh and enclosed in a box. See also **Roots Blower**.

Gear Ratio. (*a*) The ratio of the rotational speeds of the input and output shafts of a *gear-box*.

(*b*) The ratio of the greater of the two numbers of teeth in a pair of gears to the smaller number. In the case of a segment the number taken is for the corresponding complete wheel. (*a*) is the inverse of (*b*).

Gear Train. A set-up of gears on a lathe to secure a particular rate of tool traverse per revolution of the *chuck*, as required in screw cutting.

Gear-wheel. Any form of toothed wheel.

Geared Chuck. A geared form of *universal chuck*.

Geared Engine. An engine with gearing between the crankshaft and the other shafting for changing speed or for reversing.

Geared Flywheel. A *flywheel* with teeth on its periphery. The teeth are added to provide a big *gear ratio* such as between an *internal-combustion engine* and its *starter*.

Geared Headstock. A *headstock* fitted with *back gear*.

168

Geared Lathe. A lathe with a *back gear* or a multi-speed gearbox between the driving motor and the head.

Geared Locomotive. (*a*) A locomotive, working on steep inclines, having bevel pinions mounted upon their shafting and working into bevel wheels on the side of the main wheels, so that every wheel in engine and tender becomes a driving wheel.

(*b*) An electric locomotive in which the motors drive the axles through reduction gears.

Geared Pump. A power pump or series of pumps driven by a source of power through spur gearing.

Gearing. A system of gear-wheels transmitting power.

Gearing-down. A speed reduction from the driving wheel (or unit) to the driven wheel (or unit).

Gearing-up. A speed increase from the driving wheel (or unit) to the driven wheel (or unit).

Gears. For different types of gear see under: **Bevel Gear, Bevel Wheels, Differential Gear, Epicyclic Gear, Friction Gear, Helical Gear, Interchangeable Gear, Internal Spur Gear, Skew Bevel Gear, Spiral Gear, Spur Gear, Synchromesh Gear, Variable Gear, Worm Gear.**

General Gauge. A gauge designed to serve, under suitable tolerance limits, either as a *workshop gauge* or as an *inspection gauge*.

Generating Circle. Any circle in which a point on the circumference traces a curve when the circle rolls along a straight line (a cycloid) or along a curve.

Generating Line. A straight or curved line rotated about an axis in order to generate a surface.

Generator. (*a*) An apparatus for producing gases, steam, electricity, etc.

(*b*) A *generating line.*

Generator (Straight-line Generator). A straight line generating a *ruled surface.*

Geneva Wheel. See Maltese Cross Mechanism.

George. *Autopilot.*

Gib. (*a*) A metal piece which transmits the thrust of a wedge or *cotter*, as in some *connecting-rod* bearings.

(*b*) A brass bearing surface let into the working face of a steam-engine *crosshead.*

Gibbet. The triangular framework of a crane, consisting of post, *jib* and strut.

169

Gib-headed Key. (See **Key** (*a*).) A key with a head formed at right angles to its length for securing a wheel, etc., to a shaft.

Gills. Controllable flaps for varying the outlet area of the airflow through a radiator or from the cowling of an air-cooled engine. Also called ' radiator flaps ' and ' cowl flaps ' respectively.

Gimbal Joint. See **Universal Joint.**

Gimbals. *Gymbals.*

Gin. (*a*) A small *hoist* consisting of a chain or rope barrel supported in bearings and turned by a crank.

(*b*) A portable tripod carrying lifting tackle.

Gin Block (Monkey Wheel, Whip Gin). A single sheave pulley of hollow-rim section with its bearings in a skeleton frame suspended from a hook.

Gin Pulley. The pulley of a *gin block*.

Girard Turbine. See **Turbine.**

Gland. (*a*) A device to prevent leakage at a point where a shaft emerges from a vessel containing fluid under pressure or from a vacuum.

(*b*) A sleeve or nut of one-piece or two-piece design which retains and forms a means of compressing the packing in a *stuffing box*.

Screwed Gland. A gland which is adjusted by a special nut, the gland nut, to engage with a *stuffing box*.

Gland Packing. Material inserted into a gland to prevent leakage of fluid.

Gland Box. *Stuffing box.*

Glass Barrel Pump. *Acid pump.*

Glass-papering Machine. A revolving cylinder or disc or flexible moving band covered with fine ground glass brought to bear on the surface of woodwork.

Glazing. The filling of the surface of a grinding wheel with minute abraded particles so that it becomes smooth and polished and no longer grinds efficiently.

Globe Valve. A screw-down valve with the casing or body of a spherical shape. The axis of the *stem* is at right angles to the body ends which are in line with each other. Cf. **Oblique Valve** and **Angle Valve.**

Glow Plug. (*a*) An electrical igniting plug for switching on to ensure the automatic re-lighting of a gas turbine if the flame becomes unstable, as under icing conditions.

(*b*) A heater plug installed in each combustion chamber of some diesel engines. The heaters are switched on prior to cold starting and are switched off after the engine has started.

Go and No-Go Gauges. Gauges used to measure the maximum and minimum metal limits of the work and usually for each independent dimension on the work. Alternatively, ' go and not-go gauges '.

Going-barrel. A barrel in which the winding takes place from the *barrel arbor*. Power is transmitted direct by teeth on the barrel to the train of wheels in the watch or clock.

Going Fusee. A *fusee* with maintaining power.

Going Part. That part of a weaving loom known as the *sley* or *batten* which beats up the weft to the edge of the woven cloth.

Gold Spring, Passing Spring. See **Spring-detent Escapement,** under **Escapement.**

Goliath Crane. See **Crane.**

Gooseneck (Tool). (*a*) A tool for giving a finishing cut, the body of the tool having a semi-circular portion as part of the stem.

(*b*) A special press die for forming flanges in sheet metal, Fig. 80.

Fig. 80. Gooseneck, or Return, Flanging Die.

Gorge. The groove of a sheave pulley in which the rope or chain runs.

Gorge Radius of a Wormwheel. The radius of the curved throat section in a plane containing the wormwheel axis. See also **Throat Diameter.**

Governor. A mechanism for governing speed by centrifugal force or by pressure, whereby, commonly, heavy balls rotating by the motion of an engine move outwards at higher speeds, and inwards at lower speeds, to close or open a valve controlling the fuel, steam, water or hydraulic supply. See **Centre-weighted Governor, Crossed-arm Governor, Inertia Governor, Isochronous Governor, Marine Governor, Pendulum Governor, Pickering Governor, Shaft Governor, Spring-loaded Governor, Watt Governor.** (Fig. 83.)

Governor Arms. The rods connecting the *governor* sleeve to the governor balls. See also **Centrifugal Speedometer,** Fig. 167.

Governor Balls. Freely revolving masses, usually spherical, whose centrifugal force controls a throttle valve or other regulator of the speed of an engine.

Governor Cut-off. See **Automatic Expansion.**

Governor (Strapping) Motion. A mechanism regulating the revolutions of the spindle in building the *cop* bottom, as part of a *mule* cotton-spinning frame.

Governor Sleeve. The hollow cylinder which slides vertically on a *governor* spindle and carries the governor arms.

Governor Spindle. The vertical spindle which carries the *governor sleeve* and around which the sleeve revolves.

Governor Valve-gear. The governor-controlled valve-gear which regulates the opening and the closing of the induction valve in an *automatic expansion.*

Grab (Grab Bucket). A steel bucket made of two halves hinged together so that they dig out and enclose part of the material on which they rest. They are used in mechanical excavators and *dredgers.*

Grab Dredger. A *grab* suspended from the head of a *crane's jib*, with the crane raising and lowering the grab. Also called ' grapple dredger '.

Grabbing Crane. An *excavator* consisting of a crane carrying a large grab or bucket in the form of a pair of half-scoops, so hinged as to scoop or dig into the earth as they are lifted. See also **Grab.**

Grade. (*a*) The degree of slope of a road or railway. Also called ' gradient '.

(*b*) The steps of belt pulleys.

(*c*) The hardness of a grinding wheel.

(*d*) The quality of a casting.

Grade of a Feature. See **Feature.**

Grade of a Fit. That characteristic determined by the grades of the *features* of the mating pair. See also **Screw Threads.**

Grade (Class) of a Screw Thread. That characteristic determined by the relationship between the *tolerance*, the associated *allowance* and its *basic size*.

Grain Rolls. Rolls made of a tough quality of cast-iron, not chilled. Cf. **Chilled Rolls.**

Grapple Dredger. *Grab dredger.*

Grating. The perforated plate in a foot-valve or air-pump for filtering out solid matters.

Gravity Conveyor. A conveyor in which the weight of the articles is sufficient to ensure their transport from a higher to a lower level, such as sliding down an inclined runway.

Gravity Escapement. See **Escapement.**

Gravity Plane. An inclined plane on which descending full trucks pull up the ascending empty ones.

Gravity Wheel. A water wheel in which the weight of the water alone is utilized. *Overshot wheels* are gravity wheels. Cf. **Undershot Wheel.**

Grease Box. The upper portion of an *axle-box* containing the grease used for lubrication.

Grease Cock. A cup with pipe and stop-cock screwed into an engine cylinder or bearing housing to receive and regulate the supply of grease for lubricating the moving part.

Grease Cup. A cylindrical cup threaded internally and filled with grease with a nipple screwed into a bearer housing, thus feeding grease into the bearing when the cup is screwed down hard against the grease in the cylindrical reservoir.

Grease Gun. A cylinder, filled with grease, from which the grease is delivered by hand pressure on a piston, intensified by a second plunger which forms the delivery pipe and is pressed against a nipple screwed into the bearing that needs lubricating.

Great Wheel. (*a*) The first wheel in the train of a watch or clock which in *going-barrels* forms part of the *barrel*.

(*b*) The largest wheel in a watch or clock *train*. A three-train clock has a Quarter Great Wheel, a Going Great Wheel and a Striking Great Wheel.

(*c*) A very large *overshot wheel* like the one in the Isle of Man.

Grid. *Grating.*

Gridiron (Harrison's) Pendulum. See **Pendulum.**

Gridiron Valve. A *slide valve* with the ports in the valve and in the cylinder face subdivided transversely by narrow bars. The valves may be double or treble ported.

Grinder. (*a*) A machine tool for shaping pieces to exact size by means of rotating discs of emery, diamond, or other abrasive material.

(*b*) An emery wheel for grinding tools.

(*c*) A large thick circular grit-stone used in the manufacture of mechanical wood pulp. Cf. **Crusher.**

Grinding. The abrasion of metal surfaces by emery wheels, lapping, etc. See also **Honing.**

Grinding Clamps. A divided and adjustable lap used for grinding mandrels and cylindrical holes.

Grinding-in. The process of obtaining a pressure-tight seal between a conical-faced valve and its seating by grinding the two together with an abrasive mixture such as carborundum and oil.

Grinding Machine. A machine tool in which flat, cylindrical or other surfaces are finished by the abrasive action of a high-speed grinding wheel. See also **Centreless Grinding, Cylindrical Grinding, Profile Grinding, Surface Grinding, Thread Grinding.** Cf. **Honing Machine, Lapping.**

Grinding Truer. A rotating and pointed bar of steel held against a grindstone or a threaded roller of steel, clamped in a frame and allowed to rotate against the surface of the stone. See also **Dresser.**

Grinding Wheel. A wheel, composed of an abrasive powder, such as carborundum or emery, cemented with a binding agent, and fused for cutting and finishing metal. See also **Rubber Bond Grinding Wheel.**

Grip Chuck. A lathe chuck with movable jaws.

Grit. The grit, or grain, of a grinding wheel refers to the size of the abrasive particles which have standard numbers defining the number of meshes to the inch through which the grit will pass.

Grooving (Furrowing). The recessing, by revolving cutters or by a small thick circular saw with widely spaced teeth, of the edges of boards to receive the tongues of corresponding boards. Cf. **Tongueing.**

Grooving Saw. A circular saw used for cutting grooves. See also **Drunken Saw.**

Ground-effect Machine. *Air-cushion vehicle.*

Ground Resonance. Rapidly increasing oscillations of the rotor of a rotor craft when run up on the ground, due to the reaction between the dynamic frequency of the rotor and the natural frequency of the alighting gear.

Ground Wheels. (*a*) The travelling wheels of a portable crane.

(*b*) Wheels attached to seaplane hulls for maintenance when on land, instead of using trolleys.

Ground-off Saw. A saw with a very thin edge almost parallel for about 40 mm, after which one side is ground with a slightly concave taper until it runs out, leaving the thick portion of the plate in the centre of the saw.

Grub Screw. In general a screw with no *head* but with a slot across the top end for the insertion of a screwdriver. Various types are illustrated in Fig. 81.

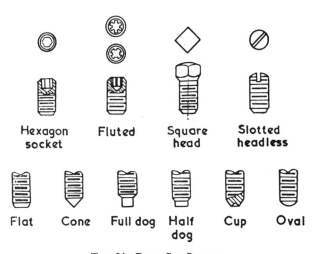

Hexagon socket   Fluted   Square head   Slotted headless

Flat   Cone   Full dog   Half dog   Cup   Oval

FIG. 81. GRUB SET SCREWS.

Grummet Washer (Gromet Washer). (*a*) A washer made of spun yarn or tar twine, etc., used to make a watertight joint under the head of a square-shouldered bolt.

(*b*) A hollow rubber or plastic washer fitted in a hole to permit electric cables, etc., to pass through a hole without chafing.

Guard. A contrivance to protect personnel from injury or accident. Cf. **Shield.**

Guard Finger. A pin used to limit the motion of a part of a mechanism. Cf. **Safety Finger.**

Guard Pin. *Safety finger.*

Guard Plate. (*a*) A fixed sheet-steel plate in front of machinery to protect personnel.

175

Guard Plate

(*b*) A curved plate in a rubber disc valve to limit the movement of the disc.

Guard Rail. *Check rail.*

Guard Straps. *Splashers.*

Gudgeon. (*a*) A pivot at the end of a beam or axle.

(*b*) A cross-shaft at right angles to a *piston rod* or pump rod, connecting the rod and the *crosshead*.

Gudgeon Pin (Gudgeon Wrist Pin). The pin connecting the piston of an internal-combustion engine with the bearing of the little end of the *connecting-rod.* Also called ' piston pin '.

A ' floating gudgeon pin ' is free to revolve in both the *connecting rod* and the piston bosses.

Guide. (*a*) An attachment or contrivance which controls the movement of any part of a mechanism along a predetermined path.

(*b*) A *check rail.*

(*c*) A chisel-shaped attachment to a *rolling mill* to lift the rail or bar off the rolls on the leaving side and thus prevent it from turning around the roll.

Guide Bars. Bars with flat or cylindrical faces for guiding the *crosshead* of a steam-engine and thus avoiding a lateral thrust on the *piston rod.*

Guide Blades. The fixed vanes within a turbine or a compressor which direct the fluid at the proper angle on to the rotating vanes. See also **Guide Vanes.**

Guide Blocks. *Slide blocks.*

Guide Plates. *Ramps.*

Guide Pulley. A loose pulley or *idler* used to guide the direction of a driving-belt wire or cable or to prevent the belt from coming into contact with an obstruction.

Guide Rail. *Check rail.*

Guide Screw. *lead screw.*

Guide Screw Stock. A *die-stock* in which the dies are divided into three portions, one being the guide and the other two the actual cutters which are placed in radial slots.

Guide Vanes. Vanes, similar in shape to aerofoils, which guide the airflow in a duct or wind tunnel. See also **Guide Blades.**

Guillotine. *Trimming machine.*

Guillotine Shears. A *shearing machine* with the shears parallel with the plane of the machine framework, which is used for the cutting up of puddled bars and slabs. Cf. **Trimming Machine.**

Gullet. A depression or gap cut in the face of a saw in front of each tooth, alternately on one side of the blade and then on the other.

Gullet Saw (Brier-tooth Saw). A saw with *gullets* cut in front of each tooth. (See Fig. 143.)

Gullet Tooth. A saw tooth with a *gullet* cut away in front of it. (See Fig. 143.)

Gulleting Machine. A machine for grinding *gullets* in front of the teeth of a saw.

Gusset (Gusset Plate). A flat plate used to stiffen the joints of a framework; frequently used in conjunction with angle-iron structures.

Guy Derrick. A *crane* operating from a mast which is held in an upright position by guy-ropes.

Gymbals (Gimbals). (*a*) A mechanical frame containing two mutually perpendicular axes of rotation. (See Fig. 82.)

(*b*) Self-aligning bearings for supporting, and keeping level, a chronometer in its box.

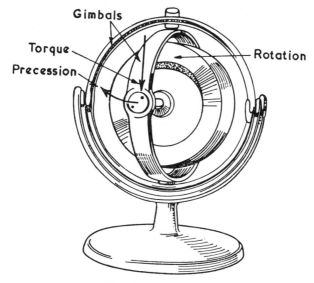

FIG. 82. GYROSCOPE.

Gyration, Centre of. *Centre of gyration.*
Gyration, Radius of. *Radius of gyration.*
Gyro. See Gyroscope.

Gyro (Gyro Compass). A type of compass which relies on the principles of rigidity of a *gyroscope* for its direction-finding ability. It embodies a spinning rotor or flywheel, driven at very high speed by an electric motor or by an air blast, with the axis of spin usually horizontal and mounted in almost frictionless bearings in a double casing. The casing allows freedom of movement about the spinning, horizontal and vertical axes so that the rotor behaves as a completely free gyroscope and thus tends to maintain its direction in space.

Gyro Integrator. A device in which the total angle of precession of a gyroscope is a measure of the time-integral of the input torque with the gyro precessing at a rate proportional to the torque applied to its gymbal.

Rate Gyro. A gyro device measuring the rate of change of direction of an axis in space and mounted in a single spring-restrained gymbal. A torque is exerted on the gymbal proportional to the rate of rotation of the unit in space about an axis at right angles to both the spin and gymbal axes.

Gyrodyne. A rotorcraft in which the rotor(s) is power-driven for take-off, climb, hovering or landing but autorotates (like an autogyro) for cruising flight. The aircraft has usually short span wings.

Gyroplane. A *rotorcraft* with a freely rotating rotor.

Gyroscope (Gyro). A small heavy wheel rotated at very high speed, usually electrically, in anti-friction bearings. Any alteration in the inclination of the axis of rotation is resisted by a powerful turning movement (gyrostatic moment) so that the axis remains in a constant direction in space, a property used in the *gyro compass* and the *artificial horizon* for guiding aircraft, ships, torpedoes and guided missiles. (See Fig. 82.)

Gyrostat. *Gyroscope.*

H

h.p. *Horse-power.*
H.I. Horizontal interval.
H.P. High pressure.
H.U.C.R. Highest useful compression ratio.
Hack-saw. (*a*) A hand-saw for cutting metal, consisting of a steel frame across which is stretched a narrow saw-blade of hardened steel.

(*b*) A larger similar reciprocating saw, power-driven through a crank and connecting-rod, usually called a ' hacksawing machine '.

Hackworth Valve-gear. A radial gear in which an eccentric opposite the crank operates a link whose other end slides along an inclined guide, the valve rod being pivoted to a point on the link.

Haigh Fatigue-testing Machine. A machine for subjecting test-pieces of materials to alternating direct stress induced by a powerful electromagnet excited by an alternating current.

Hairspring. *Balance spring.*

Half-centre. The position of the *crank pin* of an engine midway between the two dead centres.

Half Coupling. *Flange coupling.*

Half-crossed Belting. Belting which drives between two pulleys with axes at right angles to each other and can only be run in one direction.

Half-lap Coupling. The connection of two co-axial shafts by a *half-lap joint*, the two shafts being either riveted together or enclosed in a keyed-on sleeve.

Half-lap Joint. A joint formed by the process of *halving*.

Half-rip Saw. A hand-saw for cutting timber along the grain with smaller teeth than a *rip-saw*.

Half-round Bit. *Cylinder bit.*

Half-round Screws. *Button-headed screws.*

Half Shroud. A gear-wheel *shroud* extending only up to half the height of a tooth.

Half-speed Shaft. The camshaft of a piston engine which runs at half the speed of the *crankshaft*.

Halving. Cutting away half the thickness of two pieces and so that when they are joined, the outer surfaces are flush, such as in a *half-lap coupling*.

Hammer. (*a*) An instrument for beating, etc., or a metal block used for a similar purpose.

(*b*) The weight, or weighted mass, which strikes the bells, gongs, tubes or rods in striking and chiming clocks. In *repeaters* there are separate hour hammers and minute hammers, and in some clocks, quarter hammers.

Hammer Arbor. The arbor on which a clock hammer is pivoted.

Hammer Block. The steel face of a steam hammer which is attached to the *tup* by means of a dove-tailed joint.

Hammer Blow. The blow caused by the alternating force between the driving wheels of a locomotive and the rails, caused by the centrifugal force of the balance weights that balance the reciprocating masses.

Hammer Mill. A mill in which bars, hinged to discs attached to a horizontal rotating shaft, strike the material enclosed in a cage until it is fine enough to drop through the bottom openings. It is used for soft materials like coal and foodstuffs for animals.

Hammer Pallets. The pallets that fit on to the hammer arbors, two on the ' hour hammer ' and two on the ' minute hammer ' to control the striking in a *repeater*.

Hammer Rods. The vertical rods connecting the lifting cams of a turret clock to the *hammers*.

Hammer Stalk. The rod to which a *hammer* is fastened in a striking or chiming clock.

Hammer Tail. The *hammer stalk* extension which is in contact with the pins in the *pinwheel* or *pin barrel* when the hammer is lifted in a chiming or similar mechanism.

Hammer Tongs. In tool-making, tongs which bend at right angles and are used to hold the work in making hammers and similar work by hand.

Hammer Wheel (Pinwheel, Cam Wheel). In clocks, a wheel geared into the *locking wheel* which has usually eight pins fitted into the band of the wheel. These pins lift and drop the hammer which strikes a bell or gong.

Hand (of a Helical or Spiral Gear). Right hand when successive transverse sections show clockwise displacement with increasing distance from the observer; left hand when counter-clockwise. The hand of a gear may change at different parts of the face width.

Hand Expansion Gear. Variable expansion gear, usually consisting of right- and left-handed screws for adjusting by hand the positions of a pair of cut-off valves relative to the *slide valve*.

Hand Feed. The hand operation of the *feed* mechanism of a machine tool.

Hand Holes. (*a*) Holes cut in parts of machinery so that a hand can be inserted for maintenance, etc.

(*b*) A small hole, cut in the side of a pressure vessel or tank and closed by a removable cover, to provide means of access to the inside of the vessel.

Hand Lift. A lift, consisting of a *sheave* over which an endless rope passes and thus actuates a *worm* and suitable gearing, worked by hand.

Hand Traverse. The hand-operated traverse movement of a machine part.

Hand Wheel. (*a*) Any wheel turned by the hand to operate machinery or mechanisms.

(*b*) A grooved pulley with cranked handle which is mounted on a universal form of vice and used for driving a lathe or other tool by hand.

Handing. Making symmetrical work, right- and left-handed, and altering patterns from left to right hand and vice versa.

Hanging. The fixing of a pulley, etc., upon its appropriate shaft.

Hardness. Resistance to deformation, usually measured by the resistance to indentation by one of various *hardness tests*. Cf. **Toughness.** See also **Scratch Hardness.**

Hardness Numbers (Hardness Scale). An arbitrary scale of numbers determined by various *hardness tests*.

The original scale is that of Mohs as follows: talc (1), gypsum (2), calcite (3), fluorite (4), apatite (5), feldspar (6), quartz (7), topaz (8), corundum (9) and diamond (10). Ridgeway's scale inserts vitreous pure silica between (6) and (7), garnet between (7) and (8), replacing corundum by fused zirconia (11), fused alumina (12), silicon carbide (13), boroncarbide (14), thus making diamond (15). The above scales are based on the ability of each mineral to scratch the ones that come before it on the scale. On Mohs' table, lead is 1·5, iron 3·5–4·5, and case-hardened steel is 8. See **Indentation Hardness.**

Hardness Scale. *Hardness numbers.*

Hardness Tests. Tests determined either by (1) the ability of one solid to scratch another (see **Scratch Hardness**), or (2) the area of indentation formed in a given test (see **Indentation Hardness**). Dynamic tests are made to measure rebound hardness by a *Herbert pendulum* or *Shore scleroscope*. See also **Sclerometer, Scleroscope Hardness Test.**

Harmonic. A sinusoidal component of a periodic quantity with a frequency that is an integral multiple of the *fundamental frequency*. If the multiple is *n*, the harmonic is the *n*th harmonic.

Harness. Strong cords placed in position by a *comber board* as part of a *jacquard machine* by which the warp threads are operated in weaving.

Harness Cord. Varnished linen twine connecting the figuring hooks with the mails that lift the warp threads in a *jacquard loom*.

Harrison's Gridiron Pendulum. See **Pendulum.**

Hartnell Governor. A *spring-loaded governor* with vertical arms of bell-crank levers supporting heavy balls and with horizontal arms carrying rollers which push against the central spring-loaded sleeve operating the governing mechanism.   (See Fig. 83.)

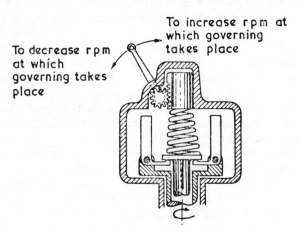

To decrease rpm at which governing takes place

To increase rpm at which governing takes place

FIG. 83. SPRING-LOADED GOVERNOR.

Harvester.   *Binder*.   See also **Thrasher.**

Harvester-thresher.   A machine which is mounted on a chassis and combines the operations of a *binder* and a *thrasher*.

Hat-leather Packing.   A leather packing ring of L-section which is gripped between discs to form a piston or attached to the ram of a hydraulic machine to prevent leakage.

Hawse Pipe (Hawser Pipe).   (*a*) A pipe which guides the barrel chain in some types of *grabs*.

(*b*) A tubular casting in a ship's bow through which the anchor chain or cable passes.

Hay Elevator.   A mechanically-driven endless band fitted with prongs to fork hay on to a rick.

Hay-loader.   A machine worked by its land wheels, which picks up hay and conveys it to the towing wagon by means of a trough and reciprocating rake-bars.

Hay-stacker.   A machine for catapulting hay up on to a rick.

Head.   (*a*) A generic term for the more important and essential part of an apparatus, a machine or a tool.

(*b*) The height of a liquid column and the pressure resulting from that height.

(*c*) The top part of a rail.   See **Rails.**

182

(*d*) The driven part of a screw or bolt.

(*e*) The removable part of an internal-combustion engine above the cylinders or the top of a cylinder.

Head Gates. The gates at the high-level end of a *lock*.

Head Race. A channel conveying water to a hydraulically-operated machine.

Head Valve. The delivery valve of a pump. Cf. **Foot Valve.**

Header. A manifold supplying fluid to a number of tubes or passages, or connecting them in parallel.

Heading Machine. A power press for producing the heads of bolts, rivets and spikes.

Headstock. (*a*) A device (Fig. 102) for supporting the head of a machine such as the fixed or poppet head of a lathe, milling machine or grinding machine; the movable head is called the *tailstock*, ' movable ' or ' loose headstock ' and the live head the ' fixed headstock '.

(*b*) The term is also used for the part of a planing machine which supports the cutters, for the supports of the gudgeons of a wheel and for the movable head of some measuring machines.

(*c*) The end timbers in the under-frame of a railway truck.

(*d*) The part of a textile machine which contains the main gearing and the drive for a beam.

Headstock Motor. A motor to drive the *headstock* of a lathe.

Heald. (*a*) That part of the mechanism which raises and lowers the warp in a *tappet* or *dobbie* loom and consists of an eye formed of twine or wire through which a thread is drawn.

(*b*) The shaft upon which a large number of healds are mounted. Also called ' stave '.

Healding. *Looming*.

Heart Cam. (*a*) A heart-shaped *cam* for the conversion of rotary into rectilinear motion.

(*b*) A heart-shaped *cam* for deriving an oscillatory motion about a pivot not coincident with the cam axis.

(*c*) A heart-shaped *cam* used in stop-watches and chronographs to bring the recording hand instantly back to zero on pressing the button. Also called a ' heart piece '.

Heart Piece. See **Heart Cam** (*c*).

Heat Engine. Any kind of engine that converts heat into mechanical energy. See also **Engine.**

Heat-engine Cycles. See **Carnot Cycle, Diesel Cycle, Internal-combustion Engine Cycle, Otto Cycle, Rankine Cycle, Steam-engine**

Heat-engine Cycles

Cycle, Wankle Engine Cycle.

Heat-exchanger. A device for transferring heat from one medium to another often through metal walls, usually to extract heat from a medium flowing between two surfaces.

Heat Pump. A machine for transferring heat from a lower grade temperature to a higher, extracting heat from flowing water by a refrigerant, which is compressed and transfers its heat to a second flow of water, and finally passes through an expander to repeat the whole process.

Heating of Bearings. *Bearings* are said to heat when their temperatures get so hot that the friction is greatly increased, or the axles or the crankshafts stick fast.

Heddle. A *heald* shaft.

Heel. The rear end of the cutting edge of the tooth of a saw or fluted drill. (See Fig. 143.)

Helical Gear. A cylindrical spur gear in which the paths traced by the teeth are helices. (See Fig. 84.)

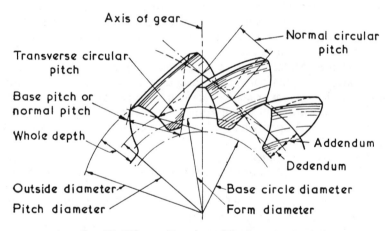

FIG. 84. HELICAL SPUR-GEAR NOMENCLATURE.

Crossed Helical Gear. Helical gears which mesh together on non-parallel axes.

Helical Gear-tooth. That portion of a helical gear which is bounded by the *root* and *tip cylinders* and by the two helicoid surfaces. See also **Double-helical Gear, Worm Thread.**

Helical Rack. (See **Rack.**) The helical rack nomenclature is shown in Fig. 138.

Helical Spring. A spring formed by winding wire into a helix along the surface of a cylinder. Cf. **Spiral Spring.**

Helicoid. The surface generated by a curve uniformly rotated about, and uniformly translated parallel to, an axis.

Helicon Gears. A *spiroid gear* without taper. It can be used at ratios less than 10 to 1. Trade name.

Helicopter. An aircraft with a main lifting *rotor* or rotors driven by power. (Fig. 85.)

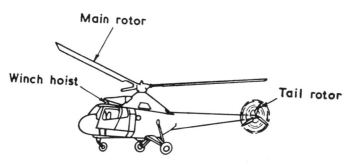

FIG. 85. HELICOPTER.

Helix. A curve traced on a cylindrical or conical surface in such a way that any tangent to the curve makes a constant angle with the intersecting cylinder or core generator. A curve like a screw thread

Helix Angle. The constant acute angle between a tangent to a helix and the intersecting generator of the cylinder, for helical and worm gears, screws, etc.

Herbert Pendulum. A massive pendulum with a 1·588 mm diameter steel ball as a pivot rocks over the surface of a specimen. The period of the pendulum and its rate of damping provide a measure of the hardness and ductility of the metal. See also **Dynamic Hardness Number, Hardness Tests.**

Hercules Crane. See **Crane.**

Herring-bone Gear. **Double-helical Gear.**

High-pressure Cylinder. The cylinder of a compound steam-engine in which the steam is first expanded. See **Compound Engines.**

High-pressure Engine. (*a*) A steam-engine which exhausts directly into the atmosphere.

(*b*) A steam-engine driven by high-pressure steam.

High-pressure Steam. Steam having a pressure considerably higher than atmospheric pressure.

High-speed Steam-engine. A vertical *steam-engine*, usually compound, with the moving parts of the piston valves totally enclosed and pressure lubricated.

Hob. A fluted, straight or helical, rotary cutter to produce spur, helical or worm gears.

Hobbing. Cutting the threads of worm wheels, dies, etc., with a *hob*.

Hobbing Machine. A machine using a *hob*.

Hodograph. A curve drawn through the ends of vectors radiating from a point, the vectors representing the velocities of a body at successive instants. The velocity along the hodograph gives the acceleration of the body along its path.

Hogger Pump. The upper portion of a deep mine pump.

Hoist. An engine with a drum used for winding up a load as in a mine shaft, travelling crane, helicopter, etc. See also **Friction Hoist, Furnace Hoist.**

Hoisting Crab. *Crab.*

Hoisting Machine or Engine. See **Crab, Differential Pulley Block, Hydraulic Lift, Jack, Jigger, Lift, Winch.**

Hole Basis Limit System. See **Limit System.**

Hole-grinding Machine. A machine in which the grinding spindle is supported plumb with the axis of the work being ground.

Hole Plate. *Division plate.*

Hollander (Beating Engine). An engine for producing paper pulp, consisting of a trough that contains a beating roll with bars set parallel to the axis of the trough.

Hollow Fusee. See **Fusee.**

Hollow Mandrel Lathe. A *lathe* with a hollow *mandrel* capable of having bar stock fed through it for repetition work.

Hollow Pinion. See **Pinion.**

Hollow Shaft. A tubular shaft.

Hollow Spindle Lathe. *Hollow mandrel lathe.*

Hone (Oilstone). A smooth stone used either dry or moistened with oil or water, to give a fine keen edge to a cutting tool.

Honey Extractor. A large cylindrical drum with centre spindle and rotating frame in which honeycombs, after uncapping of the cells, are placed for the extraction of the honey by centrifugal force.

Honing. The use of small formed abrasive stone slips mounted in a revolving holder, such as are used for finishing cylinder bores to a very high degree of accuracy; between 100 and 400 nm is the general rule with 25 nm possible. Cf. **Lapping.**

Honing Machine. A machine for finishing cylinder bores, etc., to a very high degree of accuracy. **See Honing**

Hook, Angle of. The angle between the line drawn down the front of a saw tooth and a line drawn radially from the centre of the circular saw to the tooth point. It is the same as the cutting angle of a *planing machine* cutter. (See Figs. 46, 143.)

Hook Escapement (Virgule Escapement). See **Escapement.**

Hook Tool. A tool used for vertical slotting and placed transversely to the axis of the ram of a *slotting machine.*

Hooke's Joint. Two horseshoe shaped forks, each pivoted to a single central member carrying two pins at right angles (Fig. 86). Cf. **Universal Joint.**

Hooke's Law. Strain is proportional to stress in an elastic material below the *elastic limit.*

Hopper. A container for receiving or feeding supplies of materials from, or to, machines, etc.

Hopper Dredger. A *dredger* with hopper compartments, fitted with flap-doors at the bottom, which receive the dredged material and deposit it later elsewhere.

Horizon, Artificial. *Artificial horizon.*

Horizontal Boring Machine. See **Boring Machine.**

Horizontal Crane. See **Crane.**

Horizontal Engine. Any engine with a horizontal cylinder axis.

Horizontal Escapement. *Cylinder escapement.*

Horizontal Lathe. A boring machine with a vertical axis for boring large engine cylinders and rings.

Horizontal Winch. A steam *winch* with horizontal cylinders on the side frames.

Horn. Any projecting part such as the two jaws of a *horn plate.*

Horn Balance. An extension on the outer end of an aeroplane control which gives partial aerodynamic balance and lessens the force needed to move the control.

Horn Plates. The frames whose internal edges act as guides to the axle boxes of railway rolling-stock, allowing the boxes and therefore the axles, to move up and down relative to the *chassis* in a spring *suspension.*

Hornblock. A casting which receives the axle box of railway rolling-stock and constrains it to move in a vertical plane.

Horns. The curved levers pivoted at the side of some planing machines which are controlled by the tappets to give the necessary feed to the tool and the reversing movement to the table.

Horse Gear. *Bullock gear.*

Horse-power (h.p.). The engineering unit of power equal to a rate of working of 33,000 foot-pounds per minute, 23·56 C.H.U. per minute, 42·42 Btu per minute or 745·700 watts. See also **Brake**

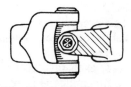

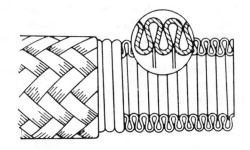

Fig. 86 (*above*). Hooke's Joint.

Fig. 87 (*right*). Flexible Metal Hose (Bellows Type).

Horse-power, Indicated Horse-power, Nominal Horse-power.

Horse-power, Transmitted. The *horse-power* from prime movers to mechanisms, via belts, wheels, shafts, etc.

Horseshoe Gauge. A fixed horseshoe-shaped gauge for gauging the outside dimensions of a piece of work.

Hose Reel. A drum, often mounted on wheels, around which flexible hose (Fig. 87) is wound.

Hot-air Engine. An engine in which the working fluid for the heat cycle is air. There have been many types of hot-air engines, but none have had a high efficiency. Cf. **Air Engine.**

Hot Iron Saw. *Hot saw.*

Hot Saw (Hot Iron Saw). A metal-cutting circular saw for cutting the ends of heated billets, steel forgings, etc. The lower portion of the saw runs in cold water.

Hot Well. The tank or pipes into which the condensate from a steam-engine or turbine condenser is pumped, and from which it is returned by the feed pump to the boiler.

Hour Rack. See **Rack.**

Hour Snail. See **Snail.**

Hour Wheel. The wheel carrying the hour hand in clocks and watches. (See Fig. 19.)

Housing. A supporting structure.

Hovercraft. *Air-cushion vehicle.*

Hub. The central part of a wheel, rotating on or with the axle and from which spokes radiate. See also **Propeller**.

Humidifier. (a) An apparatus for controlling the required humidity conditions in a room or building.

(b) An apparatus for adding moisture to the air in a room, cabin, cockpit, space suit, etc.

Humpage's Gear. An epicyclic train of wheels for the speed reduction of a machinery shaft.

Humphrey Gas Pump. A pump which acts by the periodic explosion of a gas/air mixture above an oscillating column of water in a vessel with an outlet valve. The pump is used in water-works and gives lifts up to 50 m.

Hunting. (a) An undesired variation (usually of nearly constant amplitude) from the stable running condition of a mechanism due to a feed-back in the control. Sometimes called ' cycling '.

(b) The angular oscillation of a rotorcraft's blade about its drag hinge.

(c) Abnormal time-lag between the opening or closing of the throttle and an increase or decrease in the speed of a piston-engine.

Hunting Cog. *Hunting tooth.*

Hunting Tooth. An extra tooth on a gear-wheel so that the number of its teeth shall not be an integral multiple of those in the pinion.

Hydraulic Belt. An endless belt of porous material driven at high speed with its lower end running under water, which acts like a *chain pump*.

Hydraulic Brake. (a) A motor-vehicle brake in which power is supplied by hydraulic oil pressure via small pistons to expand the brake shoes, the pressure being supplied by a pedal-operated master cylinder and piston.

(b) A piston in a cylinder filled with some liquid, frequently oil, which absorbs energy through the leakage of the fluid through small holes in the piston or otherwise. See **Liquid Spring**.

(c) An *absorption dynamometer*. See **Froude Brake**.

Hydraulic Crane. See **Crane**.

Hydraulic Cylinder. The cylinder of a hydraulic press, the pressure of the fluid in which lifts the enclosed piston. See **Hydraulic Press**.

Hydraulic Dredger. *Suction dredger.*

Hydraulic Drive. *Fluid drive.*

Hydraulic Dynamometer. A *dynamometer* which measures the

Hydraulic Dynamometer

change in angular momentum of water thrown outward by centrifugal force into stationary pockets in the casing. (See Fig. 88.)

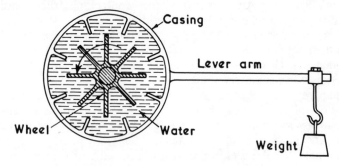

FIG. 88. HYDRAULIC DYNAMOMETER.

Hydraulic Efficiency. (*a*) The ratio between the work done by a turbine per pound of water and the available head.

(*b*) The ratio of the actual lift of a centrifugal pump to the head generated by the pump.

Hydraulic Engine. An engine driven by water under pressure from an elevated reservoir or from a loaded accumulator.

Hydraulic Forging Press. *Hydraulic press.*

Hydraulic Jack. A *jack* with the lifting head carried on a plunger working in a cylinder to which oil (or water) is supplied under pressure from a pump. (Fig. 89.)

Hydraulic Hammer. *Hydraulic press.*

Hydraulic Lift. A lift operated by water power (*a*) direct-acting by a vertical ram, or (*b*) indirect acting by a short ram whose stroke is multiplied by sheave wheels and ropes, or (*c*) by adding water to the upper of a pair of lift cars and using a brake on the connecting cable.

Hydraulic Motor. A multi-cylinder reciprocating engine driven by water under pressure and usually of radial- or swash-plate type.

Hydraulic Piston. A solid piston used in force pumps and hydraulic cylinders.

Hydraulic Press. A ram or piston carrying a *diehead* and working in a cylinder to which high-pressure fluid is admitted. The work is passed between this head and a stationary head.

Hydraulic (Force) Pump. A *force pump* delivering fluid under high pressure, usually consisting of a number of pistons arranged radially round a crank or operated by a *swash-plate.*

Hydraulic Ram. (*a*) A mechanism involving the displacement of a plunger by injecting fluid into, or withdrawing it from, a closed chamber in which the plunger moves through a gland at one end.

(*b*) A device for using the pressure head of a large moving column of water to deliver some of the water under a greater pressure.

Hydraulic Riveter. A small *ram* operated by hydraulic power to close rivets either directly or through hinged jaws.

Hydraulic Shearing Machine. A *shearing machine* operated by hydraulic power.

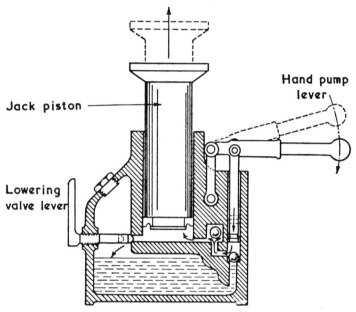

FIG. 89. HYDRAULIC CAR JACK.

Hydraulic Telemotor. A remote-operating hydraulic mechanism in which fluid displaced by the movement of the input causes a corresponding movement of the output, with the systems usually hermetically sealed to give a positive displacement.

Hydraulic Test. A test for pressure tightness and strength or for fatigue failure, by pumping water into a vessel up to a prescribed pressure.

Hydraulic Turbine (Water Turbine). A *turbine* driven by water fluid. See also **Pelton Wheel.**

Hydraulically-operated Disc Brake. A *disc brake* in which [t]
pressure is applied by hydraulic pistons. (Fig. 90.)

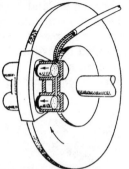

FIG. 90. HYDRAULICALLY-OPERATED
DISC BRAKE.

Hydrodynamic Governor. A small centrifugal pump wh[c]
pressure head, which varies with speed, acts on a piston connect[
to the regulating valve and thus acts as a governor. Cf. **Servomot[**

Hydrodynamic Suspension. A suspension system involving t[w]
or more fluid spring elements interconnected so that a displaceme[
at one unit is made to affect an adjacent unit. This system[
commonly used to reduce the pitching rate of road vehicles.

Hydro-extractor. *Whizzer.*

Hydrostatic Press (Bramah's Press). A machine consisting o[f]
pair of interconnected cylinders fitted with water-tight pistons, t[
cylinders having different diameters in order to obtain a *mechani[c]
advantage.*

Hypocycloid. The curve traced by a point on the circumferer[
of a circle as it rolls round the inner circumference of another circ[
Cf. **Epicycloid.**

Hypoid Bevel Gear. A *bevel gear* with the axes of the driving a[
driven shafts at right angles, but not in the same plane which cau[s
some sliding action between the teeth.

Hypoid Gear. Crossed *helical gears* which are designed to oper[a
on non-intersecting as well as non-parallel axes, that is, the axis[
the pinion is above or below that of the gear and is not parallel[
the axis of the wheel.

Hypoid Offset. The perpendicular distance between the n[c]
intersecting axes of the cones of two *hypoid gears.*

Hysteresis. The phenomenon exhibited by a system whose st[a
depends on its previous history. Cf. **Hysteresis Loop.**

Internal Hysteresis. The property of a material wh[i

192

dissipates internal energy under cyclic deformation.

Hysteresis Loop. The enclosed curve when a cycle of operations is completed, for example, the different stress/strain curves when a load is steadily increased and then decreased, making a loop in the case of some materials.

I

I. The symbol for moment of inertia.

I.C.E. Internal-combustion engine.

I.D. Induced draught.

i.g.p.m. Imperial gallons per minute.

i.h.p. *Indicated horse-power.*

i.m.e.p. *Indicated mean effective pressure.*

Ideal Efficiency. The theoretical maximum efficiency.

Idle Pulley. A pulley used in a similar manner to an *idle wheel.*

Idle Wheel (Carrier Wheel, Cock Wheel). (*a*) A wheel introduced in a *gear train* either to reverse rotation or to fill up a gap in the spacing of centres, without affecting the drive ratio. Also called a ' cock wheel '.

(*b*) An *intermediate wheel.*

Idler. *Idle wheel, idle pulley.*

Idling. The slow rate of revolution of a *piston-engine* when the throttle is in the closed position.

Igniter. A device by which the charge in a *gas-engine* or a *rocket-engine* is ignited.

Ignition. (*a*) The firing of an explosive mixture of gases in an internal-combustion engine by means of an electric spark or by a jet of gas in a gas-engine.

(*b*) The commencement of combustion in a jet-engine or a rocket-engine.

Ignition Lag. The time-interval between the passage of the spark and the resulting pressure rise in a cylinder due to combustion.

Ignition Slide. *Ignition valve.*

Ignition Timing. The crank angle relative to top *dead-centre* at which the spark occurs in a petrol- or gas-engine.

Ignition Valve (Ignition Slide). The valve of a gas-engine which opens to permit the ignition of the charge, but closes as soon as this is effected.

Impact. The sudden application or fall of a load upon a specimen, structure, etc.

For the direct impact of two elastic spheres, the ratio of the relative

Impact

velocity after impact to that before impact is constant and is calle
the *coefficient of restitution* for the material of which the spheres ar
composed. This constant is 0·95 for glass and 0·2 for lead, th
values for other solids lying between these two figures. See als
Extrusion.

Impact Loading, Safe (Impact Load Factor). The maximur
acceleration to which equipment, etc., can be subjected unde
impact or shock without mechanical damage or operational break
down. The magnitude of the acceleration is given in multiples of g
its duration and its rate of change should be specified.

Impact-testing Machine. A machine for testing the strength c
test specimens under a single blow and for measuring the amour
of energy absorbed in a fracture of the specimen. The commones
form of test-piece is a notched bar. See **Notched-bar Test.**

Impact Wheel. A water wheel driven by the impact force c
water acting at right-angles to the projecting vanes on the periphery
Turbines are impact wheels. See also **Pelton Wheel.**

Impedance. See **Mechanical Impedance.**

Impedance Wheel. The constant-speed drive *sprocket* which feed
the ciné film for each exposure.

Impeller. (*a*) The rotating member of a centrifugal pump con
pressor, or blower, which imparts kinetic energy to the fluid (water
air, etc.). (See Fig. 27.)

(*b*) A rotating member in some meters such as a *gas meter*. (Se
Fig. 75.)

Imperial Standard Wire Gauge. See **British Standard Wire Gaug**
(S.W.G.).

Impermeator. A form of self-acting lubricator used for th
cylinders of steam-engines, a two-valve contrivance by which stear
extracts oil from a small brass cylinder, then condenses to water s
that the oil floats to the top and some overflows back to the bras
cylinder.

Impulse. The force or blow by the *escape wheel* through th
escapement which is given to the pendulum or balance of a clock o
watch. Hence ' impulse post ' and ' impulse pin ' for that part c
the mechanism on which the blow falls.

Impulse-driven Clock. A clock in which the hands are driven b
electrical current impulses from a master-clock.

Impulse Pin. The vertical pin in the lever escapement *rolle*
which receives the impulse from the *pallets*, via the notch in th
lever, and which effects the unlocking on the reverse vibratior

194

See also **Escapement.**

Impulse Plane. That part of the *pallet* upon which a tooth of the *escape wheel* acts.

Impulse Reaction Turbine. See **Disc-and-drum Turbine** and Fig. 192.

Impulse Turbine. A *steam-turbine* in which steam, expanded in nozzles, is directed on to the curved blades carried by rotors, in one or more stages. No change of pressure occurs as the steam passes the blade-ring. (See Fig. 141.)

Combined-impulse Turbine. An *impulse turbine* in which the first stage consists of nozzles directing the steam on two rows of moving blades with a row of fixed blades, the *stator*, in between.

In Gear. Mechanisms and engines are said to be ' in gear ' when connected ready to be operated, or operating.

Incastar Regulator. A regulator which alters the point of contact of a watch spring and hence its length and period of vibration. (Fig. 91.)

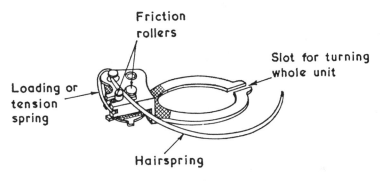

FIG. 91. INCASTAR REGULATOR.

Inching. Making adjustments by very small stages.

Inch-pound. The lifting of a pound weight a distance of one inch. *Foot-pound* is the more commonly used unit of work.

Inch-ton. The lifting of a ton weight a height of one inch. *Foot-ton* is the more commonly used unit of work.

Included Angle. The angle between the flanks of a screw thread measured in an axial plane.

Increasing Pitch. A screw is said to be of ' increasing pitch ' when the distance between each successive turn of the *helix* increases

in amount, or the *pitch* may increase in the direction of the lengt
of the blade from the centre to the circumference.

Indentation Hardness. The estimation of the *hardness* by th
permanent deformation formed in a material by an *indenter*, th
hardness being expressed in terms of the load and the area of th
indentation formed. Bell-shaped *indenters* are used to measur
Brinell hardness number and *Meyer hardness number.* Conica
indenters were introduced by Ludwik and the hardness number i
the load divided by the surface area of contact between indenter an
material. Pyramidal indenters are shaped like a square-base
pyramid and are used for determining *Vickers hardness number.*
in *Rockwell hardness tests*, in *Knoop hardness tests*, and in the *Firt.
hardometer.* Dynamic or rebound hardness is measured by th
Shore rebound scleroscope and the *Herbert pendulum.*

Indenter. An instrument used for making indentations i
materials, the depths of which give a measure of their *hardness.* Se
also **Indentation Hardness.**

Independent (Jaw) Chuck. A chuck for a lathe in which each o
the jaws is moved independently by a key to give very accurat
centring for work of irregular shape.

Independent Seconds Watch. A watch having an independen
train for driving the seconds hand.

Independent Whip Crane. *Platform crane.* See **Crane.**

Index. The lever for adjusting the rate of a watch, usually carrie
on the *balance clock*; the short end carries the *curb pins* and the lon
end moves over a scale to indicate the amount of movement give
to the curb pins. One end of the scale is marked A or F and th
other R or S: the index is moved towards A to make the watch gair
and to R to make the watch lose.

Index Centres. The centres in the *headstock* and *tailstock* a
used on *milling machines* and *gear cutters.*

Index Peg. *Division peg.*

Index Pins. *Curb pins.*

Index Plate. *Division plate.*

Indexing Head. *Dividing head.*

Indicated Horse-power (i.h.p.). The power shown by an *indicato*
being that developed by the pressure-volume changes of the workin
fluid within the cylinder of a reciprocating engine and therefor
greater than the *brake horse-power* by the power lost in friction an
pumping.

In a piston-engine it is given by ASPNn/33,000, where A is th area of the piston in square inches, S is the stroke in feet, N is the number of cylinders, n the number of strokes per minute and P is the pressure on the piston in pounds per square inch.   See also **Horse-power.**

Indicated Mean Effective Pressure (i.m.e.p.).   The average pressure exerted by the working fluid in an engine cylinder throughout the working cycle and equal to the mean height of the *indicator diagram* in Pascals (1 Pa $= 1$ N/m^2).

Indicated Thermal Efficiency.   The ratio of the heat energy equivalent to the *indicated horse-power* output and the heat energy supplied in the steam or fuel in a reciprocating engine.   See also **Thermal Efficiency** and **Mechanical Equivalent of Heat.**

Indicator.   An instrument for obtaining the pressure-volume or the pressure time changes in a steam-engine, or a piston-engine, cylinder during the working cycle.   See **Engine Indicator** and Fig. 60.

Indicator Card.   A specially prepared paper wound upon the drum of an *indicator* on which a diagram is drawn by a pencil or a metallic style.

Indicator Diagram.   The diagram drawn by the *indicator* representing to scale the work done during a cycle of operation of the engine.   See **Indicated Mean Effective Pressure** and **Light-spring Diagram.**   See also **Admission Corner** and **Admission Line, Lead Line.**

Indirect-acting Slide Valve.   A *slide valve* in a locomotive that is actuated from an intermediate rocking shaft or double-ended lever, attached at one end to the valve rod and at the other end to the die block of the slot link.   See **Link Motion.**

Indirect Action.   A motion derived through the medium of levers which are not directly related to the motion of the member which is supplying the *motive power*.

Induced Draught.   An artificial draught operated by suction, as with a fan, as opposed to *forced draught*.

Induction.   (*a*) The secondary flow of a gas (or liquid) induced by a primary flow of gas (or liquid).   Cf. **Entrainment.**

(*b*) The admission of steam into a cylinder.

(*c*) The admission of the explosive mixture into the cylinder or combustion space of an internal-combustion engine.   See **Diesel Cycle, Four-stroke Cycle, Otto Cycle, Two-stroke Cycle, Wankel Engine.**

Induction Manifold.   See **Manifold.**

197

Induction Port (Induction Valve, Inlet Port). A *port* or valve through which a charge is induced into a cylinder during the *induction stroke*.

Induction Stroke (Charging Stroke, Intake Stroke). The suction stroke during which the working charge, or air, is induced into the cylinder of an engine.

Induction Valve. *Induction port.*

Inertia. That property of a body by which it tends to resist a change in its state of rest or of uniform motion in a straight line. Inertia is measured by *mass* when linear velocities or accelerations are considered and by *moment of inertia* for rotations about an axis.

Inertia Governor. A shaft type of centrifugal governor employing an eccentrically pivoted weighted arm which responds rapidly to damped speed fluctuations by reason of its *inertia*.

Ingot. A metal casting of a suitable shape for subsequent rolling or forging.

Ingot Tilter. A machine by which *ingots* are turned between each pass of the *rolling mill*.

Injection. The process of injecting fuel into the cylinder of a *compression-ignition* or petrol engine by means of a special pump.

Injection Lag. The time-interval between the beginning of the delivery stroke of the fuel-injection pump of a *compression-ignition engine* and the beginning of the injection of the fuel into the cylinder.

Inlet Port. *Induction port.*

Inlet Valve. (*a*) The valve in a piston-engine for admitting the steam or fuel-air mixture. Cf. **Exhaust Valve.**

(*b*) A *foot valve.*

Inner Dead-centre (Top Dead-centre). The piston position of a reciprocating engine or pump at the beginning of the outstroke when the crank pin is nearest to the cylinder. See also **Dead-centres.**

Insensitive Time. *Dead time.*

Insertion Head. An automatic feed mechanism for feeding components axially into an assembly, with cutting, clinching and forming tools.

Inside Crank. A *crank* with two webs which is placed between *crankshaft* bearings and has the *big end* of the *connecting-rod* between the two webs.

Inside Cylinders. Locomotive cylinders which are fixed within the framing and smoke box.

Inside Framing. A form of locomotive framing in which the wheels are inside the main frames.

Inside Lap. *Exhaust lap.*

Inside Lead (Internal Lead). The degree of opening of the *exhaust port* of a steam-engine by the *slide valve* when the piston is at the bottom *dead-centre.*

Inside Screw Tool. *Internal screw tool.*

Inspection Gauge. A gauge used in the final inspection of a part for testing the accuracy of the finish.

Instantaneous Centre (Virtual Centre). The point about which a moving body or system of bodies, or links in a piece of mechanism, is actually turning at a given instant.

Instantaneous Grip Vice (Sudden Grip Vice). A vice operated by levers, a *toggle-joint* and a rack, instead of a screw.

Institution of Mechanical Engineers. A British engineering and scientific society incorporated by royal charter in 1930, which admits qualified mechanical engineers to full membership as Fellows and Members; it has also graduate and student members. A member of the *Council of Engineering Institutions.*

Instroke. A stroke of a gas-engine piston in a direction towards the ignition chamber. Cf. **Out Stroke.**

Intake Stroke. *Induction stroke.*

Integrator. A calculating machine such as a computer, or a mechanical machine, such as *planimeter,* which computes what is represented mathematically by an integral sign.

Ball-and-disc Integrator or Grid. A mechanism in which a disc is rotated by a variable-speed friction gear. A roller, with its axis at right angles to the disc axis, is driven through the intermediary of two balls in a cage and its angular motion is proportional to the radius of contact multiplied by the angular rotation of the disc. When the disc is driven at a constant speed the output motion is the time integral of the ball-cage displacement. (See Fig. 111(*b*).)

Gyro Integrator. (See **Gyro.**) The disc and wheel type, which is similar, is shown in Fig. 111(*a*).

Interchangeable Gears. Gears with the teeth of the wheels so designed that other gear-wheels of the same *diametral pitch,* but with any number of teeth, will mesh together correctly.

Intermediate Wheel. A wheel connecting two other toothed wheels in a clock or watch so that the latter need not be made larger and to alter the relative direction of rotation of the *followers.*

Internal-combustion Engine. An engine in which the combustion of a gaseous, liquid or pulverized solid fuel within a cylinder provides

Internal-combustion Engine

heat which is converted into mechanical work through a piston.
See **Compression-ignition Engine, Diesel Engine, Gas Engine, Petrol
Engine, Wankel Engine.**

Internal Expanding Brake. See **Brake Shoes.**

Internal Spur Gear. A *spur wheel* with teeth on the inside of the
periphery (Fig. 92). The nomenclature is given in Fig. 93.

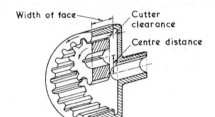

FIG. 92. INTERNAL SPUR-GEAR AND
PINION.

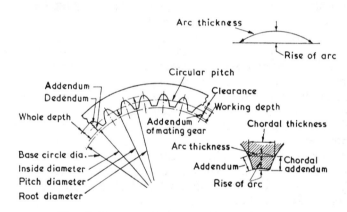

FIG. 93. INTERNAL SPUR-GEAR NOMENCLATURE.

Inverted Cylinder Engine. A vertical engine with the inverted
cylinder above the piston rod, connecting-rod and crank.

Inverted Engine. An engine with its cylinders below the crankshaft. Cf. **Vertical Engine.**

Involute. The involute of a curve is another curve of which it is
the *evolute*. The lengths $L1$, $L2$, $L3$, etc., are equal to the arc lengths
around the circumference and define the involute of a circle
(Fig. 94).

200

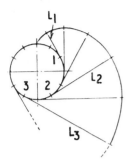

FIG. 94. INVOLUTE.

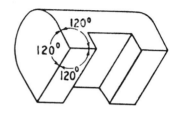

FIG. 95. ISOMETRIC PROJECTION.

Involute Gear-teeth. Wheel teeth whose flank profile is the locus of the end of a string uncoiled from a *base circle*. See also **Involute.**

Irreversible Transmission. When power can be transmitted, but not reversed.

Iris Diaphragm. A continuously variable hole forming an adjustable stop for a camera lens and usually integral with the mounting of the lens. See also **Shutter.**

Isochronism. Regular periodicity, such as a clock pendulum having the same time of swing whatever the amplitude, or the balance wheel of a watch whatever the arc of vibration.

Isochronous Governor. A governor in which the equilibrium speed is constant for all radii of rotation of the balls within the working range.

Isolator. A separate mounting, such as a heavy concrete block, to isolate a machine or instrument from external vibrations or shock. Cf. **Vibration Isolator.**

Isometric Projection. An engineering drawing projection with three mutually perpendicular axes shown equally inclined to the plane of projection in Fig. 95.

Izod Test. *Notched-bar test.*

J

J. (*a*) The symbol for Joule, the force of one *Newton* acting over the distance of one metre.

(*b*) The symbol for the *polar moment of inertia* of a shaft.

Jack. (*a*) A machine for raising a heavy weight through a short distance by means of a screw with gear or hydraulically. See **Hydraulic Jack.** Fig. 89.

(*b*) A frame of horizontal bars for supporting fixed vertica wires against which lace bobbins containing yarn can revolv freely.

Jacquard Machine. A weaving machine for operating the shed ding and controlling the figuring of a large number of warps in *loom*.

Jerk-pump. A timed fuel-injection pump with a cam-drive plunger overrunning a spill port to cause an abrupt pressure-ris which is necessary to initiate injection through the atomizer.

Jet Condenser. A *condenser* in which steam is condensed by water spray.

Jet-engine. A colloquial term for any type of engine whic produces a thrust by means of a jet of hot combustion gases. Se **Turbojet, Turboprop, Ramjet, Pulsejet** and **Rocket-engines, Rocke Motor.** Cf. **Internal-combustion Engine.**

Jet Pipe. The pipe leading the exhaust gases from a turbine t the *propelling nozzle*; sometimes called ' tail pipe '.

Jet-pipe Temperature. The temperature of the exhaust gases i a *jet pipe*.

Jet Pump. A pump which delivers large quantities of fluid at low *lift* by means of the momentum imparted to the column by th velocity of a jet of steam or compressed air.

Jib. The inclined or horizontal boom (strut member) of a *cran* or *derrick*.

Jib Crane. See **Crane.**

Jib Legs. Legs pivoted to the jib pin of a *breakdown crane* an reaching to the ground to provide a firm base for lifting.

Jig. An appliance which accurately guides and locates tool during the operations in a machine shop for producing inter changeable parts. Cf. **Fixture.** (See Fig. 96.)

Jig Borer. A precision drilling and boring machine used in th toolroom for making master jigs and prototype precision work.

Jig Grinding. The accurate location of the workpiece by jigs an the subsequent grinding of a hole to finished size by a suitabl grinding head and wheel.

Jigger. (*a*) A hydraulic lift operated by a short-stroke hydrauli ram through a system of ropes and pulleys which increase the travel

(*b*) A potter's wheel.

Jigger Saw (Jig Saw). A thin narrow-bladed saw with a reci procating vertical motion imparted by a crank and levers, now mainly replaced by a *bandsaw*. Also known as *fret saw*.

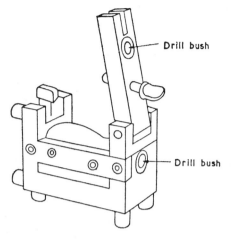

Fig. 96. Box Jig.

Drill bush

Drill bush

Jigging. Holding work in *jigs* during machining and assembly, thus facilitating interchange of components.

Jim-crow. (*a*) A swivelling tool-head, cutting during each stroke of the table of a planing machine.

(*b*) A rail-bending device.

Jinny. The travelling cab or carriage on an *overhead travelling crane*. See **Crane**.

Jockey Pulley. A small pulley wheel which is weighted so that it keeps a drive belt or chain taut as shown in Fig. 97.

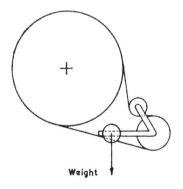

Fig. 97. Jockey Pulley.

Weight

Joggle. (*a*) A small projection on one piece of metal which fits into a corresponding recess on another piece to prevent relative movement.

203

(*b*) A lap joint so that one edge of each of two plates form a continuous surface.

Joint Pin. A pin connecting the two parts of a *knuckle joint* and prevented from moving out of place by a split-pin. See **Cotter-pin**

Jonval Turbine. See **Turbine.**

Jounce. The initial impact of an automobile wheel hitting a raised obstruction on a road surface; the opposite of rebound.

Journal. That part of a shaft supported by a *bearing* and in contact with it.

Journal Bearing. A support consisting of a box-form casting split horizontally and surrounding a shaft *journal*. The box is usually lined with a *bearing metal*.

Journal Box. An axle box. See **Journal Bearing.**

Joy's Valve-gear. A locomotive valve-gear of the *radial valve gear* type with no *eccentrics*, the valve rod being worked directly through a coupling rod or link from the *connecting-rod*. Cf **Walschaert's Valve-gear.**

Jump Joint. A *butt joint* in belting.

Jumper. A pushpiece or *click* in the form of a wedge on the end of an arm, controlled by a *jumper spring* which allows a *star wheel* in a *chiming* or *calendar mechanism* to jump forward one tooth and be reset.

Jumper Spring. A spring attached to a *jumper* for holding a *star wheel* in place.

Jumping-in. The springing-in of a ring on a piston so that it will enter the cylinder of a piston-engine.

Jumping Figure Watch. A watch in which figures on discs jumping into position in windows, indicate the time or date.

Junk Ring. (*a*) A metal ring attached to a piston of a steam engine for confining soft packing materials.

(*b*) Similar to (*a*) for holding a cast-iron piston-ring in position

(*c*) A ring for maintaining a gas-tight seal between the cylinder head and the bore of a *sleeve valve*. (See Fig. 55.)

K

k. The symbol for the radius of gyration.

kgf. A force equal to the weight of one kilogramme.

kc/s. Kilocycles per second. (SI unit: kHz.)

km.p.h. Kilometres per hour.

km.p.l. Kilometres per litre.

kt. Knot.

kWh (kwhr). Kilowatt-hour. (SI unit: 3·6 MJ.)

Kaplan Water Turbine. A propeller-type water turbine in which the pitch of the blades can be varied in accordance with the load with a consequent improvement in efficiency.

Karrusel Movement (Watch). A movement in which the revolving carriage, unlike the *tourbillon*, carries no power to the escapement and is carried, not driven, by the third pinion. The fourth pinion receives power direct from the third wheel and the fourth wheel revolves in the usual way driving the *escape pinion*. The open-faced ordinary karrusel takes $52\frac{1}{2}$ minutes to perform one revolution and the centre-seconds type takes 44 minutes.

The carriage of an ordinary karrusel revolves on a large brass pivot or boss fitted into a large hole in the pillar plate, being held in position by a flat brass wheel of 70 teeth screwed on to the brass pivot or boss of the carriage and the wheel is geared into the third pinion of usually 10 *leaves*.

The principle of the centre-seconds type is the same with a different layout, the fourth or seconds pivot being in the centre of the frame. The rim of the carriage has 136 teeth, geared into a steel wheel of 30 teeth mounted on the third pinion *arbor* instead of being turned by that pinion.

Kater's Pendulum. A bronze bar with a fixed and a movable knife-edge. The latter is adjusted until the time of oscillation about either is the same. The pendulum provides one method of determining the *acceleration due to gravity* at any place.

Keep Plate. A plate fitted in a fixed shaft to position a rotating part in an axial direction.

Keeper. A lower movable piece beneath an axle box of railway rolling-stock which limits the downward movement of the box due to track irregularities.

Kennedy Water Meter. A water meter in which the volume of flowing water continually fills and empties a cylinder of known volume, the discharge being automatically registered. It empties by tipping when the centre of gravity rises beyond a certain point.

Kentledge. Scrap material used for counterbalance on a *balance crane*. See **Crane**.

Kerf. The width of the cut made by a saw (see Fig. 153) depending on its set. See also **Drunken Saw**.

Key. (*a*) A piece of iron or steel inserted between a shaft and a hub to prevent relative rotation and fitting into a *keyway* parallel with the shaft axis. (See Fig. 98.)

Key

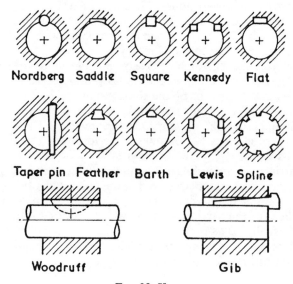

Nordberg Saddle Square Kennedy Flat

Taper pin Feather Barth Lewis Spline

Woodruff Gib

FIG. 98. KEYS.

(b) A spanner or wrench used for tightening the jaws of lath
chucks, etc.

Dovetail Key. A *parallel* key in which the part sunk in th
boss is of dovetail section, the portion on the shaft bein
of rectangular cross-section.

Gib-headed Key. A key with a head formed at right angles t
its length to facilitate withdrawal.

Round (*Norberg*) *Key.* A circular pin or bar which fits into
hole drilled half in the *boss* and half in the shaft parallel to th
shaft axis, usually for light work.

Parallel Key. A rectangular key with parallel sides used
marine tailshafts, etc., when the shafts are greater than 1 i
diameter. A square key for shafts up to 1 in. diameter or whe
it is desirable to have a greater key depth than is provided b
rectangular keys.

Saddle Key. A key with a concave face bearing on the surfac
of the shaft which its grips by friction only, being sunk in a keywa
in the *boss*.

Split Key. A key which is split at one end like a *split-pin* t
lessen the tendency to work out of its bed.

206

Sunk Key. A key which is sunk into keyways in both hub and shaft. Cf. **Feather.**

Tangential Keys. A pair of taper keys with one side largely sunk in the hub and the other in the shaft, the two keys facing in opposite directions. They are used for the transmission of reversing torque, usually under conditions of heavy loading.

Taper Key. A rectangular key with parallel sides slightly tapered in thickness along its depth for transmitting heavy torques and where periodical withdrawal of the key may be a necessity. They cannot be used in applications requiring a sliding hub member.

Woodruff Key. A key with the shape of a disc segment which is fitted in a shaft keyway milled by a cutter of the same radius, and a normal keyway in the hub. They are used for light applications or angular location of associated parts on the tapered shaft ends. See also **Feather** and **Splines,** Fig. 98.

Key Bed. *Keyway.*

Key Boss. A local thickening of a hub at the point where a *keyway* is cut to compensate for loss of strength due to the cut.

Key Chuck. A jaw chuck with jaws adjustable by screws turned by a key or spanner. See **Self-centring Chuck.**

Key Gauges. Plate gauges used for checking the width of keys and key seatings.

Key-seating. *Keyway.*

Keyboard. (*a*) The banks of keys in an instrument such as an organ or piano.

(*b*) A similar arrangement of keys in a typewriter, linotype machine, etc.

Keying. Fitting a key to the *keyways* in a *shaft* or *boss.*

Keyless Mechanism. (*a*) A winding button which may be pulled out to set the hands of a watch or operates on the depression of a special knob.

(*b*) A winding pinion which gears into the central wheel of the rocking bar (English watch type), the wheel being bevelled to take drive at one angle and transmit it at another.

(*c*) A crown and castle wheel mechanism (Swiss watch type) with a winding stem, a part of which is square and fits into a female square through the castle wheel, so called because of its shape. One end has ordinary gear-teeth and the other ratchet teeth, both set at right angles. The gear-teeth engage into the motion at will and the ratchet teeth have a counterpart in the ratchet teeth of the crown wheel fitted on a turned part of the winding stem. This keyless mechanism is of the ' shifting-sleeve ' type. When winding, the ratchet teeth of both crown and castle wheels becoming inter-locked, the motion is transmitted to the crown wheel and then through the large, or idle, crown on the barrel bar, into which the crown wheel is geared, and the watch is wound up. When the winding stem is turned in the opposite direction, the two sets of ratchet teeth slide over one another and do not wind: this is called back action and is partly a safety device.

Keyway (Key-seating). A shallow longitudinal slot cut in a shaft or a hub for receiving a *key* as shown in Fig. 98.

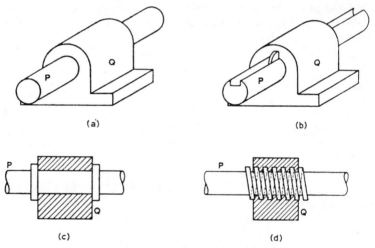

FIG. 99. KINEMATIC PAIR: (*a*) incomplete restraint, (*b*) (*c*) (*d*) complete restraint, (*b*) sliding, (*c*) turning, (*d*) screw pair.

Keyway Cutter. *Keyway tool.*

Keyway Seating Machine. A machine tool for milling *keyways* in shafts, etc., using an end mill with the work supported on a table at right-angles to the tool axis.

Keyway Tool (Keyway Cutter). A slotting machine tool used for the vertical cutting of *keyways*, the tool being equal in width to that of the keyway.

Kibble. A large unguided bucket used in shaft-sinking. Cf. **Skip** (*b*).

Kilogramme-metre. A continental European unit of work: the work done in raising a mass of one kilogramme through a vertical distance of one metre against gravity. Cf. **Foot-pound.**

Kinematic Chain. A mechanism composed of a closed chain of paired links, the movement of any link being absolutely constrained in relation to all the others in the chain. See **Kinematic Pair.**

Kinematic Pair. Two elements or links which are connected together in such a way that their relative motion is partly or completely constrained. (See Fig. 99.)

Kinetic Energy. See **Energy.**

Kinetic Friction. See **Friction.**

King Lever. A master lever in a signal-lever frame which controls the interlocking of the levers to the various railway signals.

King-pin (Swivel-pin). A pin joining the stub axle to the axle-beam of an automobile and inclined to the vertical to provide *caster* action.

Kingston Valve. See **Valve.**

Kink. (*a*) A dislocation of links in a chain.

(*b*) A twist or loop in a rope or thread.

Kite Connecting-rod. *Bow connecting-rod.*

Knee-joint. *Toggle-joint.*

Knife-edges. Hardened-steel bearing edges working on a horizontal surface or the inner circumference of a ring, permitting fine balance of the adjacent part.

Knife Tool. A lathe finishing tool with a straight lateral cutting edge used for turning right up to a *shoulder* or corner.

Knitting Frame (or Loom). A knitting machine using either *bearded needles* or *latch needles*.

Knitting Machine. An apparatus for knitting using a single thread and furnished with a number of hooked and barbed needles and also loopers for forming the meshes. See also **Looper, Needle, Sinker.**

Knocker-out. The horns of a *planing machine* against which the *tappets* strike to reverse the motion of the table.

Knocking (Pinking). (*a*) A periodic noise caused by loose parts or worn bearings.

(*b*) The noise from *pre-ignition* or *detonation*.

Knoop Hardness Test. *Hardness test* using an *indenter* in the form of a four-sided pyramid whose indentation is a parallelogram with the longer diagonal about seven times that of the shorter.

Knuckle Gearing. A gearing which has teeth with a cross-sectional

209

profile consisting of semi-circles above and below the pitch circle
This gearing is strong and is used for slow-moving rough-purpose
machinery.

Knuckle Joint. A hinged joint between two rods, a pin connecting the eye on one with a forked end on the other.

Knuckle Pin. *Wrist pin.*

Knuckle Thread. A screw thread with semi-circular cross-section
and a radius one quarter of the pitch. This is a strong thread
giving high friction, the latter necessitating generous clearance
between *male and female* parts.

Knurled. Having edges cut in a succession of ridges or ridges
forming a continuous diamond-shaped pattern.

Knurled Head. *Milled head.*

Knurling Tool (Milling Wheel). Small hard serrated steel rollers
mounted on a pin, which are pressed against circular work to make
a series of ridges on a surface to improve the finger grip on that
surface. An adjustable knurl tool-holder is shown in Fig. 100.

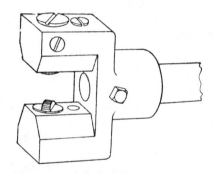

Fig. 100. Adjustable Knurl
Holder.

Kollsman Altimeter. A sensitive altimeter in which bellows-type
diaphragms expand and contract with the changes in atmospheric
pressure and control a rocking shaft whose movements are multiplied by gear-wheels to give altitude readings with three different
indicators. A bimetal bracket compensates the readings of the
instrument for changes of temperature. (Fig. 101.)

L

λ. Lambda. The symbol for wavelength.

lb. Pound (force or mass).

Labyrinth Packing or Seal. A series of grooves cut in the piston

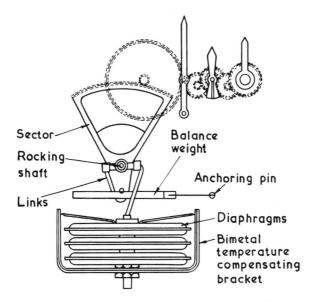

Sector

Rocking shaft

Links

Balance weight

Anchoring pin

Diaphragms

Bimetal temperature compensating bracket

FIG. 101. K.B.B.-KOLLSMAN ALTIMETER.

of a steam-engine to allow any escaping steam to expand slowly thus diminishing the leakage. Cf. **Breaking Joint.**

Lace Machine. A machine in which *bobbins*, *combs* and carriages convert two series of threads into an ornamental fabric.

Lag. (*a*) A perforated wooden strip forming part of the pattern chain for a *dobbie.*

(*b*) The delay in time between one event and the next.

(*c*) To include thermal insulation in a design.

Laminated. Made up of thin plates or sheets called ' laminates' or *shims.*

Laminated Spring. *Leaf spring.*

Land. (*a*) The portion behind the cutting edge of a tool as in *broach.* (See Figs. 17, 114.)

(*b*) The flat top of a gear-tooth or the flat bottom between two teeth. (See Fig. 172.)

Landing Gear. The *alighting gear* of a landplane designed to be strong enough to withstand a specified vertical velocity of landing.

Langström. See **Ljungström Turbine.**

211

Lantern Frame Pattern. An *inverted cylinder engine* with a hollow cylindrical standard having the *crosshead* working in the bore of the standard and the crank bearings cast in the base.

Lantern Pinion. See **Pinion.**

Lantern Ring. A spacing ring inserted in a *stuffing box* of a valve to form a pressure relief or condensing chamber.

Lantern Wheel. A wooden cog-wheel used in mills, similar in design to a ' lantern pinion ' and sometimes called a ' trundle ' or ' trundle wheel '. See **Pinion.**

Lap. (*a*) The amount by which one plate overlaps another.

(*b*) The amount by which a slide valve has to move from mid position to open the steam or exhaust port of a steam-engine. See **Angular Advance, Exhaust Lap, Outside Lap.**

(*c*) A single turn of a rope or chain around a bollard.

(*d*) The contact length of a chain around a sprocket wheel or of a belt around a pulley.

(*e*) The thin sheet in which the fibre is delivered to a *carding engine* after scutching and beating.

(*f*) A piece of soft metal, etc., or metal cylinder charged with polishing powder for *lapping*.

Lap Joint. A riveted or welded joint in which one member overlaps the other.

Lapping. The finishing and polishing of spindles, bearings, etc. to very fine limits by the use of *laps* of lead, brass, etc. See **Lapping Machine.** Cf. **Honing Machine.**

Lapping Machine. A machine tool for finishing the bores of cylinders, etc., using revolving *laps* and an abrasive powder suspended in the coolant. See also **Lapping.**

Lashlock. A *split-and-sprung gear or nut* in which relative motion of the two parts is prevented, on reversal of load, by a self-locking wedge with a light spring.

Latch Needle. A needle with a latch over the hook opening and closing while a stitch is being formed, and the completed stitch opening the latch to receive a new thread.

Lateral Traverse. The amount of end play given to locomotive trailing axles to permit the taking of sharp curves.

Lathe. A machine tool (Figs. 102, 103) consisting of a lathe bed carrying a *headstock* and *tailstock* for driving and supporting the work, and of a *saddle* which carries the slide rest for holding and traversing the tool (see Fig. 102). It produces cylindrical work, boring, facing and screw-cutting. See also **Automatic Lathe, Backing-**

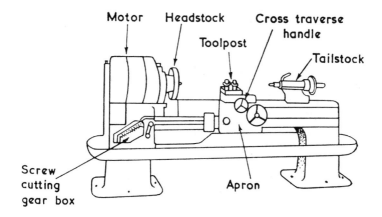

FIG. 102. LATHE.

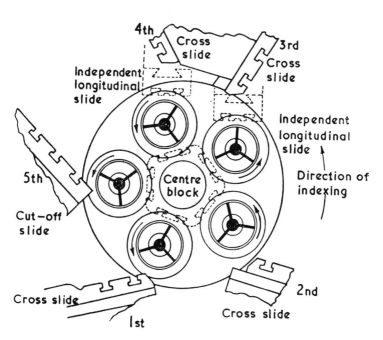

FIG. 103. OPERATION OF FIVE-SPINDLE AUTOMATIC LATHE.

213

Lathe

off Lathe, Capstan Lathe, Centre Lathe, Copying Machine, Hollow Mandrel Lathe, Slide Rest Lathe, Turret Lathe.

Lathe Bearers. The sides or cheeks of a *lathe bed*.

Lathe Bed. That part of the lathe which forms the support for the *headstock*, *tailstock* and *carriage*. It is a cast rigid box-section girder on legs with the upper surface planed to provide a true working surface. See **Ways**.

Lathe Carrier. A clamp consisting of a shank (either straight or bent) with an eye at one end and provided with a set-screw. The clamp is attached to work supported *between centres* and driven by the engagement of the *driver plate* pin with the shank of the carrier. See also **Bent-tail Carrier**.

Lathe Centre Grinder. An emery wheel with overhead drive for grinding the hardened conical points of lathe centres in place whilst mounted in the *headstock* or *tailstock*.

Lathe Cheeks. The sides of a *lathe bed*.

Lathe Dog. *Lathe carrier*.

Lathe Heads. The *headstock* and *tailstock* of a lathe.

Lathe Planer. A piece of mechanism, sometimes attached to the *saddle* of a lathe, for the surfacing of metal by rectilinear cutting, using a *milling cutter* in the *headstock*.

Lathe Standards. The supports of a *lathe bed*.

Lathe Tool. Any turning tool used in a lathe. See **Diamond-tip Turning Tool, Finishing Tool, Gooseneck Tool, Knife Tool, Parting-off Tool, Roughing Tool, Side Tool**.

Lathe Traverse. The drive mechanism for moving the saddle to perform a turning or facing operation.

Lay. (*a*) The dominant direction of tool marks or scratches in a *surface texture*. (See Fig. 65.)

(*b*) A *batten* in a *loom*.

Lay Shaft. An auxiliary, or secondary, geared shaft. See also **Gear Cluster**.

Laying-out. The marking out or setting out of work, especially plate work, to full size ready for cutting, drilling, etc. Cf. **Marking-out**.

Lazy Tongs. An arrangement of zig-zag levers for picking up objects, the picking-up grip being usually an action in a direction at right-angles to that of the applied power.

Lead. (*a*) The distance between successive intersections of a helix by a generator of the cylinder on which it lies; the distance a screw thread advances axially in one revolution.

(*b*) The lead of the helix of which the tooth trace (in a gear) forms part.

Lead Angle. (*a*) The angle of a tooth trace for a helical, spur or worm gear.

(*b*) The acute angle between a tangent to the helix and a transverse plane. The complement of the *helix angle*.

(*c*) On a parallel screw thread, the acute angle between a tangent to the helix at the *pitch point* and a plane perpendicular to the axis.

(*d*) On a taper screw thread, the angle at a given axial position between the tangent to the conical spiral at the *pitch point* and a plane perpendicular to the axis.

Lead Line. The left-hand vertical line in an *indicator diagram* representing the rise in pressure at the start of the working stroke.

Lead of Valve. The amount by which the *slide valve* of a steam-engine has uncovered the port to steam when the piston is at the beginning of its working stroke; sometimes called ' main screw '.

Lead Screw (Leading Screw, Guide Screw). The screw which runs longitudinally in front of the bed of a lathe and the master screw used for cutting a screw thread.

Leading Axle. The front axle of a locomotive.

Leading Edge. That edge of a wing, aerofoil, strut or propeller blade which first meets the air or water when the craft is in motion. Cf. **Trailing Edge.**

Leading Springs. The springs carrying the axle boxes of the leading wheels of locomotives and rolling stock.

Leading Wheels. The front wheels of a locomotive.

Leads. Lengths of thin lead wire which are inserted between a very large *journal* and the bearing cap during assembly to test the clearance.

Leaf. (*a*) A tooth of a *pinion*.

(*b*) A thin plate as part of a *laminated spring*.

Leaf Spring (Laminated Spring). A curved (sometimes flat) spring consisting of thin plates (leaves) superimposed and acting independently, to form a beam or cantilever of uniform strength (see Fig. 104). Cf. **Carriage Spring.**

Leaning Thread. *Buttress screw thread.*

Lease Rods. The two rods across a *warp* to separate the threads and keep them in their correct position in a weaving machine.

Least Angle of Traction. *Friction angle.*

215

Left-hand Engine

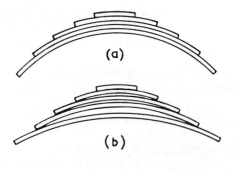

FIG. 104. LEAF SPRINGS.

(a) Pulled up (b) Free

Left-hand Engine. A horizontal engine which stands to the left of its flywheel as seen from the cylinder. Cf. **Right-hand Engine.**

Left-hand Screw. A screw which turns anti-clockwise when being inserted.

Left-hand Thread. A screw thread which, when viewed along its axis, appears to rotate counter-clockwise as it goes away from the observer, the reverse of the common wood-screw.

Left-hand Tools. Lathe side tools with the cutting edge on the right, thus cutting from left to right, that is away from the *headstock* of a conventional lathe.

Left-hand Twist Drill. A *twist drill* in which the cutting edge and *flute* run anti-clockwise up the shank.

Left-handed Engine. An aero-engine in which the propeller shaft rotates counter-clockwise when the observer is looking past the engine to the propeller.

Lentz Valve-gear. A valve-gear admitting and exhausting steam in a locomotive through two pairs of *poppet valves*, spring-controlled and operated from a camshaft rotating at engine speed.

Leveller. See **Straightening Machine.**

Level-luffing Crane. See **Crane.**

Lever. A pivoted arm carrying the *pallet* in a lever escapement. See **Escapements (Lever).**

Lever Box (Lever Bracket). A hollow casting in a crane carrying the levers connected to the motions for lifting, slewing, travelling, etc.

Lever Bracket. *Lever box.*

Lever Chuck. A concentric chuck actuated by a lever instead of a screw.

216

Lever, Differential or Floating. See **Differential Gear.**

Lever Escapement. See **Escapement.**

Lever Jack. (*a*) A simple *jack* consisting of a lever for lifting and a standard for support.

(*b*) An accessory controlling the locker carriage, located on the underside of the *combs* of a lace machine. Also called ' locker jack '.

Lever Pumps. The air and circulating pumps of marine engines worked by levers driven from the *crossheads* of the piston rods.

Lever Watch. A watch fitted with a lever escapement. See **Escapements.**

Licensed Aircraft Engineer. An engineer licensed to certify that an aircraft engine and/or components complies with current regulations.

Licker-in. A revolving cylinder covered with saw-like teeth which tears up sheets of cotton entering a *carding engine.*

Life Factor. A design factor to provide for the scatter of the results of fatigue tests, for the effects of possible deterioration in service and for the possible inadequacy of the assumed pattern of repeated loading. The estimated mean fatigue life is divided by this factor to obtain the safe fatigue life.

Lift (Elevator). (*a*) An enclosed platform working in a vertical shaft for transferring persons, goods or vehicles from one floor level to another and operating electrically, hydraulically or pneumatically with usually a *winding drum* or a *traction sheave* and a counterweight.

(*b*) The height to which a pump can raise water or other fluids.

Lift Pump. *Suction pump.*

Lift Valve. A valve in which the disc, ball, plate, etc., lifts or is lifted, vertically, to allow the passage of a fluid.

Lifting Blocks. A continuous rope passing round pulleys mounted in blocks by which a pull at the free end of the rope lifts a weight attached to the lower block heavier than the pull, thus giving a *mechanical advantage.*

Lifting Cylinder. A cylinder of a hydraulic crane used for lifting the load. Cf. **Turning Cylinder.**

Lifting Jack. *Jack.*

Lifting Piece. A cranked lever in the rack-striking mechanism of a clock, carrying the warning piece at one end and lifting the *rack hook* just before the hour or half-hour, the latter lift being appreciably less. (See Fig. 161.)

Lifting Pin. A pin on the wheel of a clock or repeater for releasing a part of the striking mechanism.

Lifting Plan. *Peg plan.*

Lifting Pump. *Suction pump.*

Lifting Ram. The smaller of the two rams in a hydraulic forging press which lifts the *crosshead* and *tup* after each stroke.

Lifting Tackle. *Tackle.*

Lifting Tongs. See Tongs (Lifting).

Lighting Chamber. *Firing chamber.*

Lighting Cock. The jet that fires the charge in a gas-engine cylinder.

Light Running. The running of mechanisms, e.g. shafting, under no load and with the minimum of friction.

Light-spring Diagram. An *indicator diagram* taken with a specially weak control spring or diaphragm to reproduce to a large scale the low-pressure part of the diagram.

Limit Gauge. A fixed gauge used for verifying that a part has been made within specified dimensional limits. It is either a ' go ' or ' no-go ' *gauge*. See also **Gauge.**

Limit Gauging. A method of measurement to ensure the fitting of two pieces together within specified clearance limits and hence permitting interchangeability.

Limit Load. The maximum load anticipated under normal conditions of operation of an aircraft.

Limit of Elasticity. *Elastic limit.*

Limit of Proportionality. The point on the stress-strain curve at which the strain ceases to be proportional to the stress. Cf. **Elastic Limit.**

Limit System. A method of classifying and selecting limits by a system of standard *allowances* and *tolerances* in graded amounts associated with specified ranges of *basic sizes*. See **Size.**

Bilateral Limit System. A limit system in which the allowed tolerance is bilateral. See **Bilateral Tolerance.**

Hole Basis Limit System. A limit system in which the *design size* for an internal feature is the *basic size* and variations in the grade of fit are obtained by varying the *allowance* and the *tolerance* on the external feature. See **Size.**

The Newall system for holes or bushes has class A for the finer grade and class B for inspection limits.

Shaft Basis Limit System. A limit system in which the *design size* for an external feature is the *basic size* and variations in the grade of fit are obtained by varying the *allowance* and *tolerance* on the internal feature.

Unilateral Limit System. A limit system in which the tolerance is unilateral. See under **Tolerance**.

Limiting Friction. See **Friction**.

Limiting Range of Stress. The greatest range of stress about a mean stress of zero that a metal can withstand for an indefinite number of cycles without failure. Also called ' endurance range ': the *fatigue (endurance) limit* is half this range.

Limits of Size. The maximum and minimum sizes allowed for a dimension. The difference between these sizes is equal to the *tolerance*. See also **Limit System**.

Limits of Tolerance. See **Tolerance**.

Limits (Metal). The greatest (maximum) and the least (minimum) allowable amount of metal or other material present on the surface of a feature.

The low limit of size for a hole and the high limit for a shaft are ' maximum metal limits ' and the high limit of size for a hole and the low limit of size for a shaft are ' minimum metal limits '. Cf. **Tolerance (Metal)**.

Lincoln Milling Machine. A horizontal spindle machine with a vertically adjustable spindle over a table of fixed height.

Line. The steel wire, gut or cord supporting a weight in a weight-driven clock.

Line of Centres. (*a*) A line passing through two or more centres in machinery or in a mechanism.

(*b*) A line joining the centres of two or more wheels in a watch or clock.

(*c*) The line joining the *balance staff* and the *pallet staff* in a lever *escapement*.

Line Shafting. The main (overhead) shafting used in factories to transmit power from the power source to individual machines.

Line Standard. A standard of length, being the distance between two fine lines on a metal bar measured under specified conditions. Line standards are the bases for checking *gauges*.

Linear Advance. The amount by which a *slide valve* is set forward for lap and lead beyond a line 90° ahead of the crank. See **Lead of Valve**.

Linear Roller-bearing. A *roller-bearing* which allows linear motion, the rollers returning along a recirculation channel. (Fig. 105.)

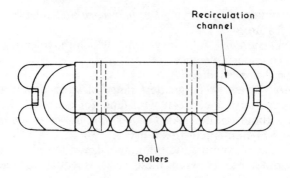

Recirculation channel

Rollers

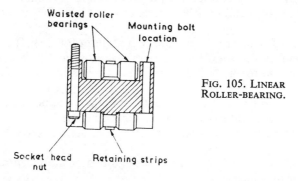

Waisted roller bearings

Mounting bolt location

Socket head nut

Retaining strips

FIG. 105. LINEAR ROLLER-BEARING.

Linear Sander. A device, with a reciprocating action, that can b
fitted to an electric drill, that is used for sandpapering or polishin
depending upon the surface attached to the reciprocating part. Se
also **Belt Polisher.**

Linear System. A motion defined by a linear differential equatio
with constant coefficients.

Linear Velocity. Velocity along a path, either straight or curvec
Cf. **Angular Velocity.**

Liner. (*a*) A separate and renewable sleeve placed within a
engine cylinder to provide a more durable rubbing surface for th
piston rings.

(*b*) Any sleeve fitted to provide a more durable surface.

Lining-out. *Marking-out.*

Lining-up. (*a*) Arranging the bearings of an engine crankshaf
etc., in perfect alignment.

(*b*) Alignment of an *assembly*.

Link. (*a*) A unit in a *chain*.

(*b*) Any connecting piece in a machine which is pivoted at both ends.

(*c*) The curved slotted member of a *link motion*.

Open Link. A link in which the slot is open, the block bearing on the inside faces of the slot. Cf. **Box Link.**

Stud Link (*Stayed Link*). A link *braced* in the centre by a short stud or by a welded connecting piece.

Link Arrangement. A link and sliding block to provide a quick-return motion to the ram in a *shaping machine*.

Link Belting. A belting (or *belt*) composed of a number of short links, arranged parallel and retained in position by pins, which permit the links to pivot freely and bend round small pulleys for transmitting power over a short distance.

Link Block. A sliding block which is pivoted to the end of a valve rod and works in the slotted link of a *link motion*.

Link Grinding Machine. A machine for grinding the curves of the slot links of *valve-gears*, having spindles of planet type and the links moved about a centre, adjustable for radius.

Link Mechanism. A system of rigid members joined together by constraints, such as *links*, so that motion can be both amplified and changed in direction.

Link Motion (Stephenson's). A valve motion for reversing and controlling the cut-off of a steam-engine, consisting of a pair of eccentrics connected to the ends of a slotted link carrying a block attached to the valve rod. The link position is varied to make either eccentric effective, an operation known as ' linking up '.

Link Reversing Motion. A reversing motion effected by a slot link operating on the two eccentrics in a *link motion*, as distinct from the direct reversal of a single eccentric on its shaft.

Link Rods. The auxiliary or articulated connecting-rods of a radial engine working on the *wrist pins* carried by the master rod.

Linkage (Computing Linkage). An assembly of rigid links, pivots and sliding members in which the motion of a particular output link is a predetermined function of the motion of the assembly.

Linkage Multiplier. A *computing linkage* in which the motion of an output link is proportional to the product of two input motions.

Linking. The process in *automation* whereby articles in a transfer

Linking

line for machining or manufacture are passed automatically, with inspection, between successive machines.

Linking-up. *Notching-up.* See also **Link Motion.**

Linotype Machine. See **Composing Machine.**

Lip. (*a*) The cutting edge or point attached to a centre bit or similar tool, which cuts the circumscribing circle during the process of boring.

(*b*) The edge of the cylinder *escapement* which receives the impulse from the *escape wheel* in a clock or watch.

Lip Drill. A drill whose cutting faces are slightly hollowed out backwards immediately above the cutting edges to give a *front rake* to the tool. (See Fig. 114.)

Liquid Spring. A piston-cylinder combination used in aeroplane suspension units, the piston forcing hydraulic fluid through a small hole.

Live Axle. *Driving axle.*

Live Head. A term sometimes applied to the *headstock* of a lathe.

Live Ring. A large roller-bearing for supporting *turntables* and revolving *cranes*. See also **Thrust Bearing.**

Live Roller. A roller free to move along its own path and rotate but which does not revolve on a spindle. Live rollers are used for the slewing motions of heavy machinery and for turntable centres.

Live Spindle. A spindle which communicates motion, such as the spindle of a headstock and the revolving mandrel of a machine tool.

Ljungström Turbine. A radial-flow double-motion reaction turbine with the groups of blades arranged in concentric rings. Alternate rings are attached to one of two discs mounted on separate contra-rotating shafts. The contra-rotation of the shafts effectively doubles the peripheral speed of all the rings except the first, thus nearly doubling the steam velocity and giving an increased efficiency for a given overall size of turbine. (Fig. 106.)

Load. The power output of an engine or power plant under given circumstances.

Load Factor. (*a*) The ratio of an average load to the maximum load.

(*b*) The ratio of the external load on an aircraft in a specified flight condition to the weight of the aircraft.

(*c*) The ratio of the number of passengers (or tons of freight) in a vehicle to the maximum number (or load).

Loader. A mechanical shovel or similar device for loading trucks.

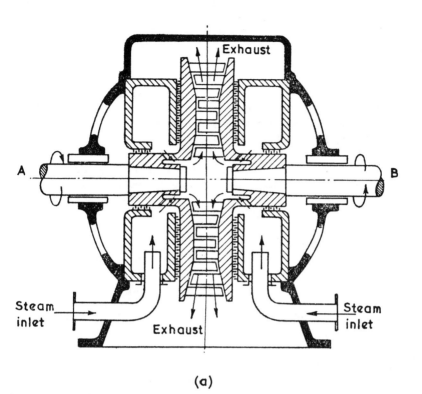

(a)

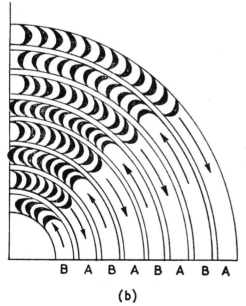

Fig. 106. Ljungström
Turbine: (a) sectional view,
(b) contra-rotation of shafts.

(b)

223

Lobe. (*a*) A rounded projection or cam.

(*b*) The projections on an ignition contact-breaker.

(*c*) The several cams formed on one ring in radial aero-engines

(*d*) The peripheral projections of a helical screw *compressor*.

Lock. (*a*) A mechanical appliance for fastening a door with a bolt that requires a special key to work it.

(*b*) A mechanism for exploding the charge in a gun or rifle.

(*c*) A communicating channel in a multi-level canal, having gates at both ends.

(*d*) An antechamber to a chamber in which engineering tests are undertaken at high pressure.

Lock. See **Steering Lock.**

Lock-nut. (*a*) An auxiliary (thin) nut used in conjunction with another to prevent it from loosening under vibration.

(*b*) A nut designed to obviate accidental loosening; it may have a plastic insert, or be of a special shape, so that one part of a nut locks against another, or a lock wire or pin can be inserted and be appropriately named.

Lock Stitch. A stitch made by a mechanism on a *sewing machine*. A lower thread is carried by a horizontally reciprocating shuttle, which passes its thread through the loop of the thread carried through the cloth by the eye-pointed needle. On the upstroke of the needle the bight of the lower thread is pulled taut to make the same stitch on both sides of the fabric.

Lock Washer. A washer made of spring steel designed to prevent the loosening of a nut.

Locker Jack. *Lever jack* (*b*).

Locker Rack. A *rack railway* with the rack centrally located and with teeth on each side in which horizontal cog-wheels work.

Locking Angle. The angle, measured from the pallet centre of a watch, through which the pallets have to move before unlocking can take place. See **Pallets.**

Locking Face. The portion of a *pallet* upon which the teeth of the *escape wheel* drop for locking.

Locking Lever (Locking Piece). The lever that locks the chiming mechanism in a clock. (See Fig. 161.)

Locking Piece. *Locking lever.*

Locking Pin. The pin on the *locking wheel*. See also **Locking Lever,** Fig. 161.

Locking Plate. A circular plate with notches cut around the periphery, the distances between the notches regulating the number of hours struck by a clock. (See Fig. 161.)

Locking Post. A steel post, screwed into the solid part of a *rack* just below the teeth, which engages the *gathering pallet* just after it has engaged the last tooth and thus locked the striking train of a clock.

Locking Springs. The two springs controlling the *locking lever* and the *hammer arbor* in a striking clock.

Locking Wheel. A wheel geared into the *warning wheel* pinion with a long and large pivot, at its pinion end, which is filed square and on to which the *gathering pallet* (see **Pallet**) is fitted. See also **Locking Lever, Locking Pin.**

Locomotive. A railway vehicle, driven by steam, electricity or oil, for hauling trucks or carriages (Fig. 107). See **Diesel Loco-**

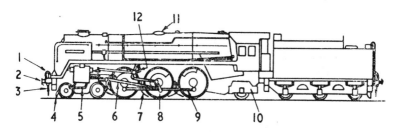

FIG. 107. LOCOMOTIVE.

1. Vacuum brake connection.
2. Buffer.
3. Coupling link.
4. Guard iron.
5. Leading bogie.
6. Connecting rod.
7. Sand pipe.
8. Driving wheels.
9. Coupling rod.
10. Trailing truck.
11. Steam dome.
12. Valve gear.

motive, Diesel-electric Locomotive, Electric Locomotive, Steam Locomotive, Traction Engine.

Locomotive Boiler. A multi-tubular *boiler.*

Logarithmic Decrement (log-dec). The natural logarithm of the ratio of the peak values of the amplitudes of a vibration in successive cycles when the amplitudes decrease exponentially. See also **Decay Factor.**

Loom. A machine for weaving yarn or thread into fabric in which two sets of yarn or thread, ' warp ' and ' weft ', are interlaced.

Looming (Healding). Drawing the threads of the warp through the eyes of the *heald* shaft, in the arranged order, prior to weaving plus knotting and twisting.

Looper. A mechanical contrivance in a sewing machine for making loops. See also **Sinker.**

Loose Centres. Heads like lathe *poppets* used for supporting work on the table of a planing or grinding machine so that the work can be rotated.

Loose Coupling. A shaft coupling capable of instant disconnection. Cf. **Fast Coupling.** See also **Claw Coupling.**

Loose Eccentric. An *eccentric* riding freely on a shaft between two stops which position and drive it for ahead and reverse-running of small steam-engines.

Loose Headstock. See **Tailstock.**

Loose Pulley. A pulley mounted freely on a shaft and generally used in conjunction with a *fast pulley* to provide means for starting and stopping a shaft by shifting a driving belt from one to the other.

Lorry. (*a*) A long low flat wagon.

(*b*) A truck used on railways and tramways.

(*c*) A motor truck for road transport.

Lost Motion. The difference between the rate of motion of driving and driven parts in a mechanism.

Low-pressure Cylinder. The largest cylinder of a multiple expansion steam-engine, in which the steam is finally expanded.

Low-pressure Engine. An engine which exhausts its steam into a condenser.

Low-pressure Steam. Steam at a pressure below or only a little above atmospheric pressure.

L-Rest. A lathe *rest* for hand turning, shaped like an inverted L.

Lubricant. A substance for reducing friction between bearing surfaces in relative motion, such as oil, graphite, air under pressure, etc. See **Lubrication.**

Lubrication. The distribution of a *lubricant* between moving surfaces in contact to reduce the friction between them. See **Boundary Lubrication, Fluid Lubrication.**

Lubricator. Any contrivance for supplying a *lubricant* to bearing surfaces.

Luff. A tackle with a double block and a single one, roped with rope 3 inches or larger.

Luffing. (*a*) Altering direction.

(*b*) Bringing a sailing boat's head closer to the wind.

226

Luffing-jib Crane. A crane with its *jib* hinged at its lower end to the crane structure to allow alteration in its radius of action. See also **Crane.**

Lug. *Ear.*

M

μ. (*a*) The symbol for the coefficient of friction.

(*b*) A micron equal to 10^{-3} mm.

(*c*) μin. one one-millionth of an inch.

M. The symbol for moment.

m. The symbol for mass.

m.e.p. Mean effective pressure.

mech. eff. Mechanical efficiency.

M.H.N. *Meyer hardness number.*

M.K.S. Unit. A unit in the metre-kilogramme-second system.

m. of i., M.I. *Moment of inertia.*

m.p.g. Miles per gallon.

m.p.h. Miles per hour. New form: mile/h.

M.R. Moment of resistance.

M.S. Maximum stress.

m.p.s. Metres per second. (SI notation: m/s.)

Machine. An apparatus consisting of an assemblage of parts, some fixed and some movable, by which mechanical power is applied at one point, to transmit a force or motion at another point. See **Mechanical Advantage, Velocity Ratio.**

Machine Centres. *Loose centres.*

Machine Moulding. The process of making moulds and cores by mechanical means, replacing hand-ramming. See also **Moulding Machine (Plastics).**

Machine Riveting. Clenching rivets with *hydraulic riveters* or *pneumatic riveters.*

Machine Shop. A shop where all the operations of engineering requiring the use of machines are carried on, and excluding fitting and erecting.

Machine Tapper. A fitting attached to drilling machines or lathes for tapping threads in holes, with a spring release to prevent breakage on completion of the tapping operation followed by reversal of the drive to remove the tap. Cf. **Diehead.**

Machine Tools. (*a*) Tools used on various machines for cutting, drilling, milling, planing, punching, shaping, shearing, slotting, turning, etc.

227

Machine Tools

(*b*) Machines which are used in manufacturing plant, including those referred to in (*a*).

Automatic Machine Tools. Machine tools with an unmanned repetitive action.

Machine Vice (Vice Chuck). A parallel-jawed vice for holding pieces of work on the tables of drilling, planing and shaping machines.

Machine Working Tap. See **Tap.**

Machining. (*a*) The operation of removing material from work by machine.

(*b*) Making or operating machines.

(*c*) The operation of printing by machine.

Magazine Feed. A mechanism by which pieces of work are fed singly to an automatic machine either by shoots or by rotating discs.

Magnetic Belting. Machinery belting with iron strips inserted at intervals along its length so that the grip is increased when passing over a magnetized pulley.

Magnetic Blocks. The movable spacers used with a *magnetic chuck*.

Magnetic Brake. A brake operated by means of an electromagnet.

Magnetic Chuck. A ' permanent magnet ' chuck, which has permanent magnets located in the base and separated from the top surface by movable spacers consisting of alternate rows of conductors, *magnetic blocks* and insulators, for holding light flat work securely on the table of a machine tool. The workpiece is released by moving the spacers with a hand lever so that the insulators are aligned and prevent the magnetic flux from reaching the surface. An ' electromagnetic ' chuck requires a source of direct-current electricity.

Magnetic Coolant Separator. A magnetic device which removes magnetic swarf and most abrasive particles from the *coolant* used in machine tools.

Magnetic Clutch. See **Clutch.**

Magnetic Damping. See **Damping.**

Magnetic Suspension. The use of a magnet to support, in part, the weight of a shaft in an instrument or meter and thus relieve the bearings of some of the load.

Magnetic Transmission. A clutch used in some small cars. Ferromagnetic power takes up the drive between two rotating members when it is drawn into an annular gap by electromagnets.

Magneto-striction. The change in dimensions produced in cer-

228

tain magnetic materials when they are magnetized, as in iron, steel and especially nickel.

Main and Tail. Rope haulage in a mine using a main rope to draw out the full wagons and a tail rope to draw back the empties.

Main Bearings. The bearings for an engine's crankshaft.

Main Cylinder. The principle or working cylinder of an engine. Cf. **Balance Cylinder, Oil Cylinder.**

Main Driving Belt. The belt from the power unit to the main driving pulley in a workshop, whose machines are belt-driven.

Main Driving Pulley. The principal pulley on a line of *shafting*.

Main Frames. The locomotive frames which carry the boiler, axle boxes, cylinders, etc.

Main Rotor(s). (*a*) The *rotor(s)* of a *rotorcraft* or *hovercraft* which provide lift, as distinct from the *tail rotor*.

(*b*) The assembly of *compressor(s)* and *turbine(s)* which comprise the rotating parts of a turbojet or turboprop engine.

Main Screw. *Leading screw.*

Main Valve. The *slide valve* proper when there is a separate expansion or cut-off valve for the steam.

Main Wheel. The *great wheel* of a clock or watch.

Mainspring. (*a*) The main source of power in a clock or watch. See also **Maintaining Spring** and **Balance Spring.**

(*b*) The spring in a typewriter that moves the *carriage* to the left after each letter has been struck.

Mainspring Hook. The means by which a *mainspring* is attached to its *barrel*.

Mainspring Winder. A tool for coiling a *mainspring* prior to its insertion or withdrawal from the *barrel*.

Maintaining Spring (Maintaining Power). A supplementary spring fitted inside the *fusee* which is wound up by the mainspring and held by the maintaining detent (see Fig. 108). The maintaining spring keeps the clock going whilst the mainspring is rewound.

Male and Female. Engineering terms applied to inner and outer members which fit together, such as threaded pieces, pipe fittings, etc.

Maltese Cross Mechanism (Geneva Wheel). A mechanism for feeding the film forward intermittently in a cinematograph projector (or camera), involving a star-wheel and cam and obviating the use of claws. (See Fig. 109.)

Manchester Principle. *Diametral pitch.*

Mandrel (Mandril)

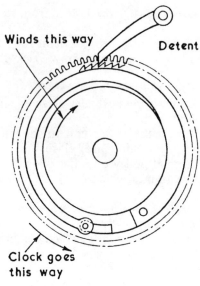

Winds this way

Detent

Clock goes
this way

Fig. 108. Maintaining
Spring

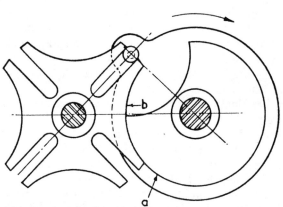

Fig. 109. Maltese Cross Mechanism: (a) star wheel, (b) cam.

Mandrel (Mandril). (a) A cylindrical rod, usually parallel an
sometimes tapered, upon which partly machined work is mounte
for turning, milling, etc.

(b) A cylindrical rod used in the opening of a die to form the in
ternal diameter when extruding tubes. When unsupported in the di
it is called a ' floating mandrel '. (See Figs. 56 and 190.)

(c) The driving or headstock spindle of a lathe.

(d) A special face-plate lathe used in watch-making.

Expanding Mandrel. One which is split and capable of expansion by a tapered plug.

Mandrel Press. A press used for the purpose of fitting mandrels to the bore of the work, the mandrel being usually slightly tapered. See also **Mandrel**.

Manifold. A chamber or pipe with many openings.

Exhaust Manifold. A pipe or chamber into which the exhaust gases of a piston-engine are led from a number of cylinders.

Fuel Manifold. A main pipe with a series of branch pipes distributing fuel to the burners of a turbojet engine or cylinders of a piston-engine.

Induction Manifold (Inlet Manifold). A branched pipe for distributing the air/fuel mixture to a number of cylinders of a *piston-engine.*

Manifold Pressure (Boost Pressure). The absolute pressure in the induction manifold of an unsupercharged piston-engine. (See also **Boost Gauge** and **Manifold**). When the engine is supercharged, it is called ' boost pressure '.

Mangle Wheel. A reciprocating gear-wheel with its teeth so arranged that it turns back and forth on its centre without making a full revolution. A pin-type mangle gear-reversing mechanism is shown in Fig. 110.

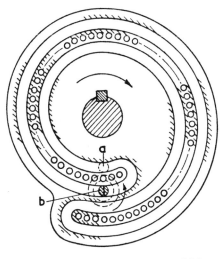

FIG. 110. PIN-TYPE MANGLE GEAR-REVERSING MECHANISM: (a) guide pin, (b) pinion spindle.

231

Margin of Safety. The difference between the *factor of safety* an the *working load*.

Marine Chronometer. See **Chronometer.**

Marine Engine. *Steam* or *compression-ignition* (*oil*) engines use for ship propulsion and directly coupled to the propeller.

Marine Governor. A *governor* to control steam admission an thus check racing of the main shaft of a marine engine, when th propeller is raised by the waves above the sea.

Marine Pattern Connecting-rod. A rod whose bearing end ha two brasses, secured by bolts to flat expansions of the wrought-iro end of the rod.

Marine Screw Propeller. A boss carrying two, three or fou helical-shaped blades which produce the thrust that drives a ship See also **Propeller.**

Marking-out. Setting out centre lines and dimensional marks o material as a guide for subsequent machining operations. C **Laying-out.**

Marlborough Wheel. An extra wide gear-wheel which enables i to mesh with other wheels mounted on different shafts which ar close to one another.

Marshall Valve-gear. A radial gear of *Hackworth* type in whic the straight guide is replaced by a curved slot to correct inequalitie in steam distribution.

Masked Valve. A *poppet valve* with its head recessed into its sea so that its outer diameter acts as a piston valve, thus allowing lower valve acceleration.

Mass. The quantity of matter in a body; the weight of a bod divided by the *acceleration due to gravity*.

Mass-balance Weight. A mass attached to an aircraft's contro surface to reduce or eliminate the inertial coupling between th angular movement of the control and some other degree of freedon of the aircraft. It may be a single lump of metal or be distribute along the span of the control surface and connected to it by a serie of links. ' Static balance ' is the condition in which the mas balance about the hinge axis is zero. It is one method of flutte prevention.

Mast. The vertical member in a derrick crane. See **Crane.**

Master. A term applied to a special *gauge*, tool, etc., or the ke member of a system.

Master Clock. A clock controlling a number of other clocks b sending out electrical impulses at predetermined time-intervals. See **Impulse-driven Clock.**

Master Connecting-rod. A specially strengthened connecting-rod, used on one cylinder of an aero *radial engine*, which carries *wrist pins* to which the other connecting-rods are articulated and thus transmits the total thrust of all cylinders to the *crank pin.*

Master Gauge. A *standard gauge* made to specially fine limits and used as the ultimate reference in the control of all products of the kind to which it relates.

Master Gear. A gear used as a reference standard.

Master Tap. An extremely accurate *tap* for use when great accuracy is required.

Master Wheel. A dividing wheel used for cutting gear-teeth.

Mating. A term which is used to describe the interlocking of two surfaces or pieces.

Maximum Material Condition (MMC). A condition implying that a finished product contains the maximum amount of material permitted by the size, dimensions and tolerances, for that product.

Maximum Value. The maximum excursion from zero, during a given time, of a non-periodic quantity (cf. **Amplitude**). A peak value is the total ' range ' of fluctuation.

Mean Effective Pressure. See **Brake Mean Effective Pressure, Indicated Mean Effective Pressure.**

Mean Line. A line from which the sum of the squares of the ordinates between it and the profile is a minimum. Cf. **Centre Line.**

Measuring Machine. A machine for the precise measurement of standard gauges to a high accuracy. One type consists of a bed supporting a sliding head carrying a micrometer spindle, and tail-stocks.

Mechanical Admittance (Mobility). The reciprocal of *mechanical impedance.*

Mechanical Advantage. The ratio of the load (or resistance) to the applied effort (or force) in a *machine.*

Mechanical Efficiency. The ratio of the *brake horse-power* of an engine to the *indicated horse-power.*

Mechanical Engineering. A branch of engineering dealing primarily with the design, production and operation of mechanisms and mechanical contrivances including prime-movers, vehicles and general engineering products.

Mechanical Equivalent of Heat. The ratio of the mechanical energy being transformed into heat to the resulting quantity of heat

Mechanical Equivalent of Heat

generated. Its value is the Joule. (1 J = 1 N.m.)

Mechanical Impedance. The ratio of the total force acting in the direction of motion to the velocity at the point, or surface, of reference and for a specified frequency; for example, when a mechanical system is vibrating with uniform amplitude and at the specified frequency. It is the reciprocal of ' mechanical admittance '.

Mechanical Integrators. Two types of *integrators* are shown in Fig. 111.

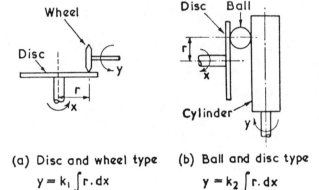

(a) Disc and wheel type

$$y = k_1 \int r \cdot dx$$

(b) Ball and disc type

$$y = k_2 \int r \cdot dx$$

FIG. 111. MECHANICAL INTEGRATORS.

Mechanical Resonance. An enhanced response at a given frequency of part of a mechanism or structure to a constant-magnitude disturbing force.

Mechanical Shovel. An excavating machine with a boom and a bucket lifting system with power for operation, such as by a diesel engine driving a pump or pumps to provide hydraulic power for the lifting system and a transmission, sometimes hydrostatic, for the vehicle drive.

Mechanical Stoker (Automatic Stoker). A device for supplying solid fuel continuously by gravity and in some cases carrying the fuel on an endless chain progressively through a furnace and depositing the ash.

Mechanics. That branch of science and technology which studies the action of forces on bodies and of the motions they produce. ' Statics ' is the section which deals with forces in equilibrium; ' dynamics ' is the section concerned with the motions in relation to

234

the forces; ' kinematics ' deals with the theory of the motion without reference to the forces. ' Kinetics ' is the science of the relations between the motions of bodies and the forces acting on them.

Mechanism. A system of mutually adapted parts working together which may not, however, constitute a complete machine.

Mechanization (Mechanisation). The change from animal to mechanical power in transport and industry.

Mechanize (Mechanise). (*a*) Make mechanical.

(*b*) Do by machinery in preference to doing anything by hand.

(*c*) Replace men or animals by machinery.

(*d*) Equip a military unit with armoured vehicles, tanks, etc.

Megadyne. A unit of force equal to one million *dynes* (10 N).

Mercurial Compensated Pendulum. See **Pendulum**.

Mesh. (*a*) The state of gears when in contact.

(*b*) The size of the openings in gratings, sieves, etc.

Meshed. A term implying that a *gear*, or system of gears, is ready for power to be transmitted through the gear.

Metal Limits. See **Limits (Metal)**.

Metal Sawing Machine. A machine used for sawing metal bars, tubes, etc., which are held in a *machine vice* whilst a reciprocating powered hacksaw cuts through them.

Metal Spinning. The shaping of sheet-metal discs into circular or moulded shapes on a lathe face-plate by the application of lateral pressure.

Metal Tolerance. See **Tolerance**.

Metallic Packing. See **Packing**.

Meter, Rotary Piston. See **Rotary Piston Meter**.

Metric Screw Thread. See **Screw Thread**.

Metronome. A clock movement with an inverted pendulum whose period of swing is regulated by a sliding weight on the pendulum; an instrument used for measuring time in the study of music.

Meyer Hardness Number. A number obtained by the same test as for the *Brinell hardness number*; it is the ratio of the load divided by the projected area of the indentation.

M.G. Machine. *Single-cylinder machine.*

Michell Bearing. A bearing in which pivoted pads support the *thrust collar* or *journal* and tilt slightly under the wedging action of the lubricant induced by their relative motion. This action gives improved lubrication conditions, a low friction coefficient and a low power loss in the bearing.

Micro-drilling

Micro-drilling. The drilling of minute holes using very small drills; for example, one of 5 micrometres diameter.

Microinch (μin.). A unit for designating surface roughness equal to one millionth of an inch.

Micrometer. An instrument with optical magnification for measuring visually small angular separations.

Micrometer Gauge. A length *gauge* using two smooth faces connected by a horseshoe-shape, the gap between the measuring face being adjustable by an accurate screw at one face. The gap is read off from a circular scale engraved under the thimble head of the screw. (See Fig. 112.)

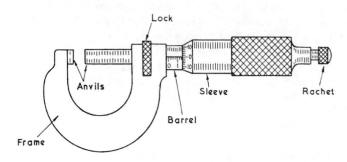

FIG. 112. MICROMETER GAUGE.

Mid (or Middle) Gear. The position of a steam-engine link motion or valve-gear when the valve motion is a minimum. Cf **Neutral Gear.**

Mil. The thousandth part of an inch. Colloquially called a ' thou '.

Mill. (*a*) A machine, or building, fitted with machinery for manufacturing processes.

(*b*) A machine for grinding, crushing or rolling.

(*c*) A grinding mill where the millstone runs round on a horizontal arbor and about a central vertical shaft.

Mill Engine. A large, low-speed horizontal steam-engine fitted with *drop valves* or *corliss valves* or a *unaflow engine*; sometimes used to drive machinery through ropes.

Mill Gearing. Gearing comprising cog-wheels, pulleys, shaft bearings and belting.

Mill Rolls (Mill Train). The rolls for making the finished product in a *rolling mill,* consisting of sets for roughing or billeting rolls and the finishing rolls.

Mill Steam-engine. *Mill engine.*

Mill Train. *Mill rolls.*

Milled Head (Knurled Head). The head of an adjusting screw roughened or cut in a succession of ridges to provide a good grip.

Milling. A shaping of metal pieces by removing metal with a revolving multi-tooth cutter to produce flat and profiled surfaces, slots and grooves. Cf. **Climb Milling.**

Milling Cutters. Rotary cutters for use on *milling machines* and sometimes used on a lathe. They are hardened-steel discs or cylinders with a great variety of slots or grooves to form the cutting teeth or alternatively separate teeth are inserted. The cutters are used for grooving, slotting and surfacing.

Milling Machine. (*a*) A machine tool with a horizontal arbor or a vertical spindle (see Fig. 113) to carry a rotating multi-tooth cutter

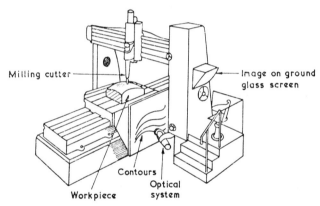

FIG. 113. CONTOUR MILLING MACHINE.

with the work supported and fed by an adjustable and power-driven table.

(*b*) A rotary machine, consisting of squeezing-rollers and a box channel called the spout, etc., fitted over a large trough, the whole being enclosed and used for the preparation of woollen fabrics for a subsequent finishing process.

Milling Wheel. *Knurling tool.*

Minus Lap. The exhaust lead on a steam valve for diminishing the amount of cushioning. **See Inside Lead.**

Minute. *Minute Hand.* The hand of a clock or watch which makes one complete turn per hour. It fits on the *centre arbor* or on the *cannon pinion pipe.*

Minute Pinion. The pinion in a clock or watch mechanism which drives the *hour wheel.*

Minute Rack. A rack in a *repeater* fitting on the pipe of the *quarter rack* with six ratchet teeth on one part of the periphery and fourteen similar teeth on another part corresponding with the fourteen minutes between each quarter.

Minute Wheel. The wheel in a clock or watch mechanism driven by the *cannon pinion.*

Minute Wheel Pin. A vertical pin in the plate on which the *minute wheel* revolves.

Misfiring. Failure of the mixture in the cylinder of an internal-combustion engine to fire normally, due to ignition failure or to an over-rich or too weak a mixture.

Mitre. A joint between two pieces of material meeting at an angle of 90° with a common surface at 45°.

Mitre Gear. A pair of bevel gear-wheels in mesh having their shafts at right angles.

Mitre Valve. A safety valve with the annular seating cut at an angle of 45°.

Mitre Wheel. See Bevel Gear, Mitre Gear.

Mitre-cut Piston-ring. A piston-ring which has the ends mitred at the joint, as distinct from steeped or square ends.

Mixed-flow (American) Water Turbine. An inward flow reaction turbine with the curved runner vanes acted on by the water as it enters radially and leaves axially.

Mixed-pressure Turbine. A steam-turbine operated from two or more sources of steam at different pressures admitted at the appropriate pressure stages.

Mixing Chamber. A chamber where fuel and air are mixed prior to ignition, especially in a gas-engine. See also **Combustion Chamber.**

Mode. Shape or form. A term used to describe the shape of a curve in a periodic oscillation. Modes are said to be ' coupled ' when motion in one mode causes motion in another mode or modes and ' uncoupled ' when motion in one mode will not cause motion in other modes. See also **Vibration.**

Modified Profile. See Profile.

Module. (*a*) The spacing of adjacent teeth of a gear divided by π.

When measured in inches, the module is the reciprocal of the *diametral pitch*.

(*b*) One of a restricted number of production items which has been standardized so as to fit together in various ways in making different articles.

(*c*) A unit standard for measuring.

Mohs' Scale. See **Hardness Numbers.**

Moment of a Force. The turning effect of a force about a given point measured by the product of the force and the perpendicular distance of the point from the line of action of the force. Generally, clockwise moments are called ' positive ' and counter-clockwise moments are called ' negative moments '. See also **Couple, Torque.**

Moment of Inertia. The sum Σmr^2, where m is the mass or a particle in the body and r its perpendicular distance from the axis. See also **Polar Moment of Inertia.**

Moment of Momentum (Angular Momentum). See **Momentum.**

Momental Ellipse. An ellipse with principal semi-axes equal to the maximum and minimum radii of gyration for a cross-section. The radius of gyration about any other axis is given by the corresponding vector line from the centre of the ellipse.

Momentum. The product of the *mass* of a body and its velocity.

Angular Momentum. The product of the *moment of inertia* and the angular velocity of a body. The sum of all the momenta remains unaltered in any one mechanical system.

Monitor. An instrument for keeping a variable quantity within definite limits by transmitting a controlling signal, as in a process plant. See also **Process Control.**

Monkey. The falling weight used in a pile driver.

Monkey Wheel. *Gin block.*

Monobloc. The integral casting of all the cylinders of an internal-combustion engine in one block.

Monocable. An *aerial railway* in which a single endless rope both supports and moves the loads.

Monotype Machine. See **Composing Machine.**

Monorail. A railway system with the carriages running along and suspended from, or mounted on, a single continuous overhead rail.

Monorail Hoist. A hoist suspended from and running on the flanges of an I-section girder.

Morse Tapers. Standard tapers for fitting the shanks of drills, etc., to machine spindles. (See Fig. 114.)

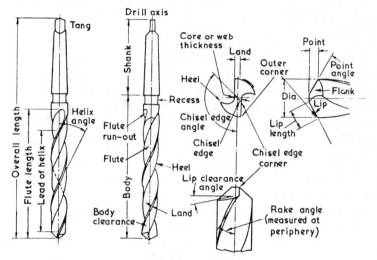

FIG. 114. MORSE TAPER TWIST DRILL NOMENCLATURE.

Mortar Mill or Mixer. A machine for crushing and mixing mechanically a mortar mix by two rollers running on the ends of a horizontal bar, which rotates about a central vertical axis and around a shallow pan containing the ingredients. Cf. **Concrete Mixer.**

Mortise Chain Cutter. See **Chain Cutter.**

Mortise Teeth. *Cogs.*

Mortise (Mortice) Wheel. A cast or machined wheel containing mortises or slots cut in it to receive wooden *cogs* instead of iron teeth. Cf. **Cog-wheel.**

Mortising Machine. A machine for cutting square or rectangular holes in wood. The reciprocating solid chisel and rotary bit in the older type machines have been replaced by the hollow chisel or the edged chain or both in combination in modern machines. The hollow chisel is particularly suitable for the automatic mortising machine.

Motion Bars. *Guide bars.*

Motion Block. A block, attached to the valve rod of some steam-engine valve-gears, which is constrained to move in a circular path by a curved slotted link. See also **Joy's Valve-gear, Walschaert's Valve-gear.**

240

Motion Disc. *Wrist plate.*

Motion Work. The auxiliary train of wheels which gives the correct relative motion to the hour and minute hands of a clock or watch.

Motor. (*a*) A *prime mover.*

(*b*) The petrol engine of a motor-car or aircraft.

Motor-car (Automobile). A private *motor-vehicle.*

Motor-vehicle (Automobile). A road vehicle powered by a petrol engine or diesel engine.

Moulding Cutter. An adjustable and specially shaped revolving cutter, often used in pairs on opposite sides, for cutting a desired moulding profile.

Moulding Machine (Plastics). Machines used for moulding plastics by compression and transfer moulding are shown in Fig. 115. The plastic may be in powder or granular form or in pellets of

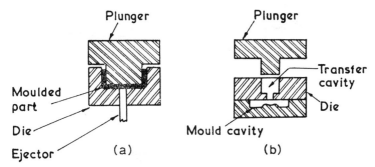

FIG. 115. MOULDING MACHINES: (*a*) compression, (*b*) transfer.

plastic powder (preforms) in the proper amount for making the moulded part. Compression moulding uses pressures of 14 MPa to 140 MPa and a mould heated to soften the plastic. In transfer moulding, the dies are closed before the plastic is added in the cylinder above. Injection moulding (Fig. 116) is similar to transfer moulding except that the soft plastic is forced into the die cavity under pressure which is maintained until the plastic has been cooled by water circulating in the walls of the die. Extrusion, blowing and laminating processes are also used for plastics. Fig. 116 shows the process of plastic extrusion, in which the granulated powder is fed into a hopper, then passes into a conveyor screw where the heat is applied to soften the plastic sufficiently to flow through the die. Blowing and vacuum-forming use a single die on a plastic sheet.

Moulding

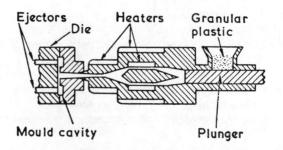

(a) INJECTION MOULDING

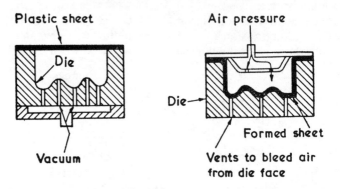

(b) BLOWING AND VACUUM FORMING

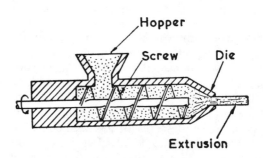

(c) PLASTIC EXTRUSION

FIG. 116. PLASTICS MOULDING.

242

Mounting. The chucking of work in a lathe. See **Chuck.**

Anti-vibration Mounting. An *isolator* to reduce the transmission of vibrations with undesirable frequencies.

Shock Mounting. An *isolator* to reduce the harmful effect of shock. See also **Vibration.**

Movable Expansion. Expansion capable of regulation by means of a second *slide valve* or other gear in a steam-engine.

Movement. (*a*) The mechanism of a clock or watch, excluding the case and dial. See **Bar Movement.**

(*b*) The essential part of the mechanism of a ciné camera or projector.

Moving-iron Instrument. An instrument depending on the movement of a piece of moving soft iron relative to a magnet as shown typically in the *air dashpot* in Fig. 1, which is an ' attraction ' type of instrument. The movement between two mutually repulsive magnetized pieces of iron, the one fixed and the other controlled by a spring is known as the ' repulsive ' type of instrument.

Moving Staircase. *Escalator.*

M Teeth. Saw-teeth, shaped like the letter M, used in some *cross-cut saws.*

Mud Bucket. The bucket or scoop of a *dredger.*

Muff Coupling. *Box coupling.*

Mule. A cotton-spinning machine which spins the yarn on the outward run and winds it on a spindle on the inward run. See also **Draw.**

Multiple Boring Machine. A *boring machine* with several *mandrels* for simultaneous boring.

Multiple Disc Clutch. See **Clutch.**

Multiple-expansion Engine. An engine in which the expansion of the working fluid is in two or more stages through cylinders of increasing size. See **Compound Engine, Quadruple-expansion Engine, Triple-expansion Engine.**

Multiple-spindle Drilling Machine. A *drilling machine* with several vertical spindles for simultaneous operation on a piece of work.

Multiple-threaded Screw (Multi-start Thread). A screw of coarse pitch with several threads to reduce the size of thread, to increase the relative size of the core and to obtain a higher *velocity ratio.* See also **Divided Pitch, Multi-start Worm.**

Multiple-tool Lathe. A heavy lathe with two large tool-posts, one on each side of the work, carrying separate tools to operate simultaneously on different parts of the work. Cf. **Duplex Lathe**

Multiplier. *Linkage multiplier.* See **Linkage.**

Quarter Squares Multiplier. A mechanism in which the product of two input shaft rotations is formed by making use of the identity $[(x+y)^2 - (x-y)^2]/4 = xy$.

Multiplier Register. The register in a calculating machine which records the number of turns of the multiplying handle (or its equivalent).

Multi-stage Pump. A centrifugal pump with two or more impellers mounted on the same shaft.

Multi-start Thread. *Multi-threaded screw.* Cf. **Single-start Thread.**

Multi-start Worm. See **Worm.**

Mushroom Follower. A cam *follower* with a flat surface, as distinct from a roller-type follower.

Mushroom Valve. *Poppet valve.*

Musical Box. A mechanical musical instrument played by causing a toothed cylinder like a *pin barrel* to work on a comb-like metal plate.

Musical Clock. A clock which plays a tune instead of chiming See also **Chiming Clock.**

Musical Watch. A *repeater* watch that plays a tune on a comb instead of striking on a gong.

Mutilated Gears. *Segmental gears.*

N

N. The symbol for a *Newton* (see Appendix: SI units).

v. The symbol for kinematic viscosity $= \mu/\rho$: viscosity $\div$ density

n. The symbol for revolutions per unit-time.

n.a. Neutral axis.

N.E.L. National Engineering Laboratory, East Kilbride, near Glasgow.

n.h.p. Nominal horse-power.

N.P.L. National Physical Laboratory, Teddington and East Kilbride, near Glasgow.

N.T.P. Normal temperature and pressure, i.e. $0°$ C and 760 mm of mercury.

Narrow Gauge. A railway gauge less than the British standard 4 ft $8\frac{1}{2}$ in.

Nave. The *hub* of a wheel.

Navvy. *Power shovel.*

Neck. A *journal* of smaller diameter than the main shafting.

Needle. See **Bearded Needle, Knitting Frame, Knitting Machine, Latch Needle, Spring Needle.**

Needle Lubricator. An inverted stoppered flask, fitted to a bearing with a wire loosely fitting in the stopper and touching the shaft.

Needle Machines. Embroidery machines of the single-needle type.

Needle Roller-bearing. See **Roller-bearing.**

Needle Valve. A screw-down stop-valve which may have the body ends in line or at right-angles to each other or may be of the oblique type with the *disc* in the form of a needle point. Needle valves are generally restricted to small sizes. See **Oblique Valve** and Fig. 24.

Negative Lead. The amount by which a steam port is closed to admission when the piston is at the bottom of the cylinder.

Negative Movement. A movement in any part of a *loom* achieved by using springs or weights. Cf. **Positive Movement.**

Netloom. *Netting machine.*

Netting Machine (Netloom). A machine producing netting with the threads knotted at their intersections.

Neutral. See **Mid Gear, Neutral Gear.**

Neutral Axis. The line of zero stress in a beam subject to bending.

Neutral Equilibrium. See **Equilibrium.**

Neutral Gear. When the gearing of a car is arranged so that no power can be transmitted, the engine or car is said to be in neutral gear. Cf. **Mid Gear.**

Newall System of Limits. See **Limit System.**

Newton. The unit of force in the *M.K.S.* system of units, equal to the force required to accelerate one kilogramme by one metre per second per second, or to 10^5 *dynes* or to about 100 grammes weight.

Nibbling Machine. A *punching machine* cutting a series of small overlapping holes to produce the rough outline of a complex sheet-metal part.

Nipple. A small drilled bush or tubular nut, or a short length of externally threaded pipe.

Nobbing. *Shingling.*

Nodal Gearing. The location of gear-wheels at a nodal point of a shaft system. See **Nodes.**

Nodes. Points of minimum displacement in a system of stationary

waves, half a wavelength apart. The term ' node ' is strictly appli(
when the displacement is zero, and when not zero is called a ' parti
node '. An ' antinode ' is a point where the displacement is
maximum.

Nominal Horse-power. An obsolete method of rating stea(
engines; for a piston-engine it is $D^2N(S)^{1/3}/15.6$, where S is tl
stroke in feet, N is the number of cylinders and D is the cylind(
diameter in inches. See also **Horse-power.**

Non-condensing Engine. An engine which exhausts its stea(
direct into the atmosphere.

Non-magnetic Watch. A watch whose performance is unaffect(
by magnetic fields with the balance, balance spring, roller and fo(
made usually of non-magnetic alloy.

Non-return Valve. *Check valve.*

Normal Helices. Two helices of the same diameter with tl
tangents at points of intersection at right-angles. The helices a(
of opposite hands.

Normal Pitch. The pitch of the traces of adjacent correspondi(
tooth flanks of helical, spur and worm gears measured along
common normal. See also **Pitch** and **Diametral Pitch.**

Normal Pitch (Parallel Helices). The distance between adjace(
intersections of a system of concentric helices with a co-cylindric(
normal helix measured along the latter. (See Fig. 84.)

Normally Aspirated Engine. A petrol or oil engine witho(
supercharge or boost.

Nose. The front part of a *spindle, mandrel* or some projecti(
part.

Nose Cap. A boss or hub fairing fitted coaxially and rotati(
with a propeller, but not extending beyond the blade roots. C
Spinner.

Nosing Motion. The increase of the speed of the tapering spind(
on which a *cop* is being wound as the diameter lessens on the mu(
spinning frame.

Notch Brittleness. The brittle property of a material causi(
fracture with small absorption of energy in an *Izod* or *Charpy* te(

Notched-bar Test (Impact Test, Izod Test). Subjecting a notch(
metal test-piece to a sudden blow by a striker which is carried by(
pendulum or falling weight by which the energy of fracture
measured. See **Impact-testing Machine, Charpy Test, Fremont Te(**

Notching. A process used to cut a configuration of indentatio(
in the edges of sheet-metal parts.

Notching-up (Linking-up). The movement towards the centre of a notched quadrant of the gear lever of a locomotive or steam-engine to decrease the valve travel and to shorten the cut-off; also called ' linking-up '.

Nozzle. (*a*) An outlet tube through which a fluid escapes from a container.

(*b*) A specially shaped passage for expanding steam in *impulse turbines.*

(*c*) An injection fuel valve for oil engines.

Nozzle Guide Vanes. A ring of radially-positioned vanes, shaped like aerofoils, which accelerate the gases from the combustion chamber of a gas-turbine type of engine and direct them on to the first rotating turbine stage.

Nuts. The mating parts of screwed members which are rotated to tighten their holds.

A compressed slotted nut is illustrated in Fig. 117, the portion above the slot locking on the *male* thread. For thread forms see **Threads.**

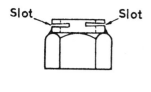

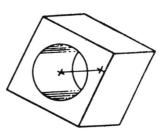

Fig. 117 (*left*). Compressed Slotted Nut.

Fig. 118 (*above*). Oblique Projection.

O

ω. Symbol for angular velocity.

Oblique Projection. A projection in which two of three mutually perpendicular edges are drawn at right-angles whilst the third is at any angle to the horizontal, usually 30° or 45°. (See Fig. 118.)

Oblique Valve. A screw-down stop-valve with the casing or body of spherical shape.　The axis of the *stem* is oblique to the body end which are in line with each other.　Cf. **Angle Valve, Globe Valve.**

Obliquity of Connecting-rod.　The angle made by the *connecting rod* with the cylinder axis of a steam-engine when the *crank pin* is at the extreme upper and lower portions of its path.

Obturator Ring.　A *gas ring*, L-shaped in cross-section, on the piston of a piston-engine to maintain a gas-tight seal between the piston and the cylinder wall.

Odontograph.　A scale to simplify the marking out of wheel teeth with appropriate numbers for wheels of various pitches by which suitable radii for the teeth of the wheels are obtained.

Odontometer.　An instrument for testing the accuracy and uniformity of gear-tooth profiles and tooth spacings in production work.　See also **Odontograph.**

Oil Cataract.　*Oil cylinder.*

Oil Cylinder (Oil Cataract).　A small cylinder which controls the amount of piston movement in a *steam reversing cylinder*, the oil pressure being regulated by a *cock*.

Oil Engine.　*Compression-ignition engine.*

Oil Feed.　(*a*) Any appliance feeding oil to a bearing or some moving part of an engine or mechanism.

(*b*) The system of pipes and manifolds used to supply diesel oil (fuel) to the fuel injector on each cylinder of a *compression-ignition engine.*

Oil Grooves.　Grooves cut in the sliding faces, bearing surfaces etc., for the distribution of lubricating oil.

Oil Pump.　A small auxiliary pump driven from an internal combustion engine crankshaft to force oil from sump or tank to the bearings.　See also **Gear Pump.**

Oil Ring.　*Scraper ring.*

Oil Sink.　The spherical recess around a *pivot* hole in a watch or clock plate to act as an oil reservoir.

Oil Sump.　The lower part of the crankcase of an internal combustion engine acting as an oil reservoir.

Oiling Ring.　A light metal ring riding loosely on a shaft located in a slot of the upper brass of a *journal bearing*.　As the ring rotates it feeds oil to the brasses from an oil reservoir in the base of the housing into which the ring dips.　Cf. **Oil Ring.**

Oils.　Neutral liquids used for lubrication with three main classes

(1) Fixed (fatty) oils from animal, vegetable and marine sources chiefly glycerides and esters of fatty acids.

(2) Mineral oils from petroleum, coal, etc., which are hydrocarbons.

(3) Essential oils, in the form of volatile products from certain plants, which are chiefly hydrocarbons.

Oil-sealing Ring. A ring outside a roller-bearing to prevent the escape of oil from the bearing.

Oldham Coupling (Double Slider Coupling). A pair of flanges, with opposed faces carrying diametral slots, between which a floating disc is supported through corresponding diametral tongues arranged at right-angles so as to connect two misaligned shafts. (Fig. 119.)

FIG. 119. OLDHAM COUPLING.

Oliver. (*a*) A simple form of power hammer used in some branches of chain-making.

(*b*) A small lift-hammer used by smiths, consisting of a horizontal shaft on end bearings with a hammer at the end of the shaft operated by a treadle underneath and a spring pole overhead.

Omnibus (Bus). Large-wheeled public vehicle plying on fixed routes with fixed places for taking up and setting down passengers.

Omtimeter (Optimeter). An optical projection measuring instrument of high precision with a comparison measuring scale magnified about 1000 times; used for comparing screw threads against a standard.

One-at-once Wheel (or Engine). A small lace machine for winding a length of yarn on a brass bobbin while maintaining uniform tension.

Open Belt. A direct-drive belt with driving and driven pulleys revolving in the same direction.

Open-end Rolls. *Mill rolls* unsupported by housing at one end.

Open-frame Connecting-rod. *Bow connecting-rod.*

Open Link. See **Link.**

Open Link Chain. See **Link, Chain.**

Open Rods. See **Crossed Rods.**

Open Shedding. The separation of warp threads so that they are only moved when required to change position.

Open Die (Self-opening Die)

Opening Die (Self-opening Die). A *die* which clears the screw thread when it comes to the end.

Opposed-cylinder Engine. An internal-combustion engine with cylinders on opposite sides of the crankcase and in the same plane with their *connecting-rods* working on a common crankshaft placed between them.

Opposed Piston-engine. An engine with a pair of pistons in a common cylinder, the explosive mixture being ignited between the two pistons. One type of opposed piston two-stroke compression-ignition engine has a triangulated form for the cylinder blocks with three crankshafts at the apices of the equilateral triangle.

Optical Flat. A surface, generally of glass or quartz, with deviations from a truly plane surface which are small in comparison with the wavelength of light. It is used to measure errors in the flatness of precision-finished surfaces by the interference fringes which can be seen when the optical flat is laid on the other surface (See Fig. 120.)

Optical Indicator. An *engine indicator* using optical methods to project an *indicator diagram* on a glass screen or for recording on a photographic plate.

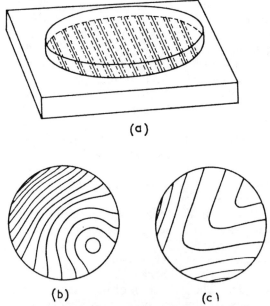

(a)

(b)   (c)

Fig. 120. Optical Flat: (a) flat, (b) peak, (c) hollow.

250

Optical Tooling. An optical method for checking the alignment of bearings, frames, etc.

Optimeter. *Omtimeter.*

Order Number. The number of vibrations or impulses which occur per revolution during the torsional oscillations of an engine crankshaft. See **Torsional Vibration.**

Oscillating Stresses. Stresses alternating between tension and compression as in a *Wöhler test.*

Oscillation, Centre of. *Centre of oscillation.*

Oscillation, Time of. The time of oscillation of a *pendulum* or *balance* is twice that of the single vibration; the time to swing from one side to the other and back again.

Otto Cycle. A working cycle of a *four-stroke* piston-engine with suction, compression, explosion at constant volume, expansion and exhaust. This involves reversible heating and reversible cooling at constant volume. Cf. **Diesel Cycle.**

Out of Gear. When the wheels of a gear train are disengaged.

Out Stroke. The stroke of a gas-engine piston in a direction away from the ignition chamber. Cf. **Instroke.**

Outer Dead-centre (Bottom Dead-centre). The piston position in a piston-engine or pump nearest the crankshaft when the piston is at the end of its out stroke.

Out-of-balance. A rotating part is said to be out-of-balance if rotation generates a resultant non-axial force.

Outside Crank. A single-web crank attached to a crankshaft outside the main bearings.

Outside Cylinders. Locomotive cylinders, carried outside the *frame*, working on to *crank pins* in the driving wheels.

Outside Lap (Steamlap). The overlap of the *slide valve* of a steam-engine beyond the edge of the *steam ports* when in mid-position.

Outside Screw Tools. *Chasers.*

Outward Flow Turbine. See **Turbine.**

Oval Chuck. A compound *chuck* in which the eccentricity is controlled by a worm wheel and a *tangent screw.*

Oval Hole Cutting. The hole made by a *cutter* controlled in a similar manner to that of an *oval chuck.*

Overcoil. The last coil of a *balance spring* raised above its plane and bent to form a terminal curve. For types of overcoil see Fig. 121.

Overdrive. A device for reducing the gear ratio in a motor-

Overdrive

Double overcoil   Single overcoil   "Duo in uno" hairsprin[g]

FIG. 121. OVERCOILS.

vehicle under optimum driving conditions to give greater speed an[d] to decrease fuel consumption.

Overhanging Cylinders. Engine cylinders bolted to the ends [of] their bed plates instead of upon their faces. This lowers the pisto[n] rod centre and shortens the foundation for the bed.

Overhanging Pulley. A pulley attached to a shaft beyond the la[st] bearing. See **Overhanging Shaft.**

Overhanging Shaft. The end portion of a shaft, overhangin[g] beyond its last bearing.

Overhead Camshaft. A *camshaft* running across the top of th[e] cylinder heads of an engine and usually driven by a *bevel*-shaft [or] *timing chain* from the *crankshaft*. The cams operate on rockers [or] directly on the valve-stems.

Overhead Gear. (*a*) Machinery working overhead.

(*b*) *Pit-head gear.*

See also **Valve-gear (Overhead Mushroom).**

Overhead Railway. A railway carried above ground level o[n] arches or viaducts. See also **Monorail.**

Overhead Tracks. Single trolley tracks hung from a ceiling [or] roof.

Overhead Traveller (Overhead Travelling Crane). See **Crane.**

Overhead Valves. Inlet and exhaust valves working in the cylind[er] head opposite the piston in a vertical petrol or oil engine. Cf. **Sid[e] Valves.**

Overlap. (*a*) The amount by which one riveted plate, etc[.] extends over another.

(*b*) The length of railway track beyond a stop signal which mu[st] be unoccupied before the previous stop signal can clear.

Overlap Ratio. (*a*) The ratio of the *face-width* to the *axial pitc[h]* of a helical gear.

(*b*) The ratio of the angle subtended at the apex of the developed *pitch cone* of a bevel gear by the tooth trace, θ, to the angle subtended at the same apex by two points on the *pitch circle* and on similar flanks of adjacent teeth, ϕ. (See Fig. 122.)

Face width

FIG. 122. OVERLAP RATIO.

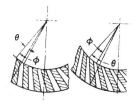

Overload Coupling. A coupling specially designed so that when a preset torque is exceeded the transmission of power is terminated.

Overriding. *Riding.*

Overshot Wheel. A water wheel which is turned by the weight of the water that runs into the buckets at the top of the periphery from the discharge flume. Cf. **Undershot Wheel.**

Overstrain. The result of stressing an elastic material beyond its *yield point.*

Overtones. Frequencies of vibration which are higher than the lowest frequency of interest in a mechanical system, but they may not be integral multiples of this lowest frequency. Cf. **Harmonic** and see **Frequency.**

P

P. The symbol for power.

p. The symbol for momentum and for pressure.

pdl. *Poundal.*

P.L. Proportionality Limit. The point on the stress-strain curve when the relation between the two quantities is no longer linear.

P.S. Proof stress.

Packing.   Material inserted in *stuffing boxes* to make engine an
pump rods pressure-tight.

   Metallic Packing.   A packing consisting of a number of so
metal rings or a helix of metallic yarn, which encircles a pistc
rod and is pressed into contact with the rod by a gland nut.

Pad Roller.   A plain roller (without sprockets) which presses th
edges of cinematograph film on to *sprockets* to ensure that a suf
cient number of teeth of the latter are engaged by the sprocket hole

Paddle Shaft.   The *paddle-wheel* shaft which is driven directly b
the engine cranks.

Paddle-wheels.   Wheels at the sides or stern of a ship, fitted wi
blades parallel to the shaft that dip into the water to propel th
vessel.   The blades (*float boards*) may be fixed or feathered.

Paddle-wheel Fan.   *Centrifugal fan.*

Pallet.   (*a*) A stand or container adapted for transportation
goods by *fork-lift truck*.

   (*b*) A part of the mechanism of a watch or clock which by i
contact with other parts releases a movement of another part.   S
Pallets.

   Gathering Pallet.   A small pallet which engages with the *lockin
piece* and releases the *rack hook* so that the rack can be gathere
up and held in the *chiming mechanism* of a clock.   See **Chimin
Mechanism,** Fig. 31.

Pallet Arbor (Verge).   The *arbor* in a clock which carries th
pallets in the centre and the *crutch* at its end.   See also **Chimir
Mechanism.**

Pallet Jewel.   The jewel in each pallet upon which the *escap
wheel* teeth act.

Pallet Staff.   The axis (which is pivoted) upon which the *palle
are mounted in a watch or clock movement.

Pallets.   (*a*) Those parts of the pendulum or *balance* to which th
teeth of the *escape wheel* impart impulses, the pallets being called th
' entering pallet ' and the ' exit pallet ', the names being sel
descriptive.   See **Escapement.**   (Figs. 62 and 123.)

   Circular Pallets.   Pallets equidistant from the *pallet staff* axi

   Exposed Pallets.   Pallets in front of a clock dial.   See als
Hammer Pallets and **Pallet.**

   (*b*) The plates of a *chain pump*.

Pantograph.   A mechanism, based on the geometry of a parallel
gram, for copying plans, etc., to a different scale.

Parabolic Governor.   *Crossed-arm governor.*

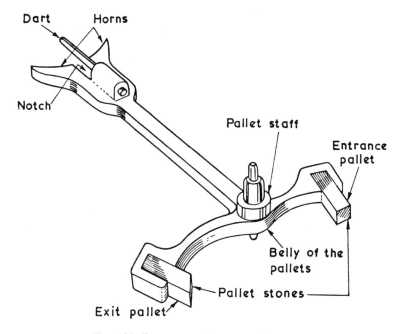

FIG. 123. ESCAPEMENT LEVER AND PALLETS.

Parallel Gate Valve. *Double disc (gate) valve.*

Parallel Key. See **Key.**

Parallel Motion. (*a*) A system of links copying a reciprocating motion to an enlarged scale, such as on piston-type engine indicators. Cf. **Pantograph.**

(*b*) A system of levers adapting a curved movement to a rectilinear reciprocating motion.

Parallel Screw Thread. A screw thread cut on the surface of a cylinder.

Parallel Slide Valve. A *gate valve* with one or two discs sliding between parallel body seats without a spreading mechanism as in a *double disc (gate) valve.* The effective closure is obtained by the pressure of the fluid forcing the downstream disc-face against its mating body seat. (See Fig. 124.) Cf. **Double-beat Valve.**

Paris (Wire) Gauge. A metric gauge with numbers for wire diameters from no. 1 (0·6 mm) to no. 30 (10 mm).

Parson's Steam Turbine. A *reaction turbine* in which rings of

255

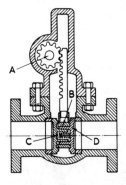

FIG. 124. PARALLEL SLIDE VALVE:
(A) actuating spindle, (B) carrier,
(C) spring for separating discs,
(D) discs.

moving blades of increasing size are arranged along the periphery
of a drum of increasing diameter. Fixed blades in the casing
alternate with these rings. Steam expands gradually through the
blading, from inlet pressure at the smallest section to condenser
pressure at the other end.

Parting-off Tool. A tool for metal and wood turning which is
narrow, deep, square across the end, and the width tapering slightly
backwards so that it will clear itself in the cut and remove the
material from the workpiece held in the chuck.

Passages. The steam-ways of a cylinder including both port
and exhaust.

Passes. The passing and repassing of bars, etc., through the
rolls of a *rolling mill*. A backward pass over the top of a two-high
mill is called a 'lost pass'.

Passing Hollow. *Crescent.*

Passing Spring. See **Spring-detent Escapement.**

Pawl (Paul). A pivoted catch or *click*, engaging by an edge or
hook with a *ratchet wheel* (Fig. 140) or a *rack* and usually spring
controlled—

(*a*) to prevent reverse motion; or

(*b*) to convert reciprocating motion into an intermittent rotary or
linear motion;

(*c*) the catch which fixes a turntable;

(*d*) the catch which prevents the winch shaft of a crane from
sliding when the gears are changed.

Pawl Feed (Ratchet Feed). The feed of a machine effected by
means of a *pawl* and *ratchet* or small cog-wheel. (See Fig. 140.)

Peaucellier Mechanism. A four-bar rhomboid mechanism

ABCD, is constrained by equal-length links *OB* and *OD* to a point, *O*, on the end of a diameter of a circle and the point *C* moves on the circumference of the circle. This mechanism constrains the point *A* to move on a line perpendicular to the diameter on the end of which is the point *O*. This mechanism gives a straight-line motion.

Pedal Feed Motion. The control of the speed at which sheets of raw cotton are fed to the *beater*, varying with the thickness of the sheets.

Pedometer. An instrument for recording the number of steps taken by a pedestrian, actuated at each step by the movement of a small weight which is balanced against a spring.

Peening. See Shot Peening.

Peg Plan (Lifting Plan). The plan indicating the order in which the *healds* are to be lifted during weaving.

Pelton Wheel. An impulse water turbine rotated by a jet of water from a *nozzle* striking specially shaped buckets attached to the periphery of the wheel, the nozzle being either deflected or valve-controlled by a governor.

Pendulum. A body suspended so as to be free to swing, especially a rod with a weight at the end for regulating the movements of a clock's works. Household clocks beat half seconds or less; long-case clocks and *regulators* have a seconds pendulum; *tower clocks* have pendulums which may beat up to two seconds. See also **Foucault's Pendulum.**

Compensation Pendulum. A pendulum so designed and constructed that its time of swing, or beat, is unaffected by change of temperature, that is, the distance between the centre of oscillation and point of suspension remains constant.

Ellicott Pendulum. A pendulum which is partly compensated by the expansion of two brass rods that lift the bob through pivoted levers.

Gridiron (Harrison's) Pendulum. Pairs of brass and steel rods arranged alternately with each pair anchored to a cross-bar, the whole being attached to a central steel control rod on which is a brass shell lead-filled bob and *rating nut*. As the control rod lengthens with heat the bob is pushed upwards by the expansion of the pairs of rods resting on the nut.

'Invar' Compensated Pendulum. A pendulum rod made of ' invar ', a nickel-steel alloy which is almost unaffected by temperature, a *bob* recessed half-way and supported in the centre so that its expansion with temperature is ineffective, and a brass tube

fitted over the rod and supported on the *rating nut*. For extreme accuracy the *zinc and steel*, and *mercurial pendulums* are more effective.

Mercurial Compensated Pendulum. A pendulum with a glass or iron jar containing mercury supported in a stirrup or cradle The pendulum rod passes through a rectangular slot in the upper part of the cradle and the *rating nut* is usually situated between the jar and the upper part of the cradle. It is used in astronomical regulators.

Zinc and Steel Pendulum. A pendulum with an internal rod of iron, covered by an easy-fitting tube of zinc and over this latter a tube of iron with a series of holes drilled through the side to permit air circulation over the zinc tube. The internal iron rod is fitted with a *rating nut* and just above this nut the zinc tube rests on a collar. The outer iron tube has at the top end a collar resting on the zinc tube and another collar at the other end, which forms the seat for the bob. This type of *compensated pendulum* is fitted in the Westminster clock ' Big Ben ' which is kept within two seconds of Mean Time.

Pendulum Bob. The weight at the bottom end of a pendulum

Pendulum Clock Mechanism. See **Barrel, Barrel Arbor, Cannon Wheel and Spring, Centre Wheel and Pinion, Chiming Mechanism, Escape Pinion, Escape Wheel, Fusee, Fusee Great Wheel, Hour Wheel, Impulse Post, Minute Wheel, Pallets, Pendulum, Striking Mechanism, Suspension Spring, Third Wheel and Pinion.**

Pendulum Cross-cutting Saw. A cross-cut *saw* mounted at the end of a swinging frame suspended from above or carried on wall brackets. The weight of the saw and its mounting is counter-balanced.

Pendulum Damper. Pivoted balance weights attached to the crank of a radial piston-engine for neutralizing the fundamental torque impulses and thus eliminating the associated *critical speed*

Pendulum Governor. An engine governor in which heavy balls swing outwards under centrifugal force, thereby lifting a weighted sleeve and progressively closing the engine throttle valve. See also **Porter Governor, Watt Governor.**

Pendulum Rod. The rod of a *pendulum* which supports the *bob*

Pendulum Rolling Mill. A rolling mill used to achieve very large reductions, up to 90% at a single pass, in the thickness of sheet-metals which are otherwise difficult to fabricate. The mill consists of two small-diameter rollers at the ends of pendulums which are synchronized to oscillate on either side of the sheet-metal being forced through by means of standard feed rollers. The pendulum rollers produce a tremendous bearing force across the width of the metal which is fed in cold, but a great deal of heat is generated beneath the rolls which oscillate at 860 strokes or more per minute.

Pendulum Spring. The thin ribbon of spring steel used for suspending the pendulum in a clock.

Perambulator. A surveying instrument for distance measurement, consisting of a large wheel supported on its axis by a long handle to wheel it along the distance to be measured, with a revolutions recorder.

Percussion, Centre of. *Centre of oscillation.*

Period. The time for one complete cycle of a periodic phenomenon. The reciprocal of the *frequency.*

Periodic Quantity. A quantity the value of which recurs at equal time-intervals.

Periodic Time. *Period.*

Periodicity. *Frequency.*

Peripheral Speed. The speed of any point on the *periphery* of a rotating wheel, cutter, etc.

Permanant Set. (*a*) An extension remaining in a test-piece after the load has been removed, the *elastic limit* of the material having been exceeded.

(*b*) A permanent deflection of any structure after being subjected to a load.

Permanent-way Crane. A *gantry crane* used for accidents on, and repair of, the permanent way of a railway.

Perpetual Screw. *Worm.*

Persian Drill. *Archimedian drill.*

Pet Cock (Priming Valve). A small plug-cock for draining condensed steam from steam-engine cylinders on the starting of the engine, or for testing the water-level in a boiler.

Petrol Engine. An *internal-combustion engine* working on the *Otto* 4-stroke or the 2-stroke cycle using a petrol spray from a *carburetter* or direct petrol injection. Ignition of the combustible petrol-air mixture is effected by a sparking plug, operated either by coil and battery or by magneto or by a.c. and transistor system.

Petrol Pump. (*a*) A small diaphragm-type pump operated either mechanically from the *camshaft* of a petrol engine, or electrically, for fuel delivery to the *carburetter.*

(*b*) A pump at a petrol station.

Phase. Two alternating quantities are said to be ' in phase ' when their maximum values occur at the same instant of time. See also **Phase Angle, Phase Spectrum**.

Phase Angle. (a) The angle between two vectors representing two harmonically varying quantities which have the same frequency, that is, the difference in *phase* measured as an angle.

(b) A quantity defining a particular stage of progress in a recurring operation.

Phase Difference. The difference between the *phase angles* of two harmonically varying quantities.

Phase Reversal. A general term for the process of obtaining a signal of identical waveform but of opposite phase to an original signal.

Phase Spectrum. The values of the *phase angles* of the components of a *vibration*, arranged in the order of the frequencies.

Photo-elasticity. A method of determining the location and direction of stress distribution in bodies under complex systems of loading by passing polarized light through a model made of a transparent plastic material, such as nitro-cellulose. The light is polarized and transmitted only on the planes of principal stress. The stress distribution is observed through a second piece of polaroid called the ' analyser '.

Pick. See Weft.

Pick-at-Will. A mechanism which picks a shuttle from any box to be thrown through a *shed* during weaving.

Pick-up. A *transducer* used in sound reproduction such as that holding the needle which follows the track on a gramophone record.

Pick-up Well. A small petrol reservoir which provides a temporarily enriched mixture during the acceleration of an automobile and is arranged between the metering jet and the spraying tube in some carburetters.

Picker. A specially designed implement for propelling the shuttle across a *loom*.

Pickering Governor. A *governor* in which the balls are connected to the centres of cambered springs of flat steel, the ends of the springs being pulled inwards as speed increases and thus reducing the opening of a throttle valve.

Picking. Throwing the shuttle across a *loom* through the *shed* formed in the warp threads; the shuttle travels along a *race board* guided by the *reed*.

Pick-off. A transducer used for monitoring or stabilizing a servo-mechanism. The input is a relative displacement of its two components and the output is generally electrical.

Pressure Pick-up. A *transducer* to provide an electrical signal proportional to the pressure to be measured.

Piercing. An operation involving the punching of a hole in a part with the interior material discarded as scrap. Cf. **Blanking.**

Pile-drawer. An appliance for extracting piles from the ground.

Pile-driver. (*a*) A power unit which raises and lets fall a weight, called a monkey, between guides to drive in a pile.

(*b*) An hydraulic and quiet version uses its own weight and its grip on adjacent piles either to drive in or pull out piles by hydraulic means.

Pile Hoop. An iron or steel band fitted around the head of a wooden pile to prevent brooming; i.e. the breaking up of the wooden head.

Pilger Mill. A *rolling mill* for rolling tubes, using a mandrel. Fig. 125 shows how a mill works as a discontinuous process.

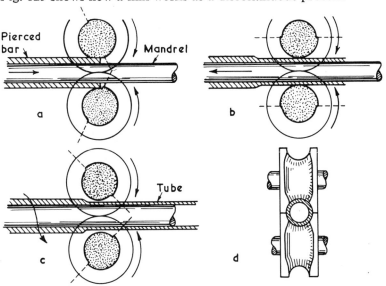

FIG. 125. PILGER MILL: (*a*) insertion, (*b*) withdrawal, (*c*) rotation and advance, (*d*) cross-section of rolls.

Pillar Drill. See **Drilling Machine** and Fig. 126.
Pillar Plate. See **Pillar.**
Pillar Pump. A *lift pump* or *force pump* attached to a base plate

Pillar Pump

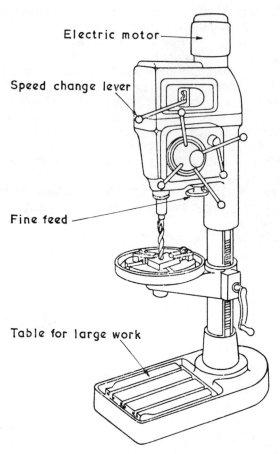

Electric motor

Speed change lever

Fine feed

Table for large work

FIG. 126. PILLAR DRILL.

carrying a pillar which forms the support for a crank, flywheel and handle for working the rod of the bucket or piston.

Pillars. (*a*) Cylindrical pieces of brass or steel which act as distance pieces between the plates of a clock or watch. The plates are called *pillar plates.*

(*b*) The lower half of a *journal bearing.*

Pillow Block (Pillow). *Plummer block.*

Pilot Engine. A separate locomotive preceding a train as a precaution against accidents to the latter.

Pilot Valve (Relay Valve). A small balanced valve controlling a supply of oil under pressure to a *servomotor* piston or to a relay for a larger control valve. It may be operated by hand or by a *governor* or by some other type of transducer.

Pin. (*a*) A small axle on which a lever oscillates or a small spindle carrying a pulley.

(*b*) A very small-diameter cylinder projecting from a surface such as those on a *pinwheel.* Cf. **Cotter-pin.**

Pin Barrel. A cylinder on the peripheral surface of which are short radial pins for lifting the hammers in a chiming clock or for lifting the *comb* in a musical clock, musical watch or musical box. When there are many pins as in a musical box it is called a toothed cylinder.

Pin Boss. The small boss of an engine crank which carries the *crank pin.*

Pin Cop. A small *cop* of a suitable size for a loom shuttle.

Pin Drill. A drill with the end in the form of a short small-diameter cylinder to fit into a previously drilled hole and thus form a concentric guide for drilling a larger hole.

Pin Pallet Escapement. See **Escapement.**

Pinion. (*a*) A small-toothed wheel, either *bevel* or *spur*, which normally has less than twelve teeth (leaves), and in a clock or watch acts as a *follower.*

(*b*) A small wheel in gear with a much larger one.

(*c*) A small gear-wheel meshing with a rack.

(*d*) Equal gears if the diameter is equal to or smaller than the width. Cf. **Wheel** (*b*).

Helical Pinion. A pinion carrying a helical gear.

Double-helical Pinion. A pinion carrying two helical gears meeting at an angle and making equal angles with a plane perpendicular to the axis.

Hollow Pinion. A pinion drilled throughout its length.

Lantern Pinion. Two coaxial discs carrying cylindrical pins parallel to the axis, the pins serving the same purpose as the teeth of (*a*) above.

Pinion Leaf. A tooth of a *pinion.*

Pin-jointed. (*a*) Said of joints in mechanisms where the only connection is a pin about which both the joined parts can turn without restriction.

(*b*) Said of joints in structural frameworks in which movements are not transmitted from one member to another.

Pinking. *Knocking.*

Pinning. Fastening a small piece of work to a larger piece.

Pintle. (*a*) The *king-pin* of a wagon.

(*b*) The pin of a hinge.

(*c*) The iron bolt on which a chassis turns.

(*d*) The upper metal brace on which a rudder swings with a *dumb-pintle* at the heel.

(*e*) The plunger of an oil engine injection valve.

Pinwheel (Pin Wheel). A wheel with pins fixed at right-angles to its plane for lifting the hammer of a striking clock or *repeater*; also called *hammer wheel.* Cf. **Pin Barrel.**

Pinwheel Escapement. An escapement with D-shaped pins, fixed on the rim at right-angles to the plane of the wheel, to give impulse to the *pallet* in *turret clocks.* It is a *dead-beat escapement* which performs best with a long heavy pendulum. Cf. **Pin Pallet Escapement.**

Pinwheel Gear. A rotary disc carrying an array of pins or teeth which mesh with a pinion that slides on a shaft at right-angles to the axis of disc's rotation. This constitutes a geared pair with a ratio that is a predetermined function of rotation. The rotation of the pinion shaft is proportional to the square of the rotation of the disc when the pins are equally spaced along an *Archimedian spiral.*

Pinwheel Principle (Odhner Wheel). A principle used in most barrel-type calculating machines and illustrated in Fig. 127. The

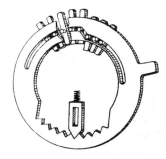

FIG. 127. PINWHEEL PRINCIPLE.

essential feature is a wheel having nine teeth or pins that can be retracted. Each pin moves in a radial slot and has a stud on one side which moves in a two-part race cut out of the setting lever. The pins are projected or retracted as the setting lever is moved, the radii of the two parts being different. The pinwheel is geared to a number wheel in the multiplier register of the calculating machine and rotates it through the required number of places.

Pipe. (*a*) A hollow spindle or shaft.

(*b*) A tubular boss or extension in a watch or clock.

Pipe-bending Machine. A machine for bending pipes or tubes between rolls operated by a lever. The pipe is often filled with short lengths of rod, slightly smaller than the bore, joined together and pulled out after the bending operation; their insertion prevents buckling of the pipe.

Pipe-threader. A machine for cutting screw threads in metal tubing.

Pirn. A small wooden bobbin which fits the *shuttle* of a loom and carries weft.

Piston. A solid or hollowed cylindrical plunger which reciprocates in a cylinder either under fluid pressure in an engine or to displace or compress a fluid in pumps and compressors. See **Bucket** (*a*), **Hat-leather Packing, Piston Ring, Plunger, Slipper Piston, Trunk Piston.** Cf. **Plunger.**

Piston Air Pump. A marine engine air pump with a solid piston and fitted with suction and delivery valves at both ends.

Piston Engine. *Internal-combustion engine.*

Piston Pin. *Gudgeon pin.*

Piston Ring. A cast-iron ring of rectangular section, cut through at one point to increase its springiness and to allow for fitting in a circumferential groove in the *piston*. The ring springs outwards against the cylinder wall to prevent leakage. See **Gas Ring, Junk Ring, Mitre-cut Piston Ring, Obturator Ring, Scraper Ring, Spring Ring.**

Piston Rod. The rod attached to the piston of an engine or pump to transmit its motion to or from the connecting-rod or crank.

Piston Rod Gland. The gland in the *stuffing box* of an engine through which the piston rod passes.

Piston Slap. The slight knock caused by a loose or worn piston slapping against the cylinder wall when the connecting-rod thrust is reversed.

Piston Speed. The speed of an engine piston measured in feet per minute which varies greatly with the type of engine from 1·3 m/s in condensing, and heavy engines, up to 4 m/s in marine oil engines and locomotive engines, from 4 to 6 m/s in large diesel engines and up to 15 m/s in aircraft engines.

Piston Travel. See **Stroke.**

Piston Valve. A *slide valve* formed by two short pistons attached to the valve-rod which slide over cylindrical ports in a close-fitting valve body, as in the steam chest of a steam locomotive.

Pit. See **Engine Pit.**

Pitch. (*a*) The uniform spacing of adjacent elements of a series of points, lines, planes, blades, teeth, etc., as in a *broach*. (See Figs. 17, 46, 143, 156, 157.)

(*b*) The distance between corresponding points on adjacent threads.

(*c*) The inclination or rake of the teeth of saws.

(*d*) The angle of setting of some tools.

(*e*) The length of two links in a *chain cutter*.

See also **Base Pitch, Circular Pitch, Diametral Pitch, Normal Pitch, Screw Pitch.**

Pitch (Helicopter). The angular setting of a helicopter blade, which is variable.

Collective Pitch Control. A control by which an equal alteration of blade angle is imposed on all the blades of a rotor independently of their position during rotation. Cf. **Cyclic Pitch Control.**

Control Advance. The phase angle by which the controlled change of cyclic pitch variation is displaced in azimuth from the direction of control-lever displacement.

Pitch (Propeller). The *pitch setting* is the blade angle measured at a standard radius, usually at 0·75 (sometimes 2/3) of the peripheral radius. (See Fig. 128.)

Braking Pitch is a pitch setting to give negative thrust, including reverse pitch. *Feathering pitch* gives the minimum drag when the engine is stopped. See also **Windmilling.**

Coarse Pitch. A large angle of pitch for high-speed flight.

Effective Pitch is the actual distance the element moves and *slip* the difference between geometric and effective pitch. (See Fig. 129.)

Experimental Mean Pitch is the distance through which a propeller advances along its axis during one revolution, when giving no thrust.

Fine Pitch. A small angle of pitch for low-speed flight. See also **Pitch Setting.**

Geometric Pitch is the distance which an element of a propeller would advance in one revolution when moving along a *helix* to which the line defining the blade angle of that element is tangential. (See Fig. 129.)

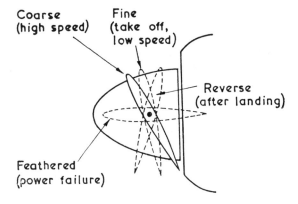

Coarse
(high speed)

Fine
(take off,
low speed)

Reverse
(after landing)

Feathered
(power failure)

FIG. 128 (*above*). PITCH SETTINGS FOR VARIABLE-PITCH PROPELLER.

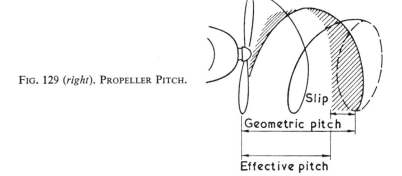

FIG. 129 (*right*). PROPELLER PITCH.

Slip

Geometric pitch

Effective pitch

Reverse Pitch. A large negative pitch setting of a propeller after an aeroplane has landed, to act as an air brake.

Pitch Angle. The angle between the axis of a *bevel gear* and the *pitch cone* generator, being the complement of the *back cone angle*.

Pitch Chain (Sprocket Chain). A chain of flat links between whose sides the projections of a *sprocket wheel* engage. See also **Silent Chain.** Fig. 130.

Pitch Circle. (*a*) The circumference of the *pitch line*. For two wheels in mesh, the pitch circles roll in contact.

(*b*) The circle of intersection of the *pitch cone* of a bevel gear and the outer end faces of the teeth. (See Fig. 12.)

267

Pitch Cones

Pitch Cones. (*a*) The contacting cones of a bevel gear on which the normal pressure angles are equal; they are coaxial with the rotation of the gears. (See Figs. 131 and 132.)

(*b*) A cone coaxial with a screw thread intersecting the surface of a *taper screw thread* so that the axial distance between the points where a generator meets the opposite flanks of the thread groove is equal to half the basic pitch of the thread. (Fig. 159.)

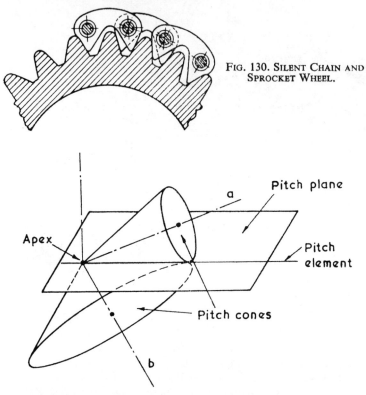

FIG. 130. SILENT CHAIN AND SPROCKET WHEEL.

FIG. 131. PITCH SURFACES, INTERSECTING AXES.

Pitch Control. The *collective* and *cyclic* pitch controls of a helicopter's main rotor(s).

Pitch Curves. The intersection of the tooth surfaces in the *pitch cone*.

Pitch Cylinder. (*a*) The cylinder of a *worm wheel* coaxial with its axis of rotation, on which the *transverse pitch* equals the *axial pitch*.

(*b*) For a screw thread, a coaxial cylinder intersecting the surface of a *parallel screw thread* so that an intercept on a generator between the points where it meets the opposite flanks of the thread groove is equal to half the basic pitch of the thread. (See Fig. 158.)

Pitch Cylinders. The contacting cylinders of helical and spur gears on which the normal pressure angles are equal; each is coaxial with the rotation of its gears. (See Figs. 132, 138, 174.)

Pitch Diameter. (*a*) The diameter of the *pitch circle* or *cylinder* of a gear. (See Figs. 84, 93.)

(*b*) The diameter of the *pitch cylinder* of a parallel screw thread or the nominal diameter of the *pitch cone* of a taper screw thread. See also **Screw Thread Diameters.**

Pitch Element. The instantaneous axis of relative motion of either of two bevel gears with respect to the other, that is, the common element of contact of the two *pitch cones* when rolling without slipping. (See Figs. 11, 12, 131, 132.)

Pitch Line. (*a*) The line or circle upon which the centres or the pitches of the wheel teeth are measured.

(*b*) In a rack, the line along which the teeth are measured.

(*c*) In a screw thread, the generator of the *pitch cylinder* or *pitch cone*. (Figs. 156, 157, 159.)

Pitch Plane. A plane in axially toothed worm wheels parallel to both the axes of the worm and worm wheel and tangential to the *pitch cylinder* of the *worm wheel.* (See Figs. 13, 131, 132.) The pitch of a *rack and pinion* is shown in Fig. 138.

Pitch Point. (*a*) The point of contact of a pair of *pitch circles* of a gear. (See Fig. 157.)

(*b*) The point where the *pitch line* intersects the flank of a screw thread. (See Figs. 158, 159.)

Pitch Setting. The blade angle of an adjustable or variable-pitch propeller. See also **Pitch.**

Pitch Surfaces. Surfaces that roll together with no sliding such as the *pitch cones* of bevel gears and the *pitch cylinders* of spur and helical gears. They have the following properties: (*a*) the *pitch element* is the instantaneous axis of relative motion, (*b*) the common tooth surface normal at any point of contact intersects the *pitch element*, and (*c*) the *pitch element* is the intersection of the surfaces of action. (See Figs. 131, 132.)

Pit-head Gear. All the machinery and its framework erected over a pit's mouth for raising and lowering the cage.

Pit Wheel

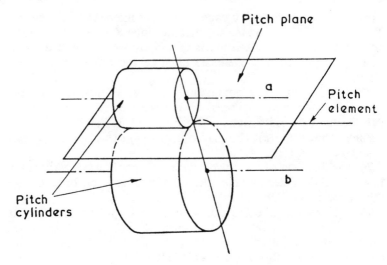

FIG. 132. PITCH SURFACES, PARALLEL AXES.

Pit Wheel. A *mortise wheel* revolving in a pit on a horizontal axis, usually the first motion wheel in a water- or wind-mill.

Pitot Tube. A tube with an opening facing into the direction of air flow and thus measuring dynamic pressure. See **Air-speed Indicator** and Fig. 2.

Pivot. (*a*) A short shaft or pin on which something turns or oscillates. Cf. **Fulcrum.**

(*b*) The end of an *arbor* (*b*) which turns in a hole, jewel or screw in a clock or watch. The end may be parallel, shouldered or conical.

Planer. *Planing machine.*

Planer Bar. A rigid steel bar carried horizontally in and standing out from the front of the tool box of some *planing machines* to assist in planing the interior of hollow workpieces.

Planer Centres. *Loose centres.*

Planer Tools. Cutting tools for a *planing machine* clamped vertically in a block pivoted in the *clapper box* on the head.

Planet Spindle. The rotating spindle of a grinding wheel which travels in a circular path.

Planet Wheel. A wheel revolving around or within the circumference of another wheel, by which it is driven. In some cases the planet wheel is driven by its own shaft.

270

Planetary Gear. (*a*) An *epicyclic gear* with a fixed annulus, a rotating sun wheel, a rotating planet carrier and planet wheels rotating about their own spindles. Cf. **Solar Gear** and **Star Gear.** (See Fig. 61.)

(*b*) Any gear-wheel whose axis describes a circular path round that of another wheel.

Planimeter. An instrument for measuring mechanically an area on a plane surface, consisting of a long arm with a tracing point that is moved round a closed curve. When the point returns to the place from which it started the area is given to scale by the reading of the revolutions of a small wheel supporting the arm.

Planing Machine (Planer). (*a*) A machine with a work-head reciprocating horizontally and having a tool-post on a *clapper box*, the work being attached to a stationary but adjustable work-table. The reciprocating motion is usually produced by a *scotch yoke* or *Whitworth quick-return motion.* See also **Plate-edge Planing Machine.**

(*b*) A machine consisting of a gear-driven reciprocating work-table sliding on a heavy bed with a stationary tool above the table on a saddle, which can be traversed across a horizontal rail carried by uprights for producing large flat surfaces. See **Clapper Box, Planer Bar, Planer Tools.**

Planoid Gears. *Hypoid*-type gearing with the pinion offset limited to one-sixth to one-third of the gear diameter. They are used in 1·5 to 1 up to 10 to 1 velocity ratio range.

Planometer. *Surface plate.*

Plant. (*a*) All the machinery requisite for carrying on business in a factory. (Also sometimes the buildings and site.)

(*b*) The rolling stock of a railway.

(*c*) To locate and set out the various centres for the *pivot* holes on the plates of a clock or watch; called ' to plant the train '.

Plate-bending Machine. A machine consisting of three rolls with bearings in housings, the top roll being adjustable to vary the curve when bending boiler and other plates between it and the other two below. See also **Straightening Machine.**

Plate Clutch. See **Clutch.**

Plate-edge Planing Machine. A metal-planing machine for truing the edges of plates, having a fixed table and a travelling tool. See also **Planing Machine** (*a*).

Plate-flattening Machine. A *straightening machine* using seven rolls, four above and three below.

271

Plate Gauge. (*a*) A thin, flat metal gauge for measuring spaces

(*b*) An external gauge formed by cutting slots of the required gauge width in a steel plate, the surfaces of which are hardened. See also **Gauges Commonly Used.**

Plate Link Chain. See **Chain.**

Plate Mill (Plate Rolls). A *rolling mill* with rolls which are plain cylinders. *Grain rolls* do the roughing and *chilled rolls* the finishing process for the plates.

Plate Rolls. *Plate mill.*

Plate-shearing Machine. A *shearing machine* furnished with specially long shears for cutting off the ragged edges of plates after leaving the rolls.

Platen. The work-table of a machine tool, usually slotted for clamping *tee-headed bolts.*

Platen Machine. A *printing machine* using a flat surface.

Plates. (*a*) Thick sheets of metal.

(*b*) The brass plates which form the framework of a clock or watch and in which holes are drilled for the *pivots* of the train of wheels.

Platform Crane. See **Crane.**

Platform Escapement. See **Escapement.**

Platform Scale. A small *weighbridge.*

Play. A limited freedom of movement in a bearing or a working part of a mechanism.

Plug. (*a*) The inner movable portion of a *cock* (*b*). (Fig. 176.) Cf. **Valve Cock.**

(*b*) A small *arbor* or *chuck* used in a larger lathe chuck.

Plug Cock. See **Plug** (*a*).

Plug Cushion Process. This method of explosive *forming* is illustrated in Fig. 133.

Plug Gauge. A gauge used to check the dimension of a hole.

Plug Limit Gauge. Two plug gauges, often on opposite ends of the same handle, which are a '*go*' and a '*no-go*' gauge. See also **Limit Gauge.**

Plug Rod. *Plug tree.*

Plug Tap (Bottoming Tap). (*a*) The final *tap*, the point of which is in no way bevelled so that it can finish an internal thread in a blind hole.

(*b*) A *plug cock.*

Plug Tree (Plug Rod). A long rod suspended from the beam of a single-acting pumping engine and provided with *tappets* for moving

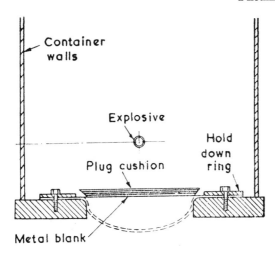

FIG. 133. PLUG CUSHION PROCESS FOR FORMING.

the handles of the equilibrium and steam exhaust valves. See also
Beam Engine.

Plummer Block (Pillow Block). A box-form casing holding the
brasses or other bearing metal for a journal bearing on line shafting
and split horizontally to take up wear. The base of the bearing is
sometimes termed the ' pillow ' and hence the term ' pillow block '.
Lubrication of the bearing is usually by an *oiling ring*. See also
Brasses.

Plunger. (*a*) The solid piston or ram of a *force pump*. A plunger
differs from a *piston* in being longer than its stroke.

(*b*) The solid piston used in moulding and extrusion. (See Figs.
115, 116.)

Plunger Bucket. A *force pump* having no valves.

Plunger Pump. *Force pump.*

Pneumatic Conveyor. See **Pneumatic Tube Conveyor.**

Pneumatic Drill. A rock drill using compressed air to reciprocate
a loose piston which hammers the shank of the bit or an intermediate
piece, or the bit is clamped to the piston rod. Provision is made to
rotate the bit by a small amount between each stroke in most
models.

Pneumatic Hoist. A light hoist lifting loads directly by the
movement of a piston in a long cylinder suspended over the work,
and operated by compressed air.

Pneumatic Pick.   A straight pick which is hammered rapidly by a reciprocating piston driven by compressed air.   Cf. **Pneumatic Drill.**

Pneumatic Riveter.   A high-speed riveting machine in which a rapidly reciprocating piston is driven by compressed air delivering up to 2,000 blows per minute.

Pneumatic Tube Conveyor.   The conveyance of small objects in suitable containers along tubes either by a vacuum or by air pressure.

Point Bar.   A horizontal bar supporting the *points* (*b*) at the back and front of a lace machine and moving with the swing of the carriages.

Point Chuck.   The point centre attached to the *headstock* of a lathe for turning work which is pivoted between centres.   See also **Lathe Carrier.**

Point Ground.   The set of a tool, such as a drill, whose normal helix angle has been ground locally at the tip to change the form of the cutting edge, as for cutting stainless steel.

Point Rail.   *Points* (*a*).

Point Tie.   An arrangement of the harness in a *jacquard machine* for the alternate tying from left to right and conversely.

Points.   (*a*) Movable tongues of metal for setting alternative routes on a railway, pivoted at the ' heel ' and with the ' toe ' locked against the *stock rail.*

Facing Points.   Points in which the toe faces the approaching train.

Trailing Points.   Points in which the heel faces the approaching train.

See also **Catch Points, Spring Points, Trap Points.**

(*b*) The parts of a lace machine carrying the twists of bobbin and warp threads to the position where the lace is made.

Poising.   The checking of a watch *balance* by supporting it horizontally with its pivots on knife-edges.   (See Fig. 20.)   It is said to be ' in poise ' when it shows no tendency to take up any particular position.   See also **Balancing.**

Poke.   *Fusee poke.*

Polar Moment of Inertia.   The moment of inertia about an axis through the centre of gravity perpendicular to the plane of the figure.   It is equal to the sum of the *moments of inertia* about two perpendicular axes through the centre of gravity in the plane of the figure.

Polishing. Making smooth and glossy, usually by friction, such as by a polishing wheel or mop to remove irregularities resulting from machining operations.

Polishing Lathe. A lathe fitted with arbor wheels, mops, etc., for *polishing*.

Poncelot Wheel. An *undershot wheel* with curved instead of the more usual flat vanes, and consequently more efficient.

Poppet (Poppet Head). The movable *headstock* of a lathe, the design of which allows the work to be revolved between centres. In America the term is used for either headstock between the centres of which the work is mounted.

Poppet Cylinder. The cylindrical poppet *mandrel* of a lathe.

Poppet Valve (Mushroom Valve). A mushroom-shaped valve, commonly used for inlet and exhaust valves of an internal-combustion engine, as shown in Figs. 137, 179, consisting of a circular head with a conical face which seats in a conical port in the cylinder and has a guided stem by which it is lifted, using a *rocker arm* and/or *tappet*. See **Valve Insert, Valve Spring** (*a*).

Port. A cylinder opening in an engine, pump, etc., by which a fluid enters or leaves, usually under the control of a valve. See **Valves.**

Portal (Jib) Crane. See **Crane.**

Porter Governor. A *pendulum governor* in which the ends of two arms are pivoted to the spindle and sleeve respectively and carry heavy balls at their pivoted joints.

Position Gauge. A gauge for checking geometrical relationships within assigned tolerances.

Positional Tolerance. See **Tolerance.**

Positive Movement. A movement in any part of a *loom* due to mechanical means. Cf. **Negative Movement.**

Pot Lid Valve. A hollow cup-shaped lift valve.

Pot Sleeper. A combined railway metal sleeper and chair used in hot climates. The gauge is maintained by cross-tie rods.

Potential Energy. See **Energy.**

Potter's Wheel. A rotating circular table, mounted on a vertical pillar and driven by human or other power, on which hollow ware is made prior to firing.

Poundal. A force that produces in a mass of one pound an acceleration of one foot per second per second, being the unit of force in the foot-pound-second system of units. 32·2 poundals equals one pound weight.

Power. The rate of doing work. For unit of power see **Horse-power and Definitions of SI Units (Appendix).**

Power Drag-line. An *excavator* comprising a large scraper pan or bucket which is dragged through the material towards the machine and below its boom or jib.

Power Hammer. Any type of *hammer* operated continuously or intermittently, by some source of power.

Power House. A building in which *power* is generated for distribution or for conversion and later distribution.

Power Plant. (*a*) A number of *power units* assembled in one place. (*b*) The complete propulsive unit for a vehicle, especially an aircraft.

Power Rating. The output power of a motor or aero-engine measured in watts under specified conditions including torque and rad/s. For aero piston engines the *manifold pressure* or *boost pressure* and the *torque*; for turbojets and turboprops the *jet-pipe temperature* and *torque*; and for *rocket motors* the pressure in the *combustion chamber*.

Power Shovel (Navvy). An *excavator* consisting of a *jib* with a radial arm along which a large bucket or scoop travels. The bucket makes a radial cut and digs above the level of the excavator.

Power Unit. An engine, or assembly of engines, complete with shafts, gears, etc., and in the case of aircraft including the propeller(s). It is often called *power plant*.

Preadmission. The admission of steam to an engine cylinder just prior to the termination of a stroke in the opposite direction. See also **Lead.** Cf. **Pre-release.**

Preloaded Ball Screw. A combination of two *ball nuts* joined together by a shimming material whose thickness is adjusted so that there is no *backlash*. (See Fig. 134.)

Pre-ignition. Premature firing of the explosive mixture in a cylinder of an internal-combustion engine. Cf. **Detonation.**

Preoptive Lathe. A *lathe* with a special headstock enabling instantaneous change of spindle speed, whilst cutting is in progress, by means of multi-disc friction clutches.

Pre-release. The opening of a steam cylinder to exhaust just before the termination of the piston stroke. Cf. **Preadmission.**

Pre-selector Gear-box. The selection of a gear-ratio in the *gear-box*, before requirement, by the movement of a small lever. The gear is afterwards engaged by pressure on a pedal. Cf. **Synchro-mesh Gear.**

Recirculating balls

Ball return tube

Precision ground screw threads

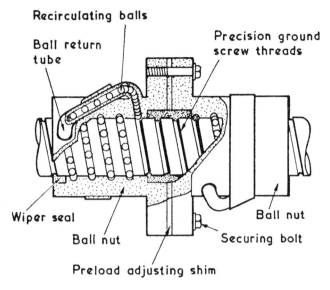

Wiper seal

Ball nut

Ball nut

Securing bolt

Preload adjusting shim

FIG. 134. PRELOADED BALL SCREW.

Press. *Hydraulic press.*

Press Forging. Forging in which pressure is applied by squeezing rather than by a hammer blow.

Pressing (Stamping). The production of forged work by pressing *dies* together under a *hammer* or *press.*

Pressure Angle (Axial). The acute angle measured in an axial plane between the axis of a helical or worm gear and a normal to the tooth profile at a point on the *reference cylinder.*

Pressure Angle (Normal). (*a*) The acute angle between a radial line passing through any point on the tooth surface of a helical, spur or worm gear and a tangent plane to the tooth surface at that point.

(*b*) The acute angle between a normal to the tooth surface at any point on the *pitch cone* of a bevel gear and the tangent plane to the pitch cone at that point.

Pressure Angle (Transverse). The acute angle between the normal to a tooth profile of a helical, spur or worm gear in a transverse plane at its point of intersection with the *reference circle* and the tangent to the reference circle at that point.

277

Pressure Angle (Working Transverse). The *transverse pressure angle* measured with reference to the *pitch circle* instead of the reference circle.

Pressure Blower. See **Blower** (*a*).

Pressure Differential. The difference in pressure between two chambers used as a driving force.

Pressure Feedback Unit. A device for feeding back a displacement or other quantity proportional to the operating pressure in an hydraulic system.

Pressure Forging. *Drop forging.*

Pressure Gauge. *Bourdon gauge.*

Pressure Jet. (*a*) A small jet-propulsion unit fitted to the tips of the rotor blades of some rotorcraft.

(*b*) A small jet nozzle operated from a gas source aboard an artificial satellite or a rocket used to control its orientation.

Pressure Ratio. (*a*) The ratio of the pressure at the beginning of a process to that on completion, as in a piston-engine for the mixture before and after compression by the piston in the cylinder.

(*b*) The absolute air pressure prior to combustion in a jet-engine divided by the ambient pressure.

(*c*) The ratio of air (gas) pressures across a compressor, a turbine or a propelling nozzle of a jet-engine. Cf. **Compression Ratio.**

Pressurized. (*a*) Maintained at a pressure greater than normal atmospheric pressure; e.g. working under water in a caisson or in special apparatus.

(*b*) Maintained at higher than ambient atmospheric pressure as in an aircraft cabin flying at high altitudes.

Primary Valve. A valve for assisting in priming and fitted on the suction side of a pump.

Prime Mover. An engine or mechanism converting a natural source of energy into mechanical power.

Prime Number. A number which has no divisors except unity and itself. Two gear- or cog-wheels should have numbers of teeth such that the two numbers have no common divisor, that is the numbers are said to be ' prime ' to each other. Such an arrangement ensures that the same two teeth on different wheels do not come into frequent contact and thus give rise to a risk of excessive wear.

Priming. (*a*) The operation of filling a pump intake with fluid to expel the air.

(*b*) The operation of injecting petrol into an engine cylinder to assist starting. See **Priming Pump.**

278

Priming Pump. A fuel pump for supplying a piston-engine with fuel during starting.

Priming Valve. *Pet cock.*

Printing Machine. A machine for transferring ink, etc., to paper, (*a*) relief type, using rotating cylinders with raised type or a flat surface; (*b*) planographic type, using prepared metal surfaces that accept or reject the ink; (*c*) intaglio and photogravure types, using surfaces with ink-carrying hollowed-out portions; (*d*) direct photographic process. Cf. **Platen Machine.**

Process Control. The automatic control of sections of an industrial plant by electronic means, including rates and accelerations of flow plus changes in temperature, pressure, etc.

Process Engineering. See **Process Control.**

Product Register. The register on a calculating machine, recording the results of multiplication and mounted on the movable carriage.

Proell Governor. A type of *Porter governor* in which the balls are attached to upward prolongations of the links to the governor sleeve.

Profile. The shape of a normal section through a surface.

Modified Profile. A profile, excluding all *texture* waviness which exceeds a certain maximum.

Profile Grinding (Form Grinding). The grinding of cylindrical work without traversing the wheel, the periphery of which has the required profile.

Profile Milling Machine. A *milling machine* (*a*) in which the rotating spindle is constrained to run over a copy of the article to be milled. (Fig. 113.)

Profile-turning Slide (Bevel Turning Slide). A tool slide mounted on the cross-slide of a *lathe*. The movement of the tool in the axial direction of the lathe is controlled by a *cam* fixed to the *tailstock*.

Profiling. The grinding and sharpening of a cutter so that it will have the correct shape for forming the material which it is going to cut.

Profilometer. An instrument for ascertaining the quality of surface finish, using a stylus and observing the surface irregularities on an oscillograph. Cf. **Surface Meter.**

Projection Gauge. An optical gauge which magnifies the profiles of screw threads about 100 to 200 times to ease inspection and comparison.

Prong Chuck. A fork-like chuck for holding and revolving wood set between lathe centres.

Prony Brake. An *absorption dynamometer* where the torque of an engine is absorbed by a pair of friction blocks bolted together across a brake drum and is balanced by weights at the end of an arm attached to the blocks; alternatively, the arm is secured to a band encircling the flywheel or pulley of the engine.

Proof Load (Test Load). (*a*) The load which a structure, or a mechanism, must be able to withstand while remaining serviceable.

(*b*) A load greater than the *working load* to which a structure or a mechanism is tested to ascertain whether it can withstand such a load without permanent distortion or damage.

(*c*) The product of the *limit load* and the ' proof factor of safety ' See **Factor of Safety.**

Proof Strain. See **Proof Stress** and **Proof Load** (*b*).

Proof Stress. In metal pieces, the stress required to produce a certain amount of extension when there is no sudden *yield point*.

Propeller. A power-driven bladed screw designed to produce thrust by its rotation in air, water or other gas or fluid. In air often called an ' airscrew ' and in water a ' marine screw propeller ' The screw produces the thrust by giving momentum to the column of air (water) which it drives backwards. Fig. 135 shows both types

Fig. 135. Propellers: (*a*) aircraft, (*b*) marine.

When the blades are fixed for an aircraft propeller it is called a ' fixed pitch propeller ', but when the setting can be changed it is called an *adjustable pitch* or a *variable-pitch propeller*. A ' tractor propeller ' produces tension and a ' pusher propeller ' produces compression in the *propeller shaft*. See also under headings **Braking,**

Coaxial, Constant-speed, Contra-rotating, Controllable Pitch, Feathering, Swivelling.

The *boss* is the central part of an integral propeller. The *hub* is the detachable part mounted on the propeller shaft or the central portion to which the roots of detachable blades are attached.

Cycloidal Propeller. A marine propeller consisting of four identical hydrofoil shapes (Fig. 136) rotating about a vertical

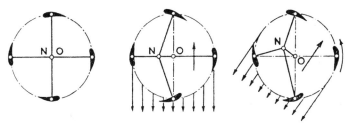

FIG. 136. CYCLOIDAL PROPELLER.

axis. With *N* and *O* coincident there is no thrust. When *N* is displaced to alter the pitch cyclically the direction of thrust is altered as shown.

Propeller Blades. The arms of a propeller with cross-sections of *aerofoil* or *hydrofoil* shape.

Propeller Efficiency. The ratio of the thrust of a propeller to the torque supplied by the engine shaft, usually 75–80% and up to 90% in special cases for aeroplanes.

Propeller Fan (Fan). An impeller or rotor usually fitted with blades of *aerofoil* form working in a cylindrical casing to provide a current of air.

Propeller Shaft. The driving shaft conveying the engine power from the gear-box to the rear or front axle of a motor-vehicle, usually connected through universal joints to allow for vertical displacement of the axle on the springs. Sometimes called ' cardan shaft '.

Propeller-type Water Turbine. A water turbine having a *runner* similar to a four-bladed marine propeller.

Propelling Nozzle. The nozzle attached to the rear end of a jet-pipe, or to an exhaust cone or to the rear end of a rocket-engine. It has usually a fixed throat of convergent-divergent type, the area of which is sometimes varied by the movement of a central bullet or by a slotted shroud called an ' ejector exhaust nozzle '. See also **Variable-area Propelling Nozzle.**

Proving Machine. *Testing machine.*

Puddling Rolls. The first set of *rolls* through which a shingled *bloom* is passed.

Pug Mill. A mill used for mixing concrete ingredients.

Pulley. A wheel on a shaft with either a cambered rim for carrying an endless belt or grooved for carrying a rope, *vee-belt* or chain. See **Differential Pulley Block, Fast Pulley, Loose Pulley, Rigger.**

Pulley Block. Pulleys placed side by side in a wooden or metal frame. See also **Differential Pulley Block, Sheave.**

Pulley Lathe. A machine for boring and facing *pulleys* with special arrangements for setting and turning.

Pull-over Mill. A *rolling mill* with a single pair of *rolls* so that, after passing through the rolls, the metal has to be pulled back over the top roll for the next feed. See also **Passes.**

Pulsator. An apparatus for causing alternate suction and pressure release fifty to sixty times a minute, as used with milking machines.

Pulsejet Engine. An air-swallowing engine, composed of a combustion chamber to which air is admitted through valves that are opened or shut by the pressure in the chamber, and of a nozzle which generates thrust by a jet of hot gases. A sparking-plug is required to ignite the fuel to start the engine, but thereafter the operation is automatic. The successful functioning of this resonating type of engine depends on the proper matching of the natural frequencies of the pressure and expansion waves within the combustion system and the mechanical properties of the valves.

Pulsometer (Pump). A steam-condensing vacuum pump, so-called from the pulsatory action of the steam, with an automatic ball-valve as the only moving part admitting steam alternately to a pair of chambers. It is sometimes used for dealing with liquids having solid matters in suspension.

Pump. A mechanism for converting mechanical energy into energy in a fluid.

Phase Pump, A pump with two opposing pistons working in each cylinder with a variable phase of operation.

Radial Pump. A pump with the cylinders radially disposed.

Replenishment Pump. A pump which delivers fluid at a suitable pressure for replenishing a system. When the pressure in the low-pressure line is raised above ambient pressure it is called ' forced replenishment '.

Swash-plate Pump. A pump with axial cylinders and pistons (or their connecting-rods) on an inclined member, the pistons reciprocating when there is relative motion between the inclined member and the cylinder system. See **Swash-plate.**

Vane Pump. A pump with its rotor axis mounted eccentrically in a cylindrical pressure chamber and with a number of vanes in radial slots maintained in contact with the chamber surface by springs or fluid pressure. The movement of the rotor causes delivery of the fluid.

Variable-stroke Pump. A *radial pump* or *swash-plate pump* with a variable crank throw or swash angle. If varied automatically as when maintaining a constant pressure it is called an ' auto variable-stroke pump'. See **Air Pump, Air-lift Pump, Centrifugal Pump, Chain Pump, Force Pump, Hydraulic Pump, Lift Pump, Pulsometer (Pump), Rotary Pump, Roots Blower, Semi-rotary Pump, Suction Pump, Vane Pump.**

Pump Barrel. The closed cylinder in which the bucket, plunger or piston of a pump moves.

Pump Bob. A rocking lever of a pump.

Pump Bucket. The piston or plunger of a *suction pump.*

Pump Duty. The overall efficiency of a pump, usually measured as the ratio of mechanical work output to heat input.

Pump Head. The head over the top of a *chain pump.*

Pump Rod. The piston rod of a pump.

Punch. (*a*) A non-rotating steel tool for shearing out a piece of material of a certain shape under pressure from plates, etc., supported underneath by a die with a slightly larger profile of the same shape.

(*b*) A tool used in *extrusion.*

Punching Bear. A portable *punching machine.*

Punching Machine. A machine for punching holes in plates using a *punch* driven by a crank and reciprocating block or by a *hydraulic ram.* See also **Nibbling Machine.**

Push Fit. See **Fit.**

Push Piece. A small cylindrical plunger on the outside of a watch-case to enable the hands or calendar to be set, or to start and stop a stop-watch, etc. See also **Slide.**

Push-rod. A rod operating the *valve rocker* of an overhead-valve engine when the camshaft is located in the crankcase (cf. **Tappet**). The vertical rod under *H* in Fig. 137 is a ' push-rod '. (See also Fig. 200.)

Pusher (Aeroplane)

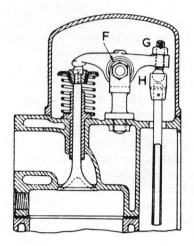

FIG. 137. PUSH-ROD AND ROCKER.
(F) Valve rocker.
(G) Tappet clearance adjusting
    screw.
(H) Spherical cup.

Pusher (Aeroplane). A piston-engined aeroplane in which the engine produces compression in the *propeller shaft*.

Q

Quadrant. A slotted guide forming a quarter circle through which an adjustable lever works as in some *link motions*.

Quadrant Plate (Wheel Plate). The plate carrying the stud wheels in the gear-box of a screw-cutting lathe.

Quadruple-expansion Engine. A *steam-engine* in which the steam is expanded successively in four cylinders of increasing size working on the same crankshaft. See also **Multiplier-expansion Engine.**

Quality Control. A statistical method of directing and testing the output of a production process.

Quarter Chiming Movement. A movement containing a train which includes a quarter rack, quarter rack tail, quarter rack hook, quarter gathering pallet, quarter snail, quarter locking wheel, quarter fly, quarter rack spring, quarter hammer and hammer springs, with an action similar to that of the same named parts of *clock* or *watch*.

Quarter Elliptic Spring. *Cantilever spring.*

Quarter Fly. See Fly (*b*).

Quarter Rack. See Quarter Chiming Movement.

Quarter Screws (Mean Time Screws). Four screws in the rim of a *compensation balance* used for rating a watch. One is placed at either end of the arms and the other two at right-angles to the arm. (See Fig. 39.) Cf. **Compensating Screws.**

Quarter Snail. See **Snail.**

Quartering. The adjustment of cranks or crank pinholes at right-angles to each other.

Quartering Machine. A double-headed machine for the accurate boring at right-angles of *crank pin* holes in locomotive driving wheels after fixing on their axles.

Quick Feed. See **Feed.**

Quick Gear. The direct lift of a crane instead of through intermediate gearing. Cf. **Slow Gear.**

Quick Return. A return made more rapidly than the cutting stroke of a machine, so as to reduce the idling (non-cutting) time of the tool. Cf. **Quick Traverse.** See also **Whitworth Quick-return Motion.** (Figs. 209, 210.)

Quick Traverse. The hand or mechanical traverse given to the slide rest of a lathe by means of a rack and pinion. Cf. **Quick Return.**

Quill. A hollow shaft or spindle of small diameter which damps out the effect of starting and compensates for misalignment with the pinion shafting due to applied load; it can be a weak link when it is necessary to prevent overspeeding of the driven part, as with an aeroplane propeller.

Quill Drive. Transmission through a quill. A drive through a hollow shaft concentric with a solid shaft, but on separate bearings. The shafts can be connected by a clutch or by a flange at one end to provide a flexible drive. In a geared quill drive the quill carries a gear-wheel which is geared to a pinion on a shaft.

R

ρ. The symbol for density.

radn.p.s. Radians per second.

ref.m.p. Referred mean pressure.

rev/min, r.p.m. Revolutions per minute.

r.p.s. Revolutions per second.

Race. (*a*) The inner or outer steel rings of a *ball-bearing* or a *roller-bearing*.

(*b*) A circular ring supporting a revolving superstructure that moves on rollers, as in some cranes.

(*c*) The channel conducting water to a water wheel.

Race Board (Shuttle Race)

Race Board (Shuttle Race). The part of the *sley* in a *loom* along which the shuttle travels.

Rack. (*a*) A straight length of toothed gearing.

(*b*) A bar or rail with teeth for engaging with a moving mechanism by means of a geared wheel; a cogged rail. A helical rack is shown in Fig. 138.

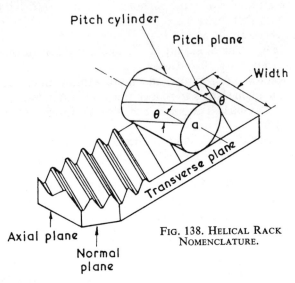

FIG. 138. HELICAL RACK
NOMENCLATURE.

(*c*) The number of motions concerned in making a piece of lace or net.

(*d*) A straight bar or a flat plate with a series of equi-distant teeth on one face (see also **Rack and Pinion**). When the pitch radius of a spur or helical gear becomes infinite, it is a rack. In clocks and watches, racks may be curved.

An 'hour rack' is a circular rack with twelve teeth corresponding to the twelve hours to be struck.

A 'quarter rack' is a straight rack with five teeth for the four quarters to be struck and one for the rack hook. (Fig. 31.)

Rack and Pinion. An arrangement of a straight-toothed *rack* and a *pinion* to convert rotary into linear and reciprocal motion, usually a pinion wheel with fixed centre actuating a movable rack. (See Figs. 139, 170.) See **Rack Railway**.

Rack and Pinion Steering-gear. A steering-gear for motor-vehicles in which a pinion on the steering-column engages with a rack attached to a divided *track rod.*

Rack, Double-sided. A rack having teeth machined on opposite sides of a straight bar.

Rack Feed. *Feed* by a *rack and pinion.*

Rack, Helical. A rack with the *pitch plane* for its *pitch surface* and its axial and transverse directions those of the mating gear.

Rack Hook. A piece on a stud which, when lifted by a *lifting piece*, releases a rack and allows it to fall until the rack tail, an extension of the arm of the rack engages with a *snail*, this snail controlling the distance that the rack can fall. See also **Rack.**

Rack Railway. A railway on a steep gradient in mountainous districts using locomotives with toothed gear to engage a rack or cogged rail, and thus obtain additional adhesion. The rack is usually laid in the centre between the rails. See also **Locker Rack.**

Rack, Spur. A *spur gear* with an infinitely large *pitch diameter.*

Rack, Straight Bevel. *Contrate gear.*

Rack Tail. See **Rack Hook.**

Racking. Moving a machine mechanism, or part thereof, forwards and backwards.

Racking Gear. The gear operating the *block carriage* of a crane.

Radial Arm. (*a*) The movable cantilever supporting the drilling saddle in a radial drilling machine. See **Drilling Machine.**

(*b*) The arm pivoting on the end of the ' leading screw ' of a screw-cutting lathe.

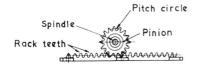

FIG. 139. RACK AND PINION.

Radial Axle. See **Bogie.**

Radial Drilling Machine. See **Drilling Machine.**

Radial Engine. A piston-engine with the cylinders arranged radially at equal angular intervals around the crankshaft. They are either in one plane (single-row) or in two planes, one behind the other (double-row). Cf. **Rotary Engine.** See **Articulated Connecting-rod, Engines (Servo-motor Types), Hydraulic Motors, Master Connecting-rod.**

Radial Paddle-wheel. A *paddle-wheel* with fixed *float boards*

287

fastened directly to the arms with their faces radiating from the centre of the wheel.

Radial Valve-gear. The valve-gear of a steam-engine in which the slide valve is given independent component motions proportional to the sine and cosine of the crank angle. See **Hackworth Valve-gear, Joy's Valve-gear, Marshall Valve-gear, Walschaert's Valve-gear.**

Radius of Gyration. The square root of the ratio of the moment of inertia of a body about a given axis divided by its mass.

Radius Rod (Brindle Rod). (*a*) A rod attached to the die or block of a *Walschaert's valve-gear* for transmitting its motion to the end of the combination lever pivoted to the valve rod.

(*b*) A rod passing from the die block of a slot link to the slide or cut-off valve of a steam-engine.

(*c*) A rod in a *parallel motion* arrangement, fixed on a pivot at one end and jointed to the back link on the other.

Rail Gauge. An iron bar with projections near each end, the distance between the faces of the projections being the gauge of the rails. See **Broad Gauge, Narrow Gauge, Standard Gauge.**

Rail Mill. A mill for rolling rails, the rolls being of the required cross-section.

Rails. Steel bars of various sections laid across *sleepers* to provide a track for rolling-stock with flanged wheels.

Rake (Rake Angle). (*a*) An angle of inclination or an angular relief. ' Front rake ', ' side rake ', ' top rake ' are positional names given to the faces of cutting tools whose angles are set to obtain the most efficient cutting angle. (Figs. 17, 114.)

(*b*) The pitch of saw teeth.

Ram. (*a*) A *hydraulic ram*. (See also Fig. 190.)

(*b*) The arm of a shaping or slotting machine carrying the tool backwards and forwards. (See Fig. 163.)

(*c*) A term used to describe the effect of the speed of an aircraft on the induced pressure in the engine air intake.

Ram-air Turbine (Wind-driven Turbine). A small turbine in an exposed position on an aircraft driven by the ram air to provide a source of auxiliary power for electric generators, fuel and hydraulic pumps, etc.

Ramjet Engine. An air-swallowing engine, composed of a diffuser, a combustion chamber and a nozzle, which generates thrust by a jet of hot gases. It has no moving parts except the fuel pump. Cf. **Pulsejet Engine.**

Ramps (Guide Plates). (*a*) Inclined surfaces.

(*b*) Appliances which clip the rails and are provided with flat helical extensions so that the wheels of rolling stock can slide up and on to the rail.

Ramsbotton Ring. See **Spring Ring.**

Ram's Horn. A symmetrically-shaped double crane hook.

Rankine Cycle. A cycle used as a standard of efficiency for composite steam plants, comprising the introduction of water at boiler pressure, evaporation, adiabatic expansion to condenser pressure, and condensation of the steam to the initial point.

Rankine Efficiency. The efficiency of an engine when compared with an ideal engine working on the *Rankine cycle* under given conditions of steam temperature and pressure.

Rapson's Slide. A mechanism which ensures a constant turning movement on a ship's rudder for all tiller positions.

Ratchet (Ratchet Wheel). (*a*) A wheel with inclined teeth for engaging with a *pawl* which allows only forward motion and arrests backward running. (See Fig. 140.)

FIG. 140. RATCHET AND DOUBLE-ENDED PAWL.

(*b*) A wheel with specially shaped pointed teeth which engage with a *click*, as used on a *barrel arbor* in watches and clocks to prevent it turning back when the spring is being wound.

Ratchet Bar. A straight bar serrated with teeth to receive the thrust of a *pawl* thus permitting movement in one direction and preventing it in the opposite direction.

Ratchet Feed. *Pawl feed.*

Ratchet Jack. A *screw jack* worked by means of a *ratchet* and *pawl.*

Ratchet Pawl. A *pawl* which engages with a *ratchet wheel* or a *spur wheel.*

Ratchet Teeth. (*a*) Fine-pointed teeth on an *escape wheel.*

(*b*) Teeth like those of *spur wheels* when the controlled motion may be in either direction.

Ratchet Teeth

(*c*) Teeth sloping in one direction only as in the case of cranes. See **Pawl.**

(*d*) In general, teeth shaped to fit the *pawl* or *click* controlling a particular motion.

Ratchet Wheel. *Ratchet.*

Rate Gyro. See **Gyro.**

Rated Altitude. The altitude at which a piston-engine gives its maximum power. With more than one stage of supercharging there will be, correspondingly, more than one rated altitude. See also **Supercharging.**

Rating Nut. A milled nut on a pendulum which supports the bob and raises or lowers it to alter the time of swing.

Ratio of Compression. *Compression ratio.*

Ratio of Expansion. *Expansion ratio.*

Reaction Turbine. A turbine in which the working gas expands through alternate rows of stator and rotor blades. The rotor blades absorb the kinetic energy after each expansion through a set of stator blades. In most cases the gas flows axially (see Fig. 192) but the *Ljungström turbine* type is an example of radial expansion. Cf. **Francis Water Turbine, Parson's Steam Turbine.**

Fig. 141 shows a simplified version of flow from a stator into a reaction rotor.

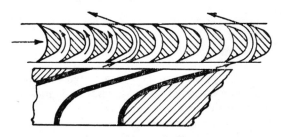

Fig. 141. Reaction Turbine.

Fig. 142 shows a turbine rotor unit as found in a *turbojet* engine.

Reaction Wheel. An enclosed wheel actuated by water pressure as in a *turbine.*

Reamer (Rimer). A tool with longitudinal or spiral flutes or separate teeth on a cylindrical or conical shank for finishing drilled holes. The tool may be solid to fit in a socket or hollow to fit on a mandrel (cf. **Broach**) and may be either a side-cutting or end-cutting type.

290

FIG. 142. TURBINE ROTOR UNIT.

Expanding Reamer. A reamer partially slit longitudinally and capable of slight adjustment in diameter by a coned internal plug.

Reaming. Enlarging a hole in metal with a *reamer.*

Reaper (Reaping Machine). An agricultural machine for cutting grain which has a protruding arm with fixed knives (fingers) and a moving scythe working between the slots of the fingers and actuated by gearing from the wheels of the machine. Usually there is a paddle-wheel to cause the cut stocks to lie all in the same direction.

Rebating. Cutting a groove into, or a shoulder on, the edge of a piece of timber, as by a machine with revolving cutters or by a thick circular saw.

Reboring. Reboring a worn cylinder and fitting a slightly larger-diameter piston.

Rebound Hardness. *Dynamic hardness.*

Receiver Gauge. A gauge for checking simultaneously all relevant features of a component.

Rechucking. Resetting a piece of work in a lathe *chuck* to complete turning, etc.

Reciprocating. Continuous alternate backward and forward motion.

Reciprocating Engine. An engine with a piston oscillating in a cylinder under the periodic pressure of the working fluid.

291

Recoil Escapement. See **Escapements.** Also called 'anchor escapement'.

Recuperator. A system of springs or compressed air which returns a gun, after firing, to its firing position.

Reducing Valve. (*a*) A valve for reducing the pressure of a fluid in a supply line.

(*b*) A valve with a weighted lever for regulating steam pressure between a boiler and its connections.

Reduction Gearing. Gearing or a system of pulleys to apply a source of power at a lower rotational speed.

Reduplication. The gain in power from a combination of pulleys in *pulley blocks*.

Reed. An arrangement of flattened steel wires, fixed in a weaving frame like a comb, to keep the *warp* threads in position, to form a guide for the shuttle and to beat up the *weft* to the fell of the cloth.

Reeling. (*a*) In general, winding on to a reel.

(*b*) Unwinding silk filaments from a number of cocoons and combing them to form a silk thread.

(*c*) Winding yarn from *cops* or bobbins on to a reel to form skeins.

(*d*) Finishing and truing tubes and bars in a machine of that name. See **Straightening Machine.**

Reference Circle. See **Circle, Reference.**

Reference Cylinder. See **Cylinder, Reference.**

Reference Gauge. A gauge used for reference to check other gauges or of the final product. It might be any of the following: *standard gauge, inspection gauge, workshop gauge* or for checking the two last.

Reflux Valve. *Check valve.*

Regulator. (*a*) A *chronometer.*

(*b*) A precision long-case clock, with a seconds pendulum and having a dial with independent hands and zones for the hours, minutes and seconds.

(*c*) The *index.*

(*d*) A *gas regulator.*

(*e*) A watch regulator. See **Incastar Regulator.**

Regulator Valve. A valve for controlling the steam supply to the cylinders of a *locomotive*, operated from the cab and located in the dome or steam space.

Reheat. The burning of additional fuel after the turbine of a *turbojet engine* to provide additional thrust. Also called 'afterburning'.

Reheating (Re-superheating). The passing of partially expanded steam in a steam-turbine back to a superheater before its further expansion.

Relative Efficiency. The ratio of the actual indicated thermal efficiency to the efficiency of an ideal cycle of an internal-combustion engine. An *air standard cycle* at the same compression ratio is one such ideal cycle.

Relay. A device for supplying additional energy, such as more pressure. See also **Servomotor.**

Hydraulic Relay. Hydraulic means by which a mechanical displacement is raised from a small power level to another displacement at a higher power level.

Jet Relay. A *hydraulic relay* in which the momentum of a fluid jet provides the necessary force.

Pneumatic Relay. Pneumatic means by which a mechanical displacement is raised from a small power level to another displacement at a higher power level.

Relay Valve. *Pilot valve.*

Release. (*a*) The opening of the steam port of a steam-engine to allow the escape of the exhaust steam.

(*b*) A trigger arrangement for releasing the shutter of a camera. See also **Antinous Release.**

Relief Angle. See **Angle of Relief.**

Relief Valve. (*a*) A *safety valve.*

(*b*) A *cylinder escape valve.*

Relieving. *Backing-off.*

Relighting. Restarting a gas-turbine engine or *reheat* in flight.

Repeater. A watch which strikes the time by blows cn gongs when a *slide* on the band of the case is pushed. A ' quarter repeater ' strikes the last hour followed by the number of quarters up to that time. A ' minute-repeater ' strikes, in addition, the number of minutes since the last quarter.

Resilience. The stored energy of a strained or elastic material, such as in a compressed spring or in rubber dampers which have inherent damping properties.

Resilient Escapement. See **Escapements.**

Resistance Welding. See **Welding** and Fig. 208.

Resonance. The synchronism of a harmonic of a forcing impulse with the natural frequency of vibration of a mechanical system, usually leading to excessive amplitudes of vibration. Cf. **Antiresonance** and see **Frequency.**

Rest, of a Lathe. The support which takes the resistance of the tool in turning operations on a lathe. See also **Floor Rest, L-rest, Slide Rest.**

Restitution, Coefficient of. See **Impact.**

Retaining Valve. An additional valve in a pump to prevent water running back when the lift is from a great depth.

Retarder (Skate, Wagon Retarder). An arrangement of braking surfaces, parallel and alongside the running rails in a shunting yard, which are operable from a signal-box.

Retractable Undercarriage. *Alighting gear.*

Return Block. *Snatch block.*

Return Crank. The short crank, fixed to the outer end of the main crank pin, that replaces an eccentric in *Walschaert's valve-gear* on locomotives with outside cylinders.

Return Valve. An overflow valve allowing the return of fluid.

Reverse Cones. Cones with bases turned away from each other, as used in some bearings of *headstocks.* Cf. **Cone Bearing.**

Reverse-flow Compressor. An *axial-flow compressor* with the axial flow in a forward direction as in some *turboprop* engines, the flow being reversed before entering the combustion chamber.

Reverse Gear. The gear by which an engine provides power for movement in a direction opposite to the normal direction.

Reverse Jaw Chuck. A *dog chuck* whose jaws are reversible, end for end.

Reverse Jaws. The jaws of a *lathe chuck* which are placed within the work when turning exterior surfaces.

Reverse Keys. A pair of steel plates, one with a projecting slip on one edge and the other with a recess of the same length, which are used as a wedge to drive two machine parts away from each other.

Reverse Pitch (of Propeller). See **Pitch Setting.**

Reversing Cam. A *cam* operating the valves of a gas-engine, which is arranged to shift along a shaft, or have its motion reversed, in order to make the engine run in the reverse direction.

Reversing Countershaft. A *countershaft* whose direction of rotation can be reversed.

Reversing Cylinder. See **Steam Reversing Gear.**

Reversing Engine. An engine whose direction of motion can be reversed. See **Link Motion.**

Reversing Gear. Gear which reverses the direction of motion of an engine, machine or mechanism. See **Joy's Valve-gear, Link Motion, Walschaert's Valve-gear.**

Reversing Link. The slotted link of an engine which alters a valve for forward or backward motion. See **Link Motion.**

Reversing Mill. A *rolling mill* in which the stock passes forwards and backwards between the same pair of rolls, which are reversed between each pass. See also **Continuous Mill, Pendulum Mill, Pull-over Mill, Three-high Rolls.**

Reversing Plate. A plate keyed on a *crankshaft* and furnished with a slot, or holes, by which an *eccentric* is moved up or down for forward or backward motion respectively. It is found on a single-cylinder engine with no *slot link* or for operating a valve opening.

Reversing Rolls. The *rolls* of a *rolling mill* whose direction of motion is reversible.

Reversing Shaft. *Weigh shaft.*

Reversing Stud. The stud or spindle carrying the *idle wheel* for reversing the motion of back-geared self-acting lathes.

Reverted Train. A train of wheels in which the axes of the first and last wheel are coincident.

Revolution Counter. *Counter.*

Revolving Escapements. See **Karrusel** and **Tourbillon Movements.**

Revolving Flats. An endless chain of metal bars extending across a revolving-flat carding engine. The bars are faced with fine teeth set in the opposite direction to those on the main cylinder which moves rapidly while the flats move slowly in the same direction carding the cotton.

Ribbon Brake. *Strap brake.*

Riding (Overriding). A term describing the action when the teeth points of one gear-wheel come into contact with the points of its engaging wheel, due to bad centring of the wheels.

Rifling. The term used for the spiral grooves in the surface of the *bore* of a gun or rifle which are engaged by the driving bands of a projectile and cause rotation of the latter.

Rigger. A fast-and-loose pair of belt pulleys.

Right-hand Engine. A horizontal engine which stands to the right of its flywheel as seen from the cylinder. Cf. **Left-hand Engine.**

Right-hand Screw. A screw having a *right-hand thread.*

Right-hand Thread. A screw thread which, when viewed along its axis, appears to rotate clockwise as it goes away from the observer, that is, as in the common wood screw.

Right-hand Tools. Lathe side-tools with the cutting edge on the left, thus cutting from right to left and putting the right-hand face on the workpiece.

Right-handed Engine. An aero-engine in which the propeller shaft rotates clockwise when the observer is looking past the engine to the propeller.

Rim Wheel (Rim Pulley). A large pulley on the rim shaft of a cotton-spinning *mule* transmitting power to the roller which drives the spindles.

Rim Width (Worm Wheel). The maximum width of the rim in the direction of the axis of the worm wheel.

Rimer. *Reamer.*

Ring Gauge. A hardened-steel ring with an internal diameter of specified size used to check the diameters of finished cylindrical work. It is manufactured to fine tolerances and used for go and no-go gauging.

Ring Lubricators. Flat metal rings, hanging from a shaft and dipping into a trough of oil, which rotate with the journal and carry oil to the bearing.

Ring Spinning. The twisting of the yarn by a positively driven bobbin and a metal *ring traveller* through which the yarn passes to the bobbin, being guided by the lifter rail.

Ring Traveller. A metal eyelet on the ring in the lifter rail. See **Ring Spinning.**

Ring Valve. A *lift valve* with a ring replacing the solid disc, the valve being guided by a central block fitting within the ring.

Rip-saw. A saw designed for cutting timber along the grain with about one tooth per centimetre (Fig. 143). Cf. **Cross-cut Saw.**

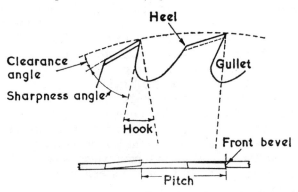

FIG. 143. RIP-SAW TEETH.

Rising and Falling Saw. A circular saw, whose spindle can be moved relative to the working table for cutting grooves of different depths. See **Rising and Falling Spindle**.

Rising and Falling Spindle. A spindle of a circular saw which rises and falls by the operation of a worm and wheel.

Rising and Falling Table. A machine table used for drilling or other purposes which is raised or dropped to suit different requirements.

Rising Rod. A rod actuating the steam and exhaust valves in a *Cornish engine* through catches, sectors and weights.

Rivet Heads. Various types of rivet heads are shown in Fig. 144.

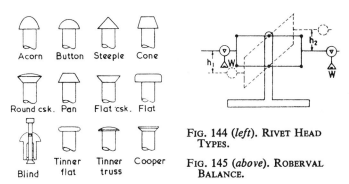

Acorn Button Steeple Cone

Round csk. Pan Flat csk. Flat

Blind Tinner flat Tinner truss Cooper

Fig. 144 (*left*). Rivet Head Types.

Fig. 145 (*above*). Roberval Balance.

Rivet Mill. A small *end mill* attached to a portable drilling machine for cutting off the stems of blind rivets which are of the break-stem variety.

Riveter. See **Hydraulic Riveter, Pneumatic Riveter.**

Road Locomotive. *Traction engine.*

Road Roller. A *traction engine* fitted with a large heavy roller for levelling road surfaces and the like. The term is also used for smaller vehicles driven by diesel engines.

Roberval Balance. A weighing balance, the original of that commonly used in retail shops. Fig. 145 shows the principle of the balance. Two horizontal members of equal length, each pivoted at its centre-point, and two vertical members of equal length joined to the ends of the horizontal members by pivots. To the mid-point of each vertical member is attached a horizontal arm on which weights are hung. Two equal weights on these arms will always balance irrespective of their positions on the arms. Machines found in shops usually have pans above the arms to avoid possible obstruction to the supports of the pans on which the weights are placed. Cf. **Beranger Balance.**

Rock Drill. A reciprocating mechanism, generally operated by steam, electricity or compressed air. The reciprocating rod has a removable cutting head, which revolves slowly and is provided with feed gear.

Rocker Arm. *Valve rocker.*

Rocket-engine. An engine which contains within itself all the chemicals necessary for producing a propulsive jet, the chemicals being in the form of liquid propellants. Cf. **Rocket Motor.**

Rocket Motor. An engine which contains within itself all the chemicals, as solid propellant(s), necessary for producing a propulsive jet. (It has no moving parts in contrast to a *rocket-engine.*)

Rocking Bar. A pivoted bar, connecting the winding stem in a keyless watch mechanism to the barrel for winding, or to the motion work for hand setting, and carrying the necessary intermediate wheels.

Rocking Disc. *Wrist plate.*

Rocking Frame. The frame carrying the gear-wheels of a self-acting lathe to reverse the direction of the back shaft.

Rocking Shaft. (*a*) A shaft with a double-ended lever to actuate the *slide valve* in an indirect-acting slide valve type of steam-engine.

(*b*) A shaft or spindle with a to-and-fro motion only, such as a *weigh shaft.*

Rockwell Hardness Test. A commercial indentation test using a conical indenter for hard metals and a spherical indenter for soft metals. A small load of 10 kg is first applied followed by a second load of 90 or 140 kg. The indentation is directly recorded on a suitable dial after the further load has been removed.

Roll Squeezers. Three rolls, two side by side with collars and a third above and adjustable, for consolidating a puddled ball into a *bloom*, whose width is fixed by the collars and its diameter by the height of the third roll.

Rolled Bars. Bars with a cross-section determined by the *rolls* through which they have been passed.

Roller. (*a*) A cylinder which rolls or turns on its axis such as the rollers in a roller-bearing and in a roller chain. See also **Road Roller.**

(*b*) **Roller, Safety.** The roller mounted on the *balance staff* of a lever escapement, against the edge of which the guard pin acts when the safety action is brought into play.

Roller-bearing. A bearing on a shaft composed of a number of steel rollers located by a cage between inner and outer steel races. Roller-bearings will carry heavier loads than *ball-bearings*. The rollers are parallel for journal loads; if tapered, both journal and thrust loads can be carried (Fig. 146). Long rollers of very small diameter are called ' needle roller-bearings '; these are located end-wise by a lip on the inner or outer race and are not positioned by a cage.

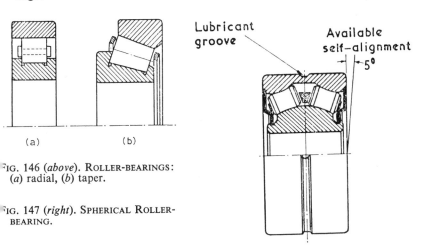

FIG. 146 (*above*). ROLLER-BEARINGS:
(*a*) radial, (*b*) taper.

FIG. 147 (*right*). SPHERICAL ROLLER-
BEARING.

Spherical Roller-bearing. A roller-bearing (Fig. 147) with two rows of barrel-shaped rollers of opposite inclination working in a spherical outer race, thus providing a measure of self-alignment. See also **Linear Roller-bearing.**

Roller Box. A cutting tool-holder used on *capstan lathes* and automatic lathes. The box holds a cutting tool and two rollers positioned so that part of the reaction force from the cutting tool is taken by the rollers, thus preventing distortion of the work.

Roller Chain. See **Chain.**

Roller Conveyor. A power-driven or gravity-operated roller-track for transporting packages or goods.

Roller Delivery Motion. The mechanism for driving the rollers of a *mule* to deliver a short length of yarn as the carriage makes its inward run.

Roller Drive. A uni-directional drive. Fig. 148 shows three

Roller Drive

balls in the driving position where the outer member goes clockwise and one ball is in its free-wheeling position.

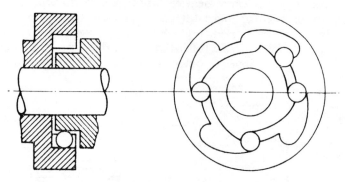

FIG. 148. ROLLER DRIVE.

Roller Path.   A smooth surface on which smooth rollers travel to carry a heavy structure or mechanism.

Roller, Safety.   A roller mounted on the balance staff of a lever escapement in a watch or clock.   A guard pin acts against the edge of the roller when the safety action is brought into play.

Roller Steady.   A support attached to the carriage of a lathe steadying the workpiece in the region of the cutting tool.   (Fig. 149.)

Rolling Circles.   *Rolling curves.*

Rolling Curves (Rolling Circles).   Templet curves used in striking out the shapes of gear-wheel *teeth.*

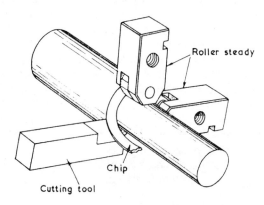

Roller steady

Chip

Cutting tool

FIG. 149. ROLLER STEADY.

300

Rolling Friction. See **Friction.**

Rolling Gate. See **Gate.**

Rolling Locker. A traverse net *lace machine* with a double tier of carriages which swing in an arc controlled by reciprocating rollers.

Rolling Mill. A set of *rolls* (Fig. 150) used for rolling metals into different shapes and sections such as *bars, billets, blooms, plates, rails, rods, slabs, strips.* See **Continuous Mill, Pilger Mill, Pull-over Mill, Rolling Mill, Reversing Mill.**

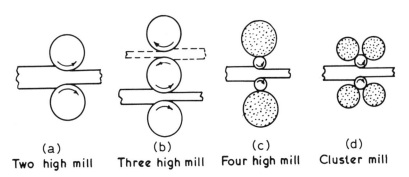

(a)	(b)	(c)	(d)
Two high mill	Three high mill	Four high mill	Cluster mill

FIG. 150. ROLLING MILL TYPES.

Rolling Stock. All coaches, trucks, etc., which carry a load and run along a railway track, but excluding rail-mounted locomotives which are not integral with the load-carrying portion of the train.

Rolls. Cylinders of steel or cast-iron used in *rolling mills.* They are called two-high or three-high according to the number of rolls one above the other. The former are for rolling iron bar or plates, and the latter when the bar or plates are passed backward through the upper rolls without the need for reversing. See also **Chilled Rolls, Grain Rolls, Mill Rolls, Puddling Rolls.**

Roman Balance. *Steelyard.*

Root. The surface of a screw thread connecting adjacent flanks at the bottom of the groove.

Root Angle. The angle between the axis and the *root cone* generator of a bevel gear.

Root Circle. The circle of intersection of the *root cylinder* of a helical, spur or worm gear by a transverse plane; the circle through the roots of the teeth.

Root Cone. The cone tangential to the bottom of the tooth spaces and co-axial with the rotational axis of a bevel gear.

Root Cylinder. The cylinder tangential to the bottom of the tooth spaces and co-axial with the gear.

Root Diameter. The diameter of the *root circle* or *root cylinder* of a gear. (See Fig. 93.)

Roots Blower. A compressor for delivering large volumes of air at relatively low pressure ratios, consisting of a pair of hour-glass shaped members rotating within a casing with but small clearance so that no valves are required. See also **Supercharger.**

Roots Supercharger. See **Supercharger.**

Rope Brake. A rope encircling a brake drum or flywheel with the one end loaded with weights and the other supported by a spring balance. The absorbed torque is equal to the difference of the tensions multiplied by the drum radius. See also **Brake.**

Rope Crane. See **Crane.**

Rope Drive. A term used to describe a method of transmitting power for a long distance by means of ropes running at a high speed.

Rope Wheel. A grooved pulley.

Rose Cutter. A small hollow milling cutter used by watch makers on a lathe for rapidly producing pivots, screws, etc.

Rose Reamer. A fluted *broach* with a conical serrated end and a sharp point (rose end).

Rotachute. A small *rotorcraft* with blades that fold into a small compass, designed as a possible alternative to a parachute for use by one man. With the blades unfolded the descent of a man with rotachute is in a steep glide, the rotachute behaving like an un-powered *gyroplane.*

Rotary Cutter. See **Cutter.**

Rotary Engine. An early type of aero-engine in which the crank-case and radially arranged cylinders revolved round a fixed crank. Cf. **Radial Engine.**

Rotary Piston Meter. A meter for measuring liquids in motion and dependent upon the motion of a rotary piston as indicated by the four positions shown in Fig. 151. The inner ring represents the piston, I and O the inlet and outlet respectively, DV the division piece and FH the fixed hub. In the first position spaces 1 and 2 are neutral and the liquid is discharged from 4. In the second position spaces 1 and 3 receive and 2 and 4 discharge liquid. In position 3, the spaces 3 and 4 are neutral, space 1 receives and space 2 dis-

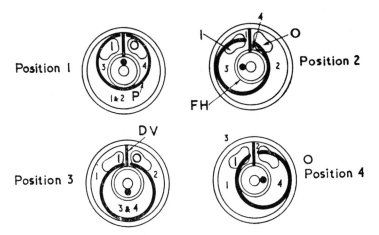

FIG. 151. ROTARY PISTON METER.

charges liquid. Finally in position 4, spaces 1 and 3 receive and 2 and 4 discharge liquid.

Rotary Planing. The work of making a surface with a vertical *milling machine*.

Rotary Pump (Drum Pump). A pump in which two specially shaped members rotate in contact. It is suited to large deliveries at low pressures. See also **Gear Pump, Roots Blower, Vane Pump.** Cf. **Semi-rotary Pump.**

Rotary Squeezer. A machine consisting of two rotating eccentric cylinders, one external and the other internal, used for the consolidation of puddled ball.

Rotary Transfer Machine. A milling machine fitted with milling cutters and multi-spindle drilling heads.

Rotary Valve (Rotating Valve). A cylinder valve, rotating on cylindrical faces in a cylinder head, which acts as a combined inlet and exhaust valve in a ported cylinder.

Rotating Valve. *Rotary valve.*

Rotative Engine. A term used in specifications to distinguish the engine with a crank and flywheel from one in which the reciprocating movement is not converted into circular motion.

Rotor. (*a*) The rotating part of a compressor or a turbine comprising the disc and its row of blades.

(*b*) A *main rotor* of a rotorcraft (Fig. 85).

303

(c) A *tail rotor* of a rotorcraft (Fig. 85).

(d) The rotating part of a steam turbine.

Rotor Blades. The aerofoil-shaped blades of a *rotor* which produce thrust or increase the pressure of the fluid.

Rotor Head. That part of the *rotor hub* to which the *rotor blades* are attached; sometimes called ' spider '.

Rotor Hub. The central rotating system in a rotorcraft.

Rotor Tip Jets. Auxiliary power units fitted to the tips of the blades of some rotorcraft; the units may be pressure jets, pulse jets, ramjets or small rockets.

Rotorcraft. An aircraft with rotating blades or wings that generate lift in flight, including *helicopters, autogiros, gyroplanes, gyrodynes*, convertiplanes and *rotachutes*.

The blades of a rotorcraft are hinged and the hinges have descriptive names as follows: drag hinge, feathering hinge, flapping hinge. One type of helicopter has its blades hinged one-third of the way out on the blade.

Roughing Down (Roughing). The removal of the bulk of the material in a machining operation.

Roughing Rolls (Billeting Rolls). The first rolls used for rolling puddled bar, squeezing and consolidating the *bloom* and bringing it roughly into shape.

Roughing Teeth. The first few teeth on a *broach*. (See Fig. 17.)

Roughing Tool. A lathe, shaping machine or planer, tool for roughing cuts, having a round-nosed or obtuse-angled cutting edge.

Roughing-out. The preliminary operation in shaping a piece of work.

Roughness. The finely spaced surface irregularities that result from cutting edges and machine tool feed used in producing the surface. It can be considered to be superimposed upon *waviness* and is measured in microns.

Round Key. See Key.

Router Cutter. A milling cutter driven through a chuck, the movement of which is controlled by a guide pin running round a former, the former being a duplicate of the desired shape of the component which is being manufactured. (Fig. 152.)

Routing Machine. (a) A machine with either (i) a fixed head in which the cutting head has only a vertical movement on vee-slides, or (ii) a radial-arm type in which a similar head is mounted on the end of a long arm which enables it to move about a centre over an

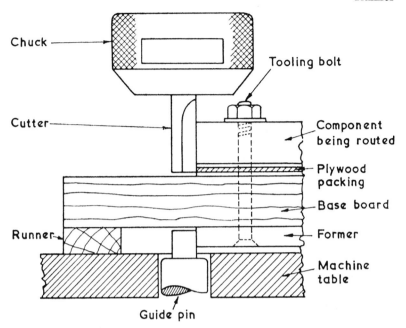

FIG. 152. ROUTER CUTTER FOR WORKPIECE.

angular area of 350°.   High-speed rotary cutters are used to cut out shapes of varied profiles determined by formers or jigs.

(*b*) A machine with a revolving point which removes unwanted metal from printing plates.

Roving (Rove).   A strand (thread) of textile fibres in one continuous length of a size suitable for spinning.   See **Fly Frames.**

Roving Frame.   See **Fly Frames.**

Rubber Bond Grinding Wheel.   A grinding wheel in which the bonding material is of rubber, which softens under the heat of grinding and acts both as a cushion for the abrasive grains and as a buff to polish out the grain marks.

Rubber Press.   A special press designed for *rubber-pad forming* (see **Forming**).

Rubbish Pulley.   *Gin block.*

Ruby Pin.   The impulse pin of a lever *escapement.*

Ruled Surface.   A surface generated by a straight line.

Runner.   The rotor (vaned member) of a water turbine.   See **Turbine.**

Runners. (*a*) Devices to assist sliding motion.

(*b*) Cylindrical sliding pieces supporting work in a pair of *turns*.

Running Centre Chuck. A *driver chuck* which revolves with the lathe mandrel. Cf. **Dead-centre Lathe.**

Running Fit. See **Fit.**

Running-in. Running a new engine, machine or mechanism under a light load and at moderate speed, to allow time for proper clearances to become established and friction-surfaces polished.

Running-out. (*a*) The working of a drill away from its centre.

(*b*) A piece of work chucked eccentrically in a lathe.

(*c*) The deviation of a cutting tool from its required path. Cf. **Run-out.**

Running Pulley. (*a*) The movable pulley in a *snatch block*.

(*b*) A *gin pulley*.

Run-out. (*a*) The total range of movement measured from a fixed point to a point on the surface of a gear or thread rotated about a fixed axis without movement along that axis.

Radial Run-out. The *run-out* measured along a perpendicular to the axis of rotation.

Axial Run-out (*Wobble*). The run-out measured parallel to the axis of rotation, at a specified distance from the axis.

(*b*) A similar movement to (*a*) in other dimensional tests of gears.

(*c*) The quantity of additional material to allow for all contingencies.

(*d*) The portion of a groove, serration, keyway, etc., where the full depth tapers away to nothing. Cf. **Running-out.**

Rymer. *Reamer.*

S

σ. Symbol for stress; also *f*.

s. & s. Spigot and socket.

scr. Screwed.

s.h.p. Shaft horse-power.

S.I. (*a*) International System of Units, derived from the six basic units; metre, kilogramme, second, ampere, kelvin, and candela. See **Definitions of SI units** (**Appendix**).

(*b*) Systeme International screw thread. See **Screw Thread.**

sp. gr. Specific gravity.

sp. ht. Specific heat.

S.V. Sluice or stop valve.

S.W.G. *Standard Wire Gauge (Imperial).*

Saddle. (*a*) That part of a machine that runs on slides.

(*b*) That part of a *lathe* which runs the length of the *lathe bed* carrying the *cross-slide, turret head* or *tailstock.* See also **Capstan Tool Head.**

(*c*) The sliding plate which carries the drill spindle and gearwheels of a radial drill.

Saddle Key. See **Key.**

Saddle Tank Engine. A steam *locomotive* with the water tank on the top and sides of the boiler. See **Tank Engine.**

Safe Fatigue Life. See **Fatigue Life.**

Safe Working Load. See **Factor of Safety.**

Safety Coupling. A friction coupling adjusted to slip at a given torque so as to protect the remainder of the system from overload.

Safety Factor. *Factor of Safety.*

Safety Finger (Guard Pin). A pin or finger which butts against the edge of the *safety roller* of a clock if the escapement is jerked or, in the case of a watch, when the hands are set back.

Safety Hoist. (*a*) A *hoist* with *differential pulley blocks.*

(*b*) A rope *hoist* with a safety catch to prevent the rope from running back.

Safety Rail. *Check rail.*

Safety Roller. See **Roller, Safety.**

Safety Stop. An arrangement of levers or springs which on the breakage of a *lift* cable, throws a *pawl* into gear with a *ratchet* on each side of the shaft to check the lift's descent.

Safety Valve. A valve, controlled by a spring or weighted with a dead-weight, fitted to a vessel or engine to allow the escape of steam, air or other gas when the internal pressure exceeds the maximum safe value.

Safety-valve Lever. The lever, loaded by a weight, as used on some *safety valves.*

Sampling Length. The profile length selected for making an individual measurement of surface *texture.* See also **Traversing Length.**

Sand Blasting. (*a*) The cleaning of a metal surface by a high-velocity jet of sand or an abrasive material, sometimes called shot-blasting. Cf. **Shot Peening.**

(*b*) The roughening of a metal surface by means similar to (*a*), prior to the application of a special finish.

Sand-papering Machines. *Glass-papering machines.*

Sand Pump. A centrifugal type of pump for extracting wet sand out of caissons, etc.

Sanding. Feeding sand on wet or frozen rails in front of the driving wheels of a locomotive.

Saw. A tool with a serrated blade. See **Bandsaw, Circular Saw, Cold-saw, Hack-saw, Hot Saw.** Cf. **Flame Cutting, Shearing.**

Saw Bench (Saw Table). The bench of a machine saw.

Saw Gullet. See **Gullet.**

Saw Gumming. Grinding the roots or *gullets* of *circular saw* teeth by emery wheels, the sections of whose edges are the counterparts of the tooth spaces.

Saw Set. An appliance for setting the teeth of saws alternately to right or left at the correct angle.

Saw-sharpening Machines. Machines fitted with abrasive wheels which automatically feed the saw tooth by tooth. See **Saw Gumming.**

Saw Spindle. The spindle of a *circular saw* furnished with bearing necks and fast and loose pulleys. The saw is clamped between two washers and prevented from turning by a *feather* fitting into a slot cut in the saw.

Saw-tooth Profiles. Figs. 46 and 153 show typical saw-tooth profiles. ' Straight-tooth ' blades have their teeth set in opposite directions, while the ' raker tooth ' has every third tooth straight. A ' wavy set ' has usually three teeth in one direction, one straight, and three teeth in the opposite direction. The purpose of the set is to permit clearance of the blade (see **Gullet).** The cut made by a saw is called a ' kerf '.

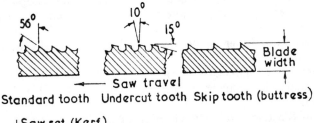

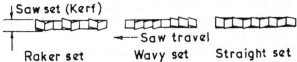

FIG. 153. SAW-TOOTH PROFILES.

Scavenging Stroke. *Exhaust stroke.*

Scavenging (or Scavenger) Pump. An oil-suction pump which returns used oil from the *crankcase* of an engine to the oil tank, using the dry sump system of lubrication.

Schiffle Machine. *Swiss machine.*

Sclerometer. An instrument for measuring hardness, consisting of a diamond at one end of a lever attached to a vertical pillar. The diamond is loaded and the pillar rotated to make a scratch of standard depth, the weight in grammes to produce this depth giving a measure of the hardness. See also **Hardness Tests.**

Scleroscope Hardness Test. The determination of the hardness of metals by measuring the rebound of a diamond-tipped hammer, weighing about two grammes, when dropped from a given height. See also **Hardness Tests.**

Scoop. A large deep shovel. See **Bucket.**

Scoop Wheel. A powered wheel fitted with *scoops* for lifting water for drainage.

Scotch Crank. A crank used on a *direct-acting pump* consisting of a square block pivoted on the over-hung crank pin and working in a slotted crosshead, which is carried by the common piston rod and ram.

Scotch Turbine. A development of a *Barker's mill* in which water is admitted through a vertical supply pipe and flows outward through horizontal curved arms.

Scotch Yoke Mechanism. A crank pin moving in a straight slot in a sliding member, the member being constrained to move at right-angles to the axis of the slot, so that uniform speed of rotation of the crank results in a simple harmonic motion of the slider.

Scragged Springs. Fig. 154 shows the stresses in the cross-sections of the wires during the process of scragging.

Scragging Machine. (*a*) A machine adapted for testing springs by impulsive loading.

(*b*) A machine for compressing springs to their solid length before use to increase their service life.

Scrap. *Capital Scrap.* Scrap arising from obsolescence and the discarding of manufactured goods.

Circulating Scrap. Scrap arising in a steel works from ingot to saleable product.

Process Scrap. Scrap arising from engineering and manufacturing operations to produce goods.

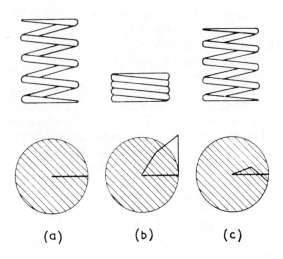

(a)   (b)   (c)

Fig. 154. Resultant Stress in Scragged Springs: (a) unstressed, (b) fully compressed; stress passes yield point, (c) resultant stress condition under no load.

Scraper Ring. An auxiliary piston ring usually fitted on the *skirt* to remove surplus oil from the cylinder walls of a piston-engine and thus reduce oil consumption, the oil being led back to the sump through holes in the piston wall.

Scratch Hardness. The hardness of a mineral determined by the ability of one solid to scratch or be scratched by another. See **Hardness Numbers.**

Another measure of scratch hardness consists of drawing a diamond stylus under a known load across the surface to be examined, the hardness being measured by the width and depth of the resulting scratch.

Screw. (a) A helix wound round a cylinder; an inclined plane wrapped round a cylinder. See **Screw Threads.**

(b) *Marine screw propeller.*

Screw Area. The area of a circle described by the blade tips of a propeller.

Screw Barrel. A *chain barrel* with a continuous spiral groove to receive the edges of alternate links in the groove with the other links lying flat on the periphery.

Screw Blade. A sector of a screw thread forming one of the blades of a marine propeller.

Screw Blade Area. The area of the oblique surfaces of the blades of a marine propeller.

Screw Chasing. See **Chaser.**

Screw Chuck. *Taper screw chuck.*

Screw Compressor. A positive displacement rotary *compressor*. The gas is progressively compressed as it is forced between two mating helical screws in an axial direction.

Screw Conveyor. *Worm conveyor.*

Screw Coupling. The coupling joining railway rolling stock using a double screw, right and left hand, by turning which the coupling is lengthened or shortened.

Screw Cutter's Gauge. *Centre gauge.*

Screw Cutting. The formation of screw threads on cylinders by *taps* and *dies*, by *chasers*, in a *screw-cutting lathe*, or by a *traversing mandrel*.

Screw-cutting Lathe. A lathe fitted with a *lead screw* and *change wheels* so that different rates of feed can be given to the slide rest relative to the rotation of the lathe mandrel. The lathe can cut screws or threads of different pitches.

Screw Die. *Die.*

Screw Feed. *Feed* effected through the medium of a screw.

Screw Gearing. (*a*) Gearing in which the teeth are not parallel with the axes of the shaft. See also **Spiral Gear.**

(*b*) The transmission of power by means of a worm and worm wheel.

Screw Heads. Fig. 155 shows a variety of types of screw heads.

Screw Jack. A *jack* consisting essentially of a vertical screw working in a nut and raised by the rotation of the nut by hand gear and a long lever, and provided with a ratchet.

Screw Machines. Machines used for turning and threading small screws, etc., from rod or bar fed through a hollow spindle.

Screw Mandrel. *Traversing mandrel.*

Screw Micrometer. *Micrometer gauge.*

Screw Pile. A pile with a wide screw or helix at the foot which is rotated to drive it into the ground.

Screw Pitch. The distance between corresponding points on adjacent thread forms, measured parallel to and on the same side of the axis.

The reciprocal of the number of threads per inch or per centimetre. See also **Cumulative Pitch.**

Screw Pitch Gauge. A small instrument containing a number of

Screw Pitch Gauge

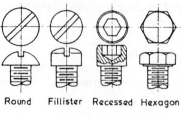

Round   Fillister   Recessed   Hexagon

FIG. 155. SCREW HEAD TYPES.

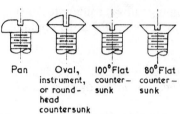

Pan   Oval,   100°Flat   80°Flat
       instrument,   counter-   counter-
       or round-   sunk   sunk
       head
       countersunk

gauges, mounted like blades on a pocket knife, for ascertaining the number of threads per centimetre of a screw.

Screw Plate. *Tap plate.* A hardened-steel plate used for cutting small screws, in which there are a number of screwing dies of different sizes.

Screw Press. *Fly press.*

Screw Propeller. *Propeller.*

Screw Rate. The number of threads per inch.

Screw Shackle. *Turnbuckle.*

Screw Stock. *Die-stock.*

Screw Surface. The surface formed by the development of a helix or of a propeller blade.

Screw Tap. *Tap.*

Screw Thread. The ridge on the surface of a cylinder or cone produced by forming a continuous helical or spiral groove of uniform section and such that the distance between two corresponding points on its contour measured parallel to the axis is proportional to their relative angular displacement about the axis. (The British Standards definition of a *perfect* screw thread.)

External threads are called *male* threads and internal threads are called *female* threads.

International Screw Thread. A metric system in which the pitch of the thread is related to the diameter, the thread having a rounded root and a flat crest.

312

Metric Screw Thread. A standard screw-thread in which the diameter and pitch are specified in millimetres.

Unified Screw Thread. An agreed system of screw-threads between Canada, the United Kingdom and the United States with an *included angle* of thread of 60°. The crest of the external (male) thread is rounded in the United Kingdom and flat in the United States, and the internal (female) has a cleared minor diameter to facilitate tapping. (See British Standard 1580.)

Multi-start Screw Thread. A thread formed by a combination of two or more helical grooves equally spaced along the axis.

Parallel Screw Thread. A thread formed on the surface of a cylinder.

Sellers' (USS) Screw Thread. The United States standard thread with a profile angle of 60° and a flat crest formed by cutting off one-eighth of the thread height.

Single-start Screw Thread. A thread formed by a single continuous helical groove.

Taper Screw Thread. A thread formed on the surface of a cone.

Screw Thread. See also **Screw Threads** and **Thread.**

Screw Thread Diameters.

Effective Diameter. Pitch diameter.

Virtual Effective Diameter. The effective diameter of an imaginary thread, with full depth of flanks but clear at the crests and roots, which would just assemble with the actual thread over a prescribed length of engagement.

Minor (Major) Diameter. The diameter of the *minor (major) cylinder* of a parallel thread or of the *minor (major) cone* of a taper thread normal to the axis.

Minor (Major) Cylinder (or Cone). An imaginary cylindrical (or conical) surface which just touches the roots (crests) of an external thread or the crests (roots) of an internal thread.

Screw Thread Milling. Cutting the threads of a *worm screw* with a revolving milling cutter.

Screw-threaded Gauge. An assemblage of thin steel plates whose edges are notched to fit screws of different pitches for the purpose of identifying the pitch of a thread.

Screw Threads. Screw threads are classified by their diameter, the number of threads N per inch or per centimetre and their form. Sometimes the pitch is given, which is the reciprocal of N.

A thread is *right-handed* if, when assembled with a stationary

313

Screw Threads

mating thread, it recedes from the observer when rotated clockwise. A *left-handed* thread recedes anti-clockwise.

The commonly-used threads in the United Kingdom are the Unified, the Whitworth, the British Association and the British Standard Cycle; in the U.S.A., the American Standard (Sellers'), the Society of Automotive Engineers (S.A.E.), the American Standard Coarse, the American Standard Fine and Unified. The oldest metric threads are the Delisle or Lowenhertz, Thury and the Système International (S.I.). Also, Model Engineers (M.E.).

The S.I. system as used today includes the French (CNM3), the German (DIN14) and the Swiss (VSM12004) for diameters over 80 mm. For diameters less than 6 mm, there are the national gauges: French (CNM132-133) and German (DIN13), the latter being the same as the Swiss (VSM12002).

Square and Acme threads are used mainly in worm wheels for power transmission. Buttress threads have special uses. See separate headings.

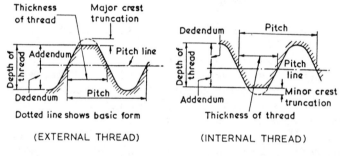

Dotted line shows basic form

(EXTERNAL THREAD) (INTERNAL THREAD)

Fig. 156. Screw Thread, Design Form.

Fig. 156 shows the design form of external and internal screw threads. See also **Thread**. Figs. 157, 158, 159 show the design forms of parallel and taper screw threads. See also **Threads, Power Transmission** and Fig. 183.

Screw Tool. A tool for cutting screws with a sectional shape the same as the interspace between two contiguous threads (cf. **Chaser**).

Screw Worm Chuck. *Taper screw chuck.*

Screw-and-nut Steering Gear. A square-threaded screw on the lower end of the steering column of an automobile engages a nut provided with trunnions, which work in blocks sliding in a short slotted arm carried by the drop-arm spindle.

314

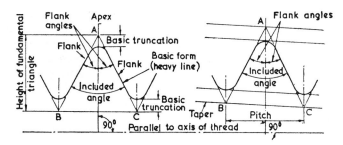

PARALLEL SCREW THREAD   TAPER SCREW THREAD

FIG. 157. SCREW THREAD FUNDAMENTAL TRIANGLE.

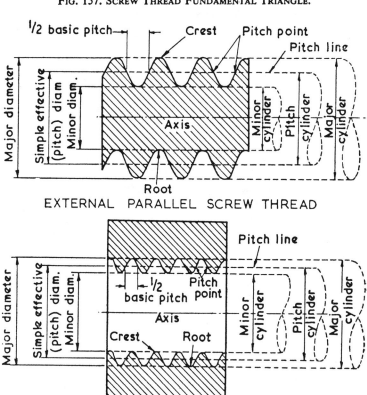

EXTERNAL PARALLEL SCREW THREAD

INTERNAL PARALLEL SCREW THREAD

FIG. 158. PARALLEL SCREW THREADS.

315

Screw-down (Stop Valve)

Screw-down (Stop) Valve. A valve in which the disc is raised or lowered by means of a screw on to the body seat by a stem perpendicular to the seat. The actuating thread on the *stem* may be within (inside screw) or exterior (outside screw) to the *bonnet*.

The following are screw-down *stop valves*: **Angle Valve, Globe Valve, Needle Valve, Oblique Valve, Tee-valve.**

Screwing Die. See **Die** (*a*).

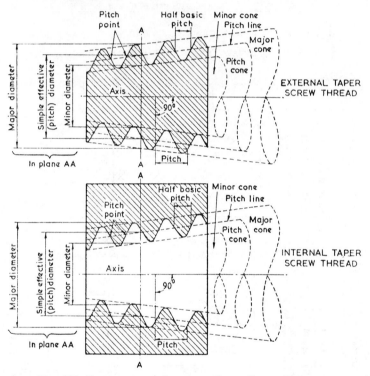

FIG. 159. EXTERNAL AND INTERNAL TAPER SCREW THREADS.

Screwing Machine. A lathe adapted for producing screws by means of *dies*.

Screwing Tackle. Appliances used for cutting screws apart from the aid of a lathe.

Scroll Chuck. A *self-centring chuck* with the jaws slotted to engage with a raised spiral on a plate which is rotated by a key to

316

advance the jaws while maintaining their concentricity. See Fig. 33.

Scroll Gear. A form of *variable gears* in which the pitch surface is in the form of a scroll, thus imparting a gradually increasing rate of motion to a shaft.

Scuffing Wear. Frictional wear due to a backward and forward rubbing between two surfaces.

Scutching. Beating to loosen fibres as of cotton and flax.

Sealing. Fig. 160 illustrates some types of sealing on shafts.

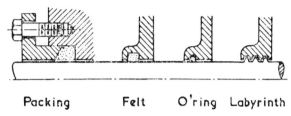

Packing   Felt   O'ring   Labyrinth

FIG. 160. SEALING ON SHAFTS.

Seat Face (of a **Valve**). *Body seat.* See **Body.**

Seating Ring. A spherical-surface ring outside a thrust bearing which renders it self-aligning.

Second Tap. A *tap* used after a *taper tap* to carry the full thread diameter further down the hole or to give the finished size of thread in a through hole.

Sector Gears. A form of toothed gearing with the wheels broken up into sectors of different curvatures, each pair of arcs transmitting different velocity ratios in an intermittent motion. See **Variable Gears.**

Segmental Gears (Mutilated Gears). Gears with teeth which are not continuous around the periphery.

Segmental Rack (Segmental Wheel). An arc of a toothed wheel imparting, through a wheel gearing into the rack, a reversible motion to a spindle.

Seismograph. An instrument in which a heavy mass is poised in such a way that a vibration of its support, together with the inertia of the mass, causes a relative motion of mass and support, that when amplified produces the record. The recording in older instruments is by a stylus on a rotating drum and in more modern instruments an electromagnetic current operates a mirror galvanometer to give a photographic trace. An observatory may have North-South and East-West horizontal instruments plus a vertical recorder.

Seizure or Seizing Up

Seizure or Seizing Up. The locking of two moving surfaces such as in a bearing due to the partial welding together of the two surfaces, caused by insufficient lubrication or insufficient clearance between the two surfaces.

Selector Forks. Forked members in an automobile gear-box with prongs that engage with grooves cut in the bosses of the gears, which they move along a splined shaft to change the gear. The forks are secured to a sliding rod which is operated by the gear-lever.

Self-acting Balance Crane. See **Crane.**

Self-acting Lathe. A lathe furnished with a slide rest whose movements are either partially or entirely self-acting; a lathe in which the tools are fed to the work by means of gearing actuated by the lathe itself, instead of being traversed by hand.

Self-aligning Ball-bearing. A *ball-bearing* with two rows of balls between an inner race and a spherical surface for an outer race, thus allowing considerable shaft deviation from the normal.

Self-centring. The automatic *centring* of a piece of work to be put into a lathe.

Self-centring Chuck (Centring Chuck, Universal Chuck). A lathe chuck for cylindrical work with the jaws maintained concentric by a scroll (see **Scroll Chuck**) or by radial screws driven by a ring gear that is operated by a key. (See Fig. 33.)

Self-correcting Mechanism for Chiming Clock. This mechanism is illustrated in Fig. 161.

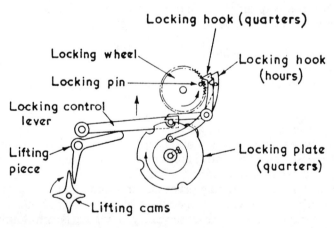

FIG. 161. SELF-CORRECTING MECHANISM FOR CHIMING CLOCK.

318

Self-induced (Self-excited) Vibration. See Vibration.

Self-opening Die. *Opening die.*

Self-winding Watch. A watch that winds itself whilst being worn by an action similar to that of a *pedometer*, or is wound by the opening and shutting of the case.

Sellers' Screw Threads. See Threads.

Semi-elliptic Spring. A *carriage spring* consisting of a pair of curved steel strips, one inverted, attached to each other at the ends, the arrangement resembling an ellipse. Cf. **Leaf Spring.**

Semi-finishing Teeth. The central teeth of a *broach.*

Semi-rotary Pump. The alternate action of a semi-rotary pump is illustrated in Fig. 162. Cf. **Rotary Pump.**

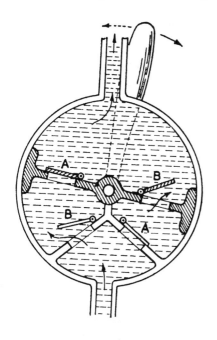

FIG. 162. SEMI-ROTARY PUMP.

Sensitive Drill. See Drilling Machine.

Servo (Servo Mechanism). A device for converting a small movement into one of greater amplitude or to exert a greater force. See also **Servo Mechanisms.**

Servo Control (in an aircraft). An additional mechanism devised to reinforce a pilot's effort by a relay.

319

Servo Mechanisms

Servo Mechanisms. Servo mechanisms are of two types, with open loop or closed loop controlling systems. In the open-loop system there is nothing in the mechanism to measure the result of the application and errors cannot be rectified. In the closed-loop system, the results of the operation are fed back into the control circuits so that they can rectify errors. In guided missiles, the servo-loop is closed by controlling impulses coming from the ground control and by error-sensing devices within the missile. See also **Servo**.

Servomotor. (a) A motor, linear or rotary, which receives the output from an *amplifier* element and drives the load, being the final control element in a servo mechanism. An example of a linear servomotor is a hydraulic ram.

(b) A device for magnifying a small effort by using hydraulic means. See also **Engine (Servomotor Types)**.

Servo System. A closed cycle automatic control system, designed so that the output follows closely the input; it usually includes power amplification and is capable of following rapid variations of input. See also **Servo Mechanisms**.

Servo Tab. A small hinged surface of an aeroplane control or flap, which is directly operated by a pilot to produce aerodynamic forces which, in turn, move the control surface (or flap).

Servomotor Type Engines. See **Engines**.

Servovalve. An hydraulic valve which enables a large flow of hydraulic fluid to be switched, by a very small initial force, from one part of a mechanism to another.

Set. The (amount of) alternate deflection of saw-teeth. (See Fig. 153.)

Set Screw (Set Bolt). A screw with a pointed, square or other shaped end used for tightening purposes, such as to secure a pulley on a shaft (Fig. 81). See also **Adjusting Screw, Binding Screw**.

Setting. The exact adjustment of a piece of mechanism such as an engine valve or the zero of an instrument.

Setting Gauge. A gauge for checking the setting of an adjustable *workshop gauge* or an *inspection gauge* or a *comparator*.

Setting-out. *Marking-out*.

Setting Over. Placing out of line the *mandrel* centre of a lathe poppet to make room for the turning of taper work between centres.

Sewing Machine. A machine for sewing fabrics with a mechanism operating a needle bar, reciprocally vertically, having the thread for the pointed-eye needle supplied from a spool on the frame, plus usually a *lock stitch* mechanism.

Shackle. (*a*) A loop to which the end of a chain is attached.

(*b*) A link or metal loop closed by a ' shackle bolt ' for connecting chains or for certain parts of the tackle in sailing boats.

Shaft. A spindle revolving in bearings and carrying pulleys, gear-wheels, etc., for the transmission of power. See also **Sealing** and Fig. 160.

Shaft Angle. (*a*) When the axes intersect as in bevel gears, the angle between these axes within which the *pitch point* lies: it is equal to the sum of the *pitch angles* of the wheel and pinion.

(*b*) When the axes do not intersect as in helical, spur and worm gears, the angle through which the axes of the nearer of the two shafts must be turned about the common perpendicular in order to bring the axes parallel with opposite directions of rotation; it is assumed that the shafts continue to rotate in their original directions. When the angle of turn is counter-clockwise, the shaft angle is positive.

Shaft Basis Limit System. See **Limit System.**

Shaft Coupling. See **Coupling** (*b*).

Shaft Governor. A *spring-loaded governor* mounted on, and rotating about, the engine crankshaft for controlling the speed of small oil engines. It is sometimes housed in the flywheel. Shaft governors are also used to control the travel of the slide valves in steam-engines.

Shaft Straightener. A machine with three rollers adjustable for uniform pressure through which a shaft is passed on to two larger rollers.

Shaft Turbine. A turbine designed for transmitting power through a shaft, such as a *free turbine* to drive a helicopter rotor or a propeller.

Shafting. *Line shafting.*

Shake. Side play in a clock or watch movement between a pivot and its hole. See also **End Shake.**

Shank. (*a*) The stem of a cutting tool, drill, broach, tap, reamer, etc., which is held in the driving member. (See Figs. 17 and 114.)

(*b*) The shaft of a tool connecting the head and the handle.

Shaper Rail. *Copping rail.*

Shaping Machine. A reciprocating ram which carries a tool horizontally in guides (Fig. 163) or vertically in a *clapper box* and is driven by a *quick return* mechanism, for producing small flat

Shaping Machine

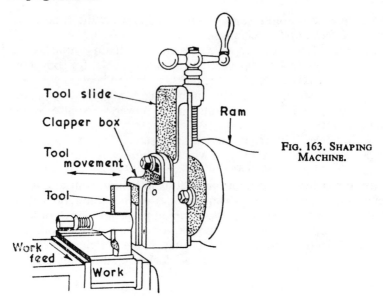

Tool slide

Clapper box

Tool movement

Tool

Work feed

Work

Ram

Fig. 163. Shaping Machine.

surfaces, slots, etc. Cf. **Planer, Slotting Machine.** See also **Table Feed, Whitworth Quick-return Motion.**

Shaving. A finishing operation in which a very small amount of metal is removed from a blanked part. See **Blanking.**

Shear. A deformation in which parallel planes in a body remain parallel, but are displaced in a direction parallel to themselves.

Shear Face. The plane of the material which is subjected to shear from external forces.

Shearing. A method of cutting metals in sheet or plate form between two straight blades without the formation of chips. Cf. **Blanking, Piercing.**

Shears. *Ways.*

Sheave. (*a*) A pulley with a grooved rim over which a rope, belt or chain passes. Sheaves are placed singly or side by side in a wooden or metal frame, which is called a *block*. Adjustable-diameter sheaves, for varying the diameter of the pulley and consequently the speed ratio of the drive, are used in small vee-belt drives. Also called a ' sheave wheel '.

(*b*) The disc of an eccentric in a steam-engine, such as that which actuates the slide valves.

(*c*) That part of an *eccentric* which is keyed directly on a shaft.

Sheave Wheel. See **Sheave** (*a*).

Shed. The horizontal opening for the passage of a shuttle carrying the weft between the warp threads in a *loom*.

Sheet Mill. A *rolling mill* for the rolling of sheet metal.

Shield. A contrivance or covering, protective plate, or screen, to protect machinery or the operator, from damage or accident.

Shims. Thin sheets of paper or slips of metal, used as spacers to regulate the distances between objects, or as wedges to make parts of it. The use of shims is illustrated in Fig. 134.

Shimmy. The violent oscillation of a castoring wheel of an aeroplane's undercarriage, or of a vehicle, about its castor axis, which can be suppressed by a damping device called a ' shimmy damper ' or sometimes by redesign of the tyre tread to increase the resistance to lateral movement. The phenomenon occurs when the coefficient of friction of the surface exceeds a critical value. In the case of cars it is called ' wheel wobble '.

Shingling. Hammering or rolling puddled ball to convert it into bar or sheet iron.

Shock (Motion). (*a*) A motion of a *transient* nature caused by a very rapid change of displacement, velocity, acceleration or force.

Shock Spectrum. The spectrum of an ' applied shock ' defined as an excitation of a sudden character and of significant magnitude.

(*b*) The sudden application of a load to a member, or a sudden change of pressure, etc.

Shock Absorber (Shock Damper). (*a*) An energy-absorbing device, such as a frictional spring, rubber or hydraulic damper, in any part of a mechanism or moving vehicle.

(*b*) A bumper on a motor-vehicle.

(*c*) An ' oleo leg ' in an aircraft's undercarriage where air is compressed by the flow of oil through valves with a restrictor valve to prevent rapid return flow on the rebound. See also **Damper**.

Shock Mounting. See **Mounting**.

Shockproof Watch. A watch with a flexible mounting for the *balance staff* to avoid damage to its pivots.

Shoe. See **Brake Shoes**.

Shoot. See **Weft**.

Shop Traveller. *Overhead travelling crane.*

Shore Scleroscope. A machine in which a small indenter is dropped from a specified height on to the specimen and the height of rebound, read on a graduated scale, is a measure of the dynamic hardness. See **Hardness Tests**.

Short Link Chain. *Close link chain.*

Shot. See Weft.

Shot Peening. Blasting the surface of a metal with small hard steel balls driven by an air blast to harden the surface layers. Cf. **Sand Blasting.**

Shoulder. That portion of a shaft or of a flanged structure where a sharp increase of diameter or other dimension occurs.

Shoulder Bolt (Stripper Bolt). A bolt in which the shank is hardened and ground to provide a good fit in a bearing or bush.

Shrinkage Allowance. The difference in diameter between two parts when both are cold, to be united by *shrinking on.*

Shrinking on. Fastening together two parts by heating the outer member until it expands sufficiently to pass over the inner, and on cooling grips it tightly; a process used for attaching steel tyres to locomotive wheels.

Shroud. (*a*) Circular webs for stiffening the sides of the gear-teeth on a gear-wheel; called a ' shrouded wheel '.

(*b*) A peripheral strip to strengthen turbine blading.

(*c*) A peripheral ring fitted to prevent the escape of gas past the tips of the blades in an axial-flow turbine; sometimes called a ' turbine shroud ring '. See also **Shroud Plate.**

(*d*) A deflecting vane on one side of an inlet port of some piston-engines to promote swirl in the cylinder.

(*e*) The circular flanges around the edges of a water wheel between which the buckets extend; the wheel is called a ' shrouded wheel '.

Full Shroud. A *gear-wheel* which has shrouding extending to the tips of the teeth.

Shroud Plate. A side plate of a water-wheel bucket.

Shroud Ring. See **Shroud.**

Shrouded Wheel. See **Shroud.**

Shutter. A mechanical device in a camera or cinema camera for exposing photographic film for a known time and at will, using a roller blind with a slot, a rotating disc with a slot, an *iris diaphragm* or a circular eyelid device.

Shuttle. (*a*) A weaving device shaped like a cigar with two pointed ends by which the weft thread is shot across between the warp threads.

(*b*) The sliding or rotating device that carries the lower thread in a *lock-stitch* sewing machine.

Shuttle Box. A box-like extension at each side of a *loom* from which the *shuttle* is thrown, to and fro, when a loom is working.

Shuttle Guard. A guard fixed to the *sley* of a loom to deflect or keep low a shuttle which accidentally flies out of the loom.

Shuttle Machine. *Swiss machine.*

Shuttle Race. See **Race Board.**

Side Dresser. A saw tool carrying a pair of dies which are squeezed one on either side of the tooth with a means of adjustment to locate the same spot on each tooth and to ensure that each tooth is set an equal amount on each side of the saw blade.

Side Frames. The main frames carrying the bearings for the shafts of engines, cranes, pumps, etc.

Side Hooks. The coupling hooks of railway wagons placed at the side of the *draw hook.*

Side Rail. *Check rail.*

Side Rake. See **Rake.**

Side Tank. See **Tank Engine.**

Side Tool. A tool, with the cutting face at the side, which is fed laterally along the work.

Side Valves. *Poppet valves* mounted in the side wall of an engine cylinder block. Cf. **Overhead Valves.**

Sieving (Sifting). The operation of shaking loose materials in a sieve so that the smaller particles pass through the mesh. Cf. **Agitators.**

Sight-feed Lubricator. (*a*) A *drip-feed lubricator* with a small glass tube supplying oil drops from a reservoir so that the drops can be seen.

(*b*) A small glass tube filled with water so that oil from a pump rises in visible drops en route to the oil pipe.

Silent Chain. See **Chain.**

Simple Harmonic Motion (S.H.M.). A motion represented by projecting on to a diameter the uniform motion of a point round a circle.

Simple Steam-engine. A *steam-engine* in which the steam expands from the initial to the exhaust pressure in a single stage.

Sine Bar. A tool used for the accurate setting out of angles by arranging to convert angular measurements to linear ones, thus depending on the accuracy of the bar length and the raising of one end by gauge blocks (Fig. 164).

Sine Wave. A wave in which the particles (or points) move in transverse vibrations of a simple harmonic type. See **Simple Harmonic Motion.**

Single-acting Engine. A reciprocating engine in which the

Single-acting Engine

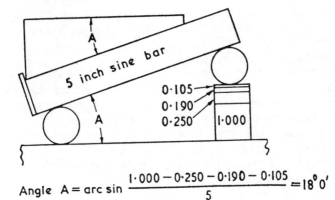

$$\text{Angle} \quad A = \text{arc sin} \; \frac{1\cdot000 - 0\cdot250 - 0\cdot190 - 0\cdot105}{5} = 18°0'$$

FIG. 164. SINE BAR.

working fluid acts on one side only of the piston, as in most piston-engines, in an early form of steam-engine and in steam-hammers.

Single-acting Piston. A piston in contact with the working fluid on one side only.

Single-acting Pump. A pump delivering liquid on alternate strokes only.

Single-beat Escapement. An *escapement* in which the balance receives impulses on alternate vibrations, as in chronometer and duplex *escapements*.

Single Curve Teeth. *Involute gear-teeth.*

Single-cylinder Engine. (*a*) An engine with only one steam cylinder, plus a *flywheel* to carry it over dead centres.

(*b*) An internal-combustion engine with only one cylinder, used for research purposes, for motor bicycles and for ' bubble ' cars.

Single-cylinder Machine (Yankee Machine). A machine in which wet paper is pressed on a polished heated cylinder and dried during one revolution to impart a high glaze on one side. It is a common process in the production of high-class photographic prints.

Single-entry Compressor. See **Centrifugal Compressor.**

Single Gear. A gear consisting of one pinion and one wheel only in combination.

Single-plate Clutch. Fig. 165 shows a single-plate dry clutch operated through lever *E* which pushes the collar and disc *B* forward with the assistance of springs *D*. This brings the clutch plate *F* along the splines *G, H* to complete the contact between parts *A* and

326

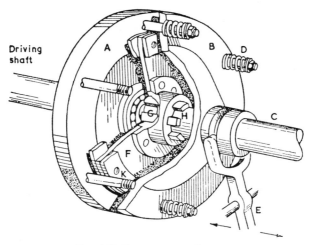

FIG. 165. SINGLE-PLATE CLUTCH.

B via the clutch lining material *K*. The driving shaft is then transmitting its torque to shaft *C*.

Single-ported Slide Valve. See **Slide Valve.**

Single Purchase. A lifting arrangement consisting of only a *single gear* or pulley. Cf. **Double Purchase.**

Single-start Thread. A *screw thread* with a single helical crest winding around its body. Cf. **Multi-start Thread.**

Single-threaded Screw. A screw with a *single-start thread.*

Sink. The hemispherical depression around a pivot hole in a watch or clock for holding the lubricating oil.

Sinker. A mechanism in a *knitting machine* for pushing a length of thread over the spring needles so as to form a new course of loops.

Sinking Engines. Engines used for excavating the shafts of mines.

Sinking Pump. A pump consisting of two suction barrels with the pistons worked alternately with a bob lever, by hand or power, to clear water from foundations, etc.

Sinusoidal. The trace of an alternating quantity plotted to a time base in the form of a sine wave; hence ' sinusoidally '.

Size. A generic term denoting magnitude, especially geometrical magnitudes but also including weights, capacities, horse-powers, ratings, etc.

Actual Size. The measured dimension of a part assumed correct at 68° F (20° C).

Basic Size. (q.v.)

Design Size. That size which defines the design requirements for a particular dimension. See also **Limits of Tolerance.**

Limits of Size. (q.v,)

Nominal Size. A convenient size for specifying an object.

Skate. *Retarder.*

Skew Bevels (Skew Bevel Gears). *Bevel gears* whose axes are not in the same plane.

Skew Lines. Two straight lines which are neither parallel nor intersecting: they have one common perpendicular.

Ski Lift. An *aerial cable railway* up the side of a mountain for transporting skiers to the tops of snow slopes. They can be of the cable car, chair or ski-tow types.

Skip (Skep). (*a*) A bucket on a crane or cableway used for the transport of material.

(*b*) A guided metal box for hoisting coal or mineral up a shaft. Cf. **Kibble.**

Skip Tooth. See **Saw-tooth Profiles.**

Slabbing. Drawing down steel ingots into slabs using a steam-hammer.

Slabbing Machines. Large *milling machines* (*planer* type) which take wide and deep cuts off heavy work.

Slack Side. The lower side of a driving belt.

Slackness. *Play.*

Slat Conveyor. *Apron conveyor.*

Sleepers. Beams of wood, reinforced concrete, etc., passing across and beneath *rails* which they support and prevent from spreading apart. Also called ' cross-tie ', ' cross-sill '.

Sleeve. A tubular piece, usually machined externally and internally, into which a rod, another tube or a piston is inserted.

Sleeve Bearing. A bearing which is long in proportion to its diameter.

Sleeve Coupling. *Box coupling,*

Sleeve, Follow-up. A sleeve in a *servo* consisting of a valve, sleeve and controlled member, the valve working in the sleeve and the controlled member causing the sleeve to follow the valve or vice versa.

Sleeve Valve. A thin steel sleeve fitted within the cylinder of a petrol or oil engine with a reciprocating rotary motion so that ports cut in it register alternately with inlet and exhaust ports in the cylinder wall. See also **Double Sleeve Valve.** Fig. 55.

Sleeve Wheel. A wheel with a long hollow boss which fits over and slides along a shaft.

Slewing. Circular movement of a *crane* or other revolving machine.

Slewing Cylinders. *Turning cylinders.*

Slewing Gear. The gear for *slewing* a *travelling crane*, normally consisting of friction cones which actuate a vertical shaft; this last revolves a pinion in a curb ring on the crane's truck.

Sley (Slay). (*a*) (Noun) the *reed* which guides the warp threads in a loom. (Verb) to pass the warp threads through a reed for dressing or weaving.

(*b*) A guideway in a knitting machine.

(*c*) That part of a lace machine between the beams and the thread guides which keep the threads properly arranged.

Sley Sword. One of two metal arms on a rocking shaft which support the *sley* in a *loom*.

Slide. (*a*) A piece of a mechanism which moves in a linear direction over a flat or curved smooth face between guides.

(*b*) The operating piece on the outside of a *repeater*. Cf. **Push Piece.**

Slide Bars. *Guide bars.*

Slide Blocks (Guide Blocks, Slipper Blocks). Blocks, attached to an engine or pump *crosshead* and moving between *guide bars*, which ensure that the path of the *piston rod* is truly rectilinear.

Slide Rest Lathe. A slotted table which carries the tool-post of a lathe, is mounted on the carriage and is capable of longitudinal and cross-traverse. (See also **Rest.**) A 'compound slide rest' has sliding, surfacing and swivelling motions.

Slide Rod (Slide Valve Spindle). The rod of a *slide valve* connecting the eccentric to the valve, with or without a slot link.

Slide Valve. (*a*) A valve which slides in contrast to one that rotates or lifts, referring usually to the valves of steam- and gas-engines.

(*b*) A steam-engine inlet and exhaust valve shaped like a rectangular lid, alternately admitting steam to the cylinder and connecting the ports to exhaust through the valve cavity. See **D Slide Valve, Parallel Slide Valve, Piston Valve.**

The simplest slide valve has a single port and its travel is the same as the throw of an eccentric. Hence 'single-ported slide valve'.

Slide-valve Spindle. *Slide rod.*

Slider-crank Chain. A chain consisting of three turning pairs and one sliding pair. The cylinder and crank combination is the most common.

Sliding. The motion of a *slide rest* on its bed.

Sliding Friction. See Friction.

Sliding-head Automatic Lathe. A lathe with a sliding *headstock* the cutting tools being held radially in a tool bracket fixed to the *lathe bed* each tool being brought into operation in the correct preset order by means of *cams* and *levers*.

Sliding Mesh Gear-box. A gear-box in which the gear is changed by sliding one pair of wheels out of engagement and sliding another pair in.

Sliding Pulley. A guide pulley free to move along its shaft.

Slip. *Belt slip.*

Slip (Screw Propeller). The amount by which the product of the *pitch* and rad/s exceeds the actual forward velocity, expressed as a percentage of the former. (See Fig. 129.)

Slip (Pump). The difference between the volume swept through by the plunger of a reciprocating pump and the actual discharge. If the discharge is greater due to pressure rise in the cylinder, the slip is negative.

Slip Gauge. A very accurate gauge block, a set including lengths up to 10 cm. Three or four slip gauges can be 'rung' together to give a combined thickness accurate to $0.25 \mu m$ and then used as a 'go' or 'no-go' gauge. See also **Block Gauge.**

Slip Jaws. Temporary facings on the jaws of lathe chucks or machine vices.

Slip Table. A slip table consists of a slab of slate, marble or light alloy maintained in position by hydrostatic suspension, the oil under high pressure leaking out around the periphery of the slab into a trough for return to the hydraulic pump. It is used in conjunction with a vibrator enabling vibration of heavy objects in the horizontal plane. The vibration test specimen is rigidly attached to the slab which is in turn affixed to the driving shaft of the vibrator. A slip table is required when the object under test is too heavy to be hung up on a resilient chord.

Slipper (Slipper Block). *Slide block.*

Slipper Brake. (*a*) An electromechanical brake acting directly on the rails of a tramway.

(*b*) Iron blocks on a truck, which with screws and levers can be forced down on to the rails.

Slipper Guides. *Guide bars.*

Slipper Piston. A light piston with its skirt cut away between the thrust faces, thus saving weight and reducing friction.

Slipper. *Retarder.*

Slip-winding Engine. A machine for winding yarn from a hank to flanged *bobbins* in lace manufacture, using drums and *swifts*.

Slit Bar Motion. A name for the mechanism of a rod or link with an adjustable throw, depending upon the limits of the slits in the bar and the position of a tightening stud.

Sliver. A continuous strand of fibres formed after carding. See **Carding Engine** and **Draw Box**.

Slot Drill. A flat-ended double-cutting or ball-nosed drill with no centre point which is traversed as it operates. Cf. **End Mill**.

Slot Link. See **Link Motion**.

Slot Machine. A retail dispensing or amusement device in which a mechanism operates upon the insertion of a coin. Cf. **Slotting Machine**.

Slotted Crosshead. A *crosshead* slotted to receive a sliding block which is attached to the end of the connecting-rod of a steam pump thus combining rectilinear with rotational motion. See **Scotch Yoke**.

Slotting Machine (Vertical Shaping Machine). A machine tool, having a ram with a vertical motion balanced by a counterweight and cutting on the downstroke towards the table. It is similar to a *shaping machine*. Cf. **Slot Machine**.

Slotting Tools. Tools used for cutting *keyways*, etc., in a *slotting machine*, having a narrow edge and deep, stiff section, with top and side clearance but little rake.

Slow Gear. (*a*) The double or multiple purchase gear of cranes and hoisting machinery. Cf. **Quick Gear**.

(*b*) The back gear of lathes, drilling and other machines giving an increase of power with slower movement. See **Back Gear**.

Slug. (*a*) A unit of mass in the foot-pound system of units. See **Foot-pound**.

(*b*) A solid line of type as cast by the *linotype* process. See **Composing Machine**.

Sluice. (*a*) A water channel equipped with means for controlling the flow by a *sluice gate*.

(*b*) The opening through which the water is admitted to a water wheel or turbine.

Sluice Gate. A barrier plate free to slide vertically across a water channel or an opening in a lock gate. Cf. **Sluice Valve**.

Sluice Valve. A solid *wedge gate valve* used for waterworks purposes.

Snail. A cam used in a clock for the gradual lifting (on a smooth curve), or the discharge (by a step), of a lever to control the striking mechanism. The size of the step controls the number of teeth picked up on a rack and hence the number of strikes on a gong. An ' hour snail ' has twelve steps of gradually increasing height on the one wheel and a 'quarter snail' has four steps and is mounted on the minute wheel (Fig. 166). A *repeater* has similar snails on a

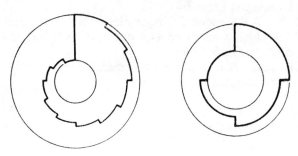

FIG. 166. HOUR AND QUARTER SNAILS.

smaller scale, but the 'quarter snail' is fitted directly on to the *cannon pinion* arbor on which it moves freely with a *surprise piece* which is screwed to it. A ' minute snail ' in a minute repeater is fitted or riveted on to the cannon pinion and has four curved sections each with fourteen steps.

Snailed. (*a*) A surface finished with eccentric curves. Cf. **Snailing.**

(*b*) A *barrel arbor* shaped so that the inner coil of the mainspring passes over the hook without forming a kink.

Snailing. Putting a very high polish on a wheel or arbor for a watch or clock.

Snap. (*a*) A *punch* with a hemispherically recessed end, used to form rivet heads.

(*b*) To hold parts together by springing one part over or into another.

Snap Gauge. An adjustable go- no-go gauge with the body in the form of a C-frame casting. One leg has a fixed anvil and the other has two adjustable anvils to set the limits, the front as the go portion and the back anvil has the no-go portion. Cf. **Gap Gauge.**

Snatch Block (Return Block). The suspended block, in lifting tackle, which contains the pulleys resting in the lower bight of the chain.

Snift Valves. Valves which open for the passage of air but close if liquids attempt to pass.

Snifting Valve. A *relief valve* to release fluid or gas when a certain pressure is exceeded.

Snout Boring Machine. A boring machine with the cutters carried on an unsupported extension of the spindle, used for boring engine cylinders.

Snubber (Snub). A device for limiting the movement of a system and not coming into action until the displacement of the system reaches a predetermined amount; for example, the checking of a cable or rope suddenly when running out.

Socket. A hollow for something to fit into, or stand firm, or revolve in.

Solar Gear. An *epicyclic gear* with a rotating annulus, a fixed sun wheel, a rotating planet carrier and planet wheels rotating about their own spindles. Cf. **Planetary Gear** and **Star Gear.**

Sole Plate (Sole). The *bedplate* of a marine engine.

Solid Coupling. A coupling forged in a solid piece with its shaft, as in a propeller shaft.

Solid Head. A cylinder or cylinder block cast in one piece to distinguish it from the detachable head type.

Spacing. (*a*) The average distance between prominent irregularities in a profile.

(*b*) The dividing and setting out of holes, etc., in a piece of work.

Sparking-plug (Spark). A plug screwed into the cylinder head of a petrol engine to obtain ignition of the fuel-air mixture in the cylinder by a spark discharge between the insulated central electrode and one or more earthed points.

Specific Speed. (*a*) The speed at which a water turbine will run to produce one horse-power under a head of one foot.

(*b*) The speed at which a pump will deliver one gallon of water per minute under a head of one foot.

Speed. The ratio of the distance, in a straight line or in a continuous curve, covered by a moving body to the time taken. Rotational speed is measured by the number of complete revolutions in a given time.

Speed Cones. *Speed pulleys.*

Speed Indicator. See **Speedometer, Tachometer.**

Speed Governing. See **Governors.**

Speed of Rotation. See **Speed.**

Speed Pulleys (Speed Cones). Pairs of pulleys with steps of different diameters, mounted head to tail on two shafts so that by shifting the belts connecting them, different speed ratios are obtainable.

Speedometer. A *tachometer* fitted to the gear-box or propeller shaft of a motor or other road vehicle and usually graduated to indicate the speed in miles per hour. The centrifugal type is specially described (see **Speedometer, Centrifugal**) and reference is made there to other mechanical types; in addition, there are electrical air-vane types and magnetic ' drag cup ' types.

Speedometer, Centrifugal. Mechanical speedometers usually employ one of three principles, namely, centrifugal, chronometric or magnetic. Fig. 167 shows the first, the second has a clockwork

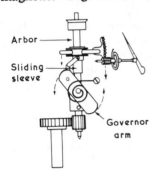

Arbor

Sliding
sleeve

Governor
arm

Fig. 167. Centrifugal Speedometer.

movement to determine the quotient of distance divided by time and the third depends on a rotating magnetic field in an air gap. In the illustration the weights are thrown outwards on the *governor* principle, the arm of which is hinged to a rotating arbor. The arm is linked to a sliding sleeve on the arbor and through this the position of the weights controls the instrument pointer.

Speeds. (*a*) The number of steps of a cone, driving pulley or drum.

(*b*) The rotation speeds of a lathe chuck or machine tool. Cf. **Feed.**

Spherical Roller-bearing. See **Roller-bearing** and Fig. 147.

Spherometer. An instrument with a central micrometer screw for measuring the height of convex or concave surfaces above or below a zero mark.

Spider. (*a*) **Cathead.**

(*b*) A spoked central member, fitting on a propeller shaft which transmits the drive to the blades mounted one on each spoke. The pitch of the blade can be altered by rotation on the spoke.

Spigot Bearing. A bearing carrying two shafts in line while allowing them to turn independently.

Spin Drier. A rapidly rotating cylinder for drying wet clothes by centrifugal force. Cf. **Tumbler Drier, Whizzer.**

Spindle. (*a*) A slender metal rod or pin on which something turns.

(*b*) The long slender rod in spinning wheels by which thread is twisted and on which, when twisted, it is wound.

(*c*) That component in a screw-down stop valve on which the actuating thread is formed and by which control of the *disc* is effected.

Spindle Valve. *Screw-down stop valve.*

Spindle Wheel. The toothed wheel on the *leading screw* of a screw-cutting lathe.

Spindles. *Breaking pieces.*

Spinner. A streamline fairing, enclosing the hub or boss of a propeller, fitted coaxially and rotating with the propeller. Cf. **Nose Cap.**

Spinning. (*a*) The twisting together of short fibres to form a continuous length of thread.

(*b*) The moulding of circular articles in thin sheet metal by pressure applied during rotation in a lathe. (See Fig. 168.)

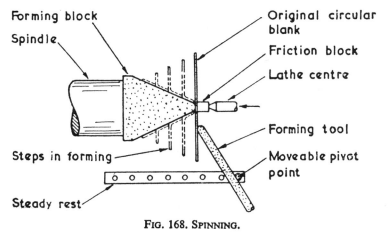

Forming block
Spindle
Original circular blank
Friction block
Lathe centre
Forming tool
Steps in forming
Moveable pivot point
Steady rest

FIG. 168. SPINNING.

Spinning-jenny. A mechanism for spinning more than one strand at a time.

Spinning Machine. A machine that spins fibres continuously.

Spinning Wheel. A household instrument for spinning yarn or thread with a flywheel driven by crank or treadle.

Spiral. A curve whose radius vector increases or decreases progressively with the polar angle. The curve need not be in one plane.

Spiral Angle. The angle between the *pitch cone* generator of a bevel gear and the tangent to the *tooth trace* at that point. The angle is positive for a right-hand gear.

Spiral Bevel Crown Gear. A gear whose *pitch curves* are inclined to the *pitch element* at the *spiral angle* and are usually circular. Cf. **Straight Bevel Crown Gear.**

Spiral Gear. *Crossed helical gear.* See **Bevel Gear, Helical Gear, Spiral Bevel Crown Gear.**

Spiral Spring. A spring formed by coiling a steel ribbon into a helix of increasing diameter, which, when compressed flat, forms a true spiral. Cf. **Helical Spring.**

Spiral Winged Valve. A lift valve with wings arranged as sections of a spiral of very long pitch so that the valve turns round on its seating to a slight degree at each lift thus ensuring uniform wear.

Spiroid* Gears. A type of *worm gear* which is conical in shape with the mating member a face-type gear. A large number of teeth are in simultaneous contact, very high ratios can be accomplished in a single step, and these gears can be manufactured by standard hobbing machines. [* Trade name.]

Splashers (Guard Straps). The sheet-iron strips which arch over the tops of locomotive wheels as a protection against injury to the drivers.

Splice Piece. *Fish-plate.*

Splined Shaft. A shaft provided with several long *feather* ways. See also **Splines.**

Splines. Relatively narrow keys integral with a shaft and resembling long gear-teeth. (See Fig. 98.) Cf. **Feather.**

External Splines. Splines produced by milling longitudinal grooves in a shaft. (Fig. 169.)

Internal Splines. Grooves formed in a hole into which a splined shaft fits.

Split Bearing (Divided Bearing). A shaft bearing with split housing and the bearing bush (or brasses) clamped between the two parts. See also **Bearings.**

Reduced diameter
for bearing

Splines

FIG. 169. SPLINED SHAFT.

Split Compressor. See **Spool**.

Split Crankcase. A common type of engine crankcase split horizontally at about the centre line of the crankshaft. Cf. **Barrel-type Crankcase**.

Split Gear (Split-and-sprung Gear). A fixed and a loose gear-wheel mounted alongside on the same shaft and coupled together by springs, pre-loaded so that backlash is eliminated when both gears mesh with a common rack, gear-wheel or pinion. Sometimes called ' anti-backlash gear '. See under **Gears**.

Split Key. See **Key**.

Split-pin. See **Cotter-pin**.

Split Pulley. A belt pulley made in halves which are bolted together to grip the shaft.

Split Ring. A ring divided either diagonally or with a lapped joint, such as a piston ring.

Split Ring Clutch. See **Clutch**.

Split Wheel. A split *cog-wheel*, split for a use similar to a *split pulley*.

Split-and-sprung Nut. A device for eliminating backlash between a lead-screw and a nut, by mounting a second nut alongside the first and keying it to the first to prevent relative rotation, but sprung apart from it axially so as to take up backlash.

Split-seconds Chronograph. See **Chronograph**.

Spoke. (*a*) A radial bar of a wheel or of a steering-wheel of a vehicle.

(*b*) A rung of a ladder.

(*c*) An arm of a running wheel of a locomotive or rolling stock.

Spoke Machine. A copying machine or lathe with two sets of centres having a *template* in the one and the piece of work in the other, a roller moving on the template and a cutter following, being set in the same slide rest. Cf. **Routing Machine**.

Spool. (*a*) A gas-turbine engine which has two compressors driven separately by two turbines is called a two-spool jet-engine: hence the terms single-spool compressor and two-spool compressor (cf. **Compound Engine**). The highest practicable compression ratio of a single-shaft engine is about 7 to 1, but a two-spool engine may be more than 9 to 1.　See **Axial Compressor, Compression Ratio.**

(*b*) A *pirn* or reel for winding yarn, etc.　See also **Bobbin.**

(*c*) A reel for photographic film.

Spreadboard.　An endless belt upon which handfuls of sorted flax fibre are laid for conversion into *sliver.*

Spreader.　In a *double disc gate valve* the ' upper spreader ' is the component attached to, or engaging, the actual thread of the stem and the ' lower spreader ' is its complementary component. Together in conjunction with the stop in the body, they constitute the spreading mechanism to force the discs apart against the body seats when the valve is closed.

Spring.　A piece of bent or coiled metal with elastic properties. See below and **Carriage Spring, Helical Spring, Relay Spring, Scragged Springs, Spiral Spring.**

Spring Balance, Circular.　A weighing machine dependent on the extension of a helical wire spring the extension being proportional to the load, provided the wire is not stretched beyond its elastic limit.　In circular spring balances the springs are used in one or more pairs, as indicated in Fig. 170.　The extension of the springs moves the rack and thus actuates the pinion carrying the pointer. A small spring presses the rack into contact with the pinion.

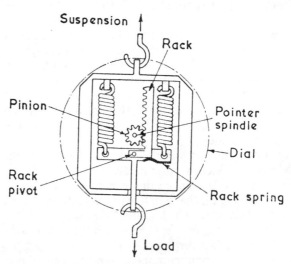

FIG. 170. SPRING BALANCE.

Spring Balance Valve. A form of safety valve in which the lever is attached to the end of a spring balance.

Spring Box. A hollow box containing a *helical spring* which is thus constrained to exert a force along one axis only.

Spring Chuck. A wooden *chuck* with a hold on the work by a cone with slits, the closing of which is controlled by a sliding ring.

Spring-detent Escapement. See **Escapements.**

Spring Hammer. A power hammer in which the *tup* is suspended from laminated springs to lessen the jar.

Spring-loaded Governor. An engine *governor* in which the rotating masses, which move outwards under centrifugal force, are controlled by a spring. See **Hartnell Governor** and Fig. 83.

Spring Needle (Bearded Needle). A knitting *needle* for producing a close and even texture.

Spring Pin. A hollow split dowel pin made of spring steel.

Spring Points. Points normally closed by springs which can act as either facing points or trailing points. See **Points.**

Spring Rate. The ratio of load to deflection for a spring. For many types of metallic springs this rate is approximately a constant.

Spring Ring. A metallic ring cut diagonally and bigger than the bore to render the close fitting of a piston when sprung into an engine cylinder. A Ramsbottom-ring is a spiral of two or three coils or separate cut rings pressed into grooves in the periphery of a solid piston. See also **Piston Ring.**

Spring Safety Valve. *Spring balance valve.*

Springs, Disc. Springs in the form of shallow dished discs. Fig. 171(*a*) shows a combination which gives a lower *spring rate* than Fig. 171(*b*).

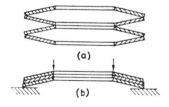

(a)

(b)

FIG. 171. DISC SPRINGS.

Sprocket. (*a*) Each of several teeth on a wheel engaging with the links of a chain or a toothed belt.

(*b*) A cylindrical wheel with protruding pins on one or both rims for pulling a film by means of the perforations on its edges.

339

Sprocket

Sprocket Holes. The accurately punched holes (perforations) on one or both edges of cinematograph film for pulling it at the correct speed through cameras and projectors.

Sprocket (Wheel). A wheel with spikes on its periphery with a pitch between spikes equal to the lengths of chain links with which the spikes engage. The wheel may drive or be driven by the chain such as in the case of a pedal bicycle. Also called ' bracket chain wheel '.

Sprocket Chain. *Pitch chain.*

Spur Gear (Spur Gearing). (*a*) Gearing composed of combinations of *spur wheels.*

(*b*) A system of *gear-wheels* connecting two parallel shafts.

(*c*) A cylindrical gear with teeth parallel to the axis and on the outer diameter; a *spur wheel.* (See Figs. 92, 93, 172, 173(*a*), 173(*b*).)

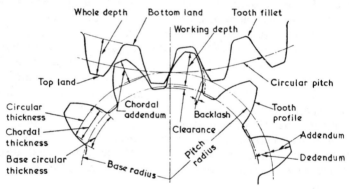

Fig. 172. Spur-, Helical-, and Bevel-gear Nomenclature.

Spur-gear Cutting. The cutting is done with single rotary disc cutters, or with hobs or with a pinion-shaped cutter which is rotated with the blank, thus generating the correct tooth form.

Spur-gear Nomenclature. This nomenclature is shown in Figs. 172, 173(*a*), 173(*b*). The reference planes are given in Fig. 174.

Spur Rack. See **Rack, Spur.**

Spur Wheel. A toothed wheel with its teeth on the outer diameter parallel to the axis. The teeth may be straight, helical or double helical.

Square Centre. A pyramidal-shaped lathe centre which is placed in the *poppet* mandrel.

340

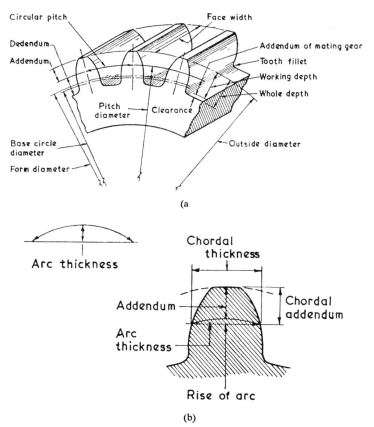

(a

(b)

FIG. 173. SPUR-GEAR NOMENCLATURE.

Square-nose Tool. A finishing tool for turning with its cutting edge at right-angles to the edges of the shank.

Square Thread. A *screw thread*, with a basically square section, twice the pitch of a similar vee-thread and slightly rounded corners, which is used for transmitting a thrust. (See Fig. 183(*a*).)

Squaring Bands. *Steadying bands.*

Squeezing Machine. A machine in which bars or rails are bent or straightened.

Stable Equilibrium. See Equilibrium.

Stacking Truck. *Fork-lift truck.*

Staff. An arbor or axis especially that of the *balance* or *pallets*.

341

Staggered

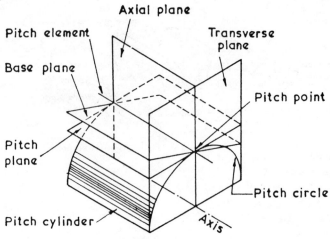

FIG. 174. SPUR-GEAR REFERENCE PLANES.

Staggered. A term to describe the alternate inclination in opposite directions of the *spokes* of a *wheel*.

Staking On. The fitting of wheels on shafts solely with keys, as in millwright's work.

Stamp. A die for moulding sheet metal by means of a screw or hydraulic pressure.

Stamping. *Pressing*.

Standard Gauge. (*a*) A *gauge* whose size has been determined in relation to an ultimate standard of length.

(*b*) The distance 4 ft $8\frac{1}{2}$ in. (1·44 m) between the inner edges of locomotive rails in Great Britain, North America, China, and most of Europe.

Standard Specification. A specification to which a machine, material, etc., must conform as laid down in the United Kingdom by the British Standards Institution and in the U.S.A. by the Bureau of Standards.

Standard Wire Gauge (S.W.G.). The legal wire gauge in the United Kingdom. See **British Standard Wire Gauge** and **Gauges Commonly Used.** (No SI Wire Gauge has been defined. See **Gauges Commonly Used.**)

Standing Pressure. The value at which the pressures in the two steam outlets are equal, usually with the steam-engine piston at the geometrical central position.

Star Gear. An *epicyclic gear* with a rotating annulus, a rotating

342

sun wheel, a fixed planet carrier and planet wheels rotating about their own spindles. Cf. **Planetary Gear** and **Solar Gear**.

Star Wheel. (*a*) A small manually operated wheel, on *capstan* and *turret lathes* furnished with radial spokes and attached to the end of a feed screw, which engages with a *pawl* once per revolution and thus feeds the screw round a definite amount each time to actuate the slide carrying the capstan or turret.

(*b*) A wheel in a watch or clock with pointed triangular teeth. The number of these teeth on the wheels of a calendar watch varies according to what it records, such as date, month, etc.

Star-wheel Motion. A feed motion, using a *star wheel*, which is employed for surfacing in lathes and boring machines.

Starter (Starter Motor). The device for starting the engine of an automobile, consisting essentially of a battery, electric motor and suitable gearing for starting the flywheel of the engine.

Starting Engine. (*a*) A small engine for starting large diesel and steam-engines.

(*b*) A small auxiliary gas-turbine engine carried by some aircraft to provide ground power supplies, including high-pressure air for starting the main engines.

Starting Lever. The lever attached to a starting *slide valve*.

Starting Valve. A valve for admitting steam from the boiler to the cylinders of a steam-engine.

Static Balance. See **Balancing (Static).**

Static Friction. *Stiction.*

Static Line. A tube connected to the ambient pressure. See **Air-speed Indicator.**

Static Stiffness. (See also **Stiffness**.) The change in deflection with change in dead load.

Stationary Link. A *slot link* in which the pin or die block only is moved for the purpose of reversing.

Stator. A row of radially disposed *stator blades*.

Stator Blades. Fixed blades of aerofoil shape mounted on the casing of an axial compressor to direct the airflow between the several rotary stages. The first row in many engines is adjustable. The same term is used for the stationary blades performing a similar function between the rows of turbine rotor blades.

Stator Blades (Exhaust). An assembly of stator blades situated behind the axial compressor or turbine discharge to remove residual swirl from the exhaust gases and usually in sections to allow for thermal expansion.

Stave. (*a*) The projecting end by which a crankshaft is held during forging.

(*b*) The shaft upon which a large number of *healds* are mounted.

Stay Tap. A very long *tap*.

Stayed Link Chain. *Studded chain.* See **Chain.**

Steady (Back Rest, Back Stay). A support attached either to the bed or the carriage of a lathe for backing up slender work. It may consist of slotted radial rollers adjusted to bear on the rough-turned work or in the form of a ring encircling the work or some similar steadying device. See also **Roller Steady.**

Steady Pin. (*a*) A pin for fitting mechanical parts together accurately with one fixing screw. Cf. **Dowel.**

(*b*) A pin for accurately fixing together two parts of a clock or watch; it is fixed to one part and a close fit in a hole in the other part.

Steady State. See **Vibration.**

Steadying Bands (Squaring Bands). Bands at each end of a *mule* carriage to ensure that the ends and centre make the same movement during the outward and inward run.

Steadying Roller. (*a*) A small roller, just in front of the cutter, for pressing down a wooden plank in a wood *planing machine.*

(*b*) The small roller used in a *steady* which can precede or follow up on a turned diameter if the *steady* is saddle or turret mounted.

Steam Chest. The chamber to which the steam pipe is connected and in which the *slide valve* works.

Steam Crane. See **Crane.**

Steam Dome. See **Dome.**

Steam Donkey Pump. A small pump for supplying steam boilers with feed water, having the piston rod of the steam cylinder and the ram of the pump in line, and a flywheel to carry the momentum through the *dead-centres.*

Steam-engine Cycle. Steam is admitted to a cylinder on one side of a steam-tight fitting piston. The piston is connected by a piston rod to a crosshead which slides between guides and the crosshead moves the crank through a connecting-rod. After expansion in the cylinder the exhaust steam either passes to a condenser or through two or more cylinders of increasing size. The admission and exhaustion of the steam from each cylinder is controlled by a *slide valve.*

Steam-engines. Engines in which the motive power is supplied by the elasticity and expansion or rapid condensation of steam. See **Compound Engine, High-speed Steam-engine, Mill Steam-engine,**

Multiple-expansion Engine, Quadruple-expansion Engine, Simple Steam-engine, Triple-expansion (Steam) Engine, Steam Turbine, Unaflow Steam-engine.

Steam Gauge. *Bourdon gauge.*

Steam Hammer. A heavy hammer used for forging which slides between vertical guides. Steam is used to raise and sometimes to accelerate the descent of the hammer which otherwise falls freely under gravity. See also **Double-acting Steam Hammer.**

Steam Hammer Type. *Inverted cylinder engine*, marine type.

Steam Jacket. A jacket, round a steam-engine cylinder, which is supplied with live steam to prevent excessive condensation of the working steam in the cylinder.

Steam Lap. *Outside lap.*

Steam Line. The upper approximately straight portion of an *indicator diagram* which records the entrance of the steam into an engine cylinder.

Steam Locomotive. A steam-engine and boiler integrally mounted on a frame for service on a railway. Cf. **Traction Engine.**

Steam Ports. Passages supplying, and exhausting steam from the *valve face* to the cylinder of a steam-engine.

Steam Pump. A *force pump* driven by a steam-engine.

Steam Reversing Gear (Steam Reversing Cylinder). A power reversing gear of a steam locomotive operated by a reversing lever to admit steam to an auxiliary cylinder whose piston operates the reversing links of the valve gear.

Steam Trap. An automatic device for ejecting condensed steam from steam pipes, etc., without permitting the escape of steam.

Steam Turbine. A *turbine* in which the steam does work by expanding, thus creating kinetic energy, as the steam passes over the blades of the turbine disc (or drum). See **Combined-impulse Turbine, Disc-and-drum Turbine, Impulse Turbine, Reaction Turbine.**

Steam Valve. A valve for regulating the supply of steam. See **Lift Valve, Slide Valve, Starting Valve, Stop Valve.**

Steelyard (Roman Balance). A weighing arm with a fixed weight at one end and marked positions on the other side of the point of support to indicate, when balanced, the weight of the object being weighed at that position.

Steering Head. The gear for steering a ship.

Steering-arm. *Drop arm.*

Steering-gear. The two geared members transmitting motion from the steering-wheel of an automobile to the stub axles through the *drop arm, drag link,* steering-arms and *track rod.* See **Cam-type Steering-gear, Rack-and-pinion Steering-gear, Screw-and-nut Steering-gear, Worm-and-wheel Steering-gear.**

Steering-lock. The maximum amount that the steered wheels of a vehicle can turn from side to side. See also **Turning Circle.**

Steering-wheel. The spoked handwheel at the top of the steering column of an automobile which actuates the *steering-gear.*

Stem. (*a*) The shaft of any component. See **Spindle.**

(*b*) The shaft to which the button is attached in a keyless watch.

Stephenson's Link Motion. See **Link Motion.**

Stepped Pulleys. *Speed pulleys.*

Stepping-in. Fitting a shaft or spindle with a shouldered shank into a hole in a support.

Stern Tube. The tube which carries the bearings of a ship's propeller shaft.

Stiction. A tendency of two surfaces to adhere to one another unless kept in relative vibratory or rotatory motion. See also **Friction.**

Stiffness. The restoring force per unit displacement. The reciprocal of *compliance.*

Stiffness Ratio. The ratio of stiffness along two defined principal axes in *isolators* and in rigid bodies with resilient supports. See also **Dynamic Stiffness, Static Stiffness.**

Stirrup. (*a*) A strap or loop supporting a spindle, etc., vertically.

(*b*) Generally, a support.

Stock. The principal part or body of a tool or instrument to which the working part is attached. See also **Die-stock.**

Stock Rail. The outer fixed rail on a railway against which a *point* works. See **Points.**

Stocking Cutter. A milling gear-cutter for roughing out the material between gear-teeth, to be followed by a finishing cutter.

Stockless Anchor. An anchor with no cross-piece on the shank and the arms pivoted so that both can engage at the same time. (See Fig. 175.) Cf. **C.Q.R. Anchor.**

Stocks and Dies. *Dies* and their holders for cutting screws. Cf. **Stock.**

Stone Tongs (Nippers). A large pair of scissors with the points curving inwards to clip into the sides of stone blocks while chains connect the loops of the tongs to a hoisting ring. They are similar to *lifting tongs.*

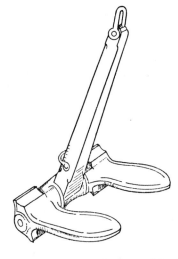

Fig. 175. Stockless Anchor.

Stop. A metal pin or block to stop or reverse the action of a machine or mechanism.

Stop Cock. (*a*) A *cock*. (Fig. 176.)

(*b*) A *plug*.

Stop Drill. A drill provided with a collar to prevent penetration beyond a certain depth.

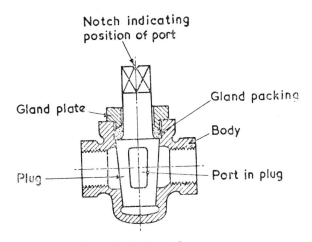

Fig. 176. Common Stopcock.

Stop Motion. A lever arrangement for disengaging a machine or mechanism without stopping it. Cf. **Clutch.**

Stop Piece. A piece which limits the movement of a hammer in a striking mechanism of a clock or repeater.

Stop Slide. *All-or-nothing piece.*

Stop Valve. (*a*) A valve which, when shut, prevents the flow of liquid or gas. See **Angle Valve, Globe Valve, Needle Valve, Oblique Valve, Screw-down Stop Valve, Tee-valve.**

(*b*) The main steam valve on a boiler for controlling the steam supply and isolating the boiler from the main steam pipe.

Stop-watch. A watch that is started and stopped by pressure on a winding or special knob. The commonest type records fifths of a second, but others record one-hundredth and one-tenth seconds. The dial has usually a sweep-seconds hand and a minute-recording hand. Cf. **Chronograph.**

Stop Work. A mechanism to prevent the overwinding of a mainspring in a watch or clock or a hanging weight in a clock.

Straddle Milling. Using side-cutting *milling cutters* fitted on an arbor to machine, for example, both side-faces of a piece of work at one operation and sometimes the top face as well. Cf. **Gang Milling.**

Straight Bevel Crown Gear. A gear whose *pitch curves* are straight lines, intersecting the *apex*. The *spiral angle* at any cone distance is zero. Cf. **Spiral Bevel Crown Gear.** (See Fig. 13.)

Straight-flute Drill. A conical pointed drill with straight longitudinal flutes in the shank.

Straight-line Edger. A sawbench with the saw carried on a spindle above the table of the machine. A travelling endless chain feeds the material through the saw in the nature of a travelling bed, assisted by a power-driven pressure roller or rollers above the bed.

Straight-line Lever Escapement. See **Swiss Lever** under **Escapements.**

Straight Link. A *slot link* with the reversal movement divided between the link and the block.

Straight Shank Drill. A drill with a parallel shank for use with self-centring chucks.

Straightening Machine. (*a*) A machine for straightening rails and bars as they leave the rolls after forging.

(*b*) A machine for straightening or bending channel, angle and bar iron in boiler and smith shops, by squeezing; also called a ' bending machine '.

(*c*) A machine for flattening metal sheet after it has been coiled up.

Strain Gauge. A device for converting strain into an electrical signal, consisting essentially of a flattened and closely coiled fine insulated wire attached to the specimen or structure under investigation. Strain in the test-piece stretches the wire and alters its electrical resistance which can be calibrated in relation to the strain.

Strap. (*a*) A *belt*.

(*b*) An *eccentric strap*.

Strap Bar. A sliding bar which carries a *belt fork*.

Strap Brake. A brake consisting of a strap or belt, sometimes fitted with shoes, enveloping a flywheel.

Strap End. A loose strap on the end of some *connecting-rods* which encloses the *brasses* and is held in place by a *gib* and *cotter*.

Strap Fork. *Belt fork*.

Stretch Forming. See **Forming.**

Striking Gear. The lever, fork and essential fittings for shifting belts on and off the pulleys of rotating shafts.

Striking Mechanism. See Fig. 36; **Gathering Pallet; Jumper; Jumper Spring; Lifting Piece; Lifting Pin; Locking Post; Minute, Cannon** and **Hour Wheels; Rack; Snail; Star Wheel; Warning Piece.**

Striking Screws. The starting of screw threads with chasing tools. See **Chaser.**

Striking Work. The mechanism of a clock which causes the striking.

String Warp Machines. Lace machines supplied by spools and by beams when making warp lace.

Stripper and Worker. A pair of wire-covered rollers which are part of a roller *carding engine*.

Stripper Bolt. *Shoulder bolt*.

Stripping. Damaging the teeth of wheels due to too sudden starting and stopping.

Stripping Plate. Plates with ends sloped to fit the grooves of mill rolls and bevelled to lift or clear the bars from the grooves and thus prevent *collaring*.

Stroboscope. A device for measuring the rotational speed of a rapidly rotating object; the object appears to rotate at a speed equal to the actual speed difference of the device and the object and appears fixed when the two are synchronized. The object is usually observed (*a*) through a slot in a rotating disc or drum, or (*b*) with the aid of a flashing light of pre-fixed frequency.

The device can be used for the detection of wear, distortion or chatter of moving parts and mechanisms.

Stroke. (*a*) The total length of the movement of a piston in the cylinder of a piston-engine.

(*b*) The length of travel of a reciprocating part of an engine or mechanism.

Stub Axle.   See **Axle.**

Stub-tooth Gear.   A gear with robust teeth of short height as used in the manufacture of automobile gears.

Stud.   (*a*) A headless bolt screwed from both ends and plain in the middle.   One end is usually tapped into a screwed hole of one piece of work and the other left to receive a tightening nut for attachment of a second piece.

(*b*) The part of a watch to which the outer coil of a *balance spring* is attached.

(*c*) A short vertical pin in a watch or clock.

(*d*) A single dowel.

(*e*) A projecting pin used as a pivot for lever, rod or wheel.

Stud Link.   See **Link.**

Stud Wheel.   The idle wheel on the intermediate pin of a screw-cutting lathe.

Studded Chain (Stayed Link Chain).   See **Link Chain.**

Stuffing Box.   An annular space through which a machinery part moves, and in which packing is compressed by a *gland* to make a pressure-tight joint, such as the rod of a pump or the stem of a valve.

Stuffing Box Spacer.   *Lantern ring.*

Suction Box.   The lower chamber of a *suction pump* into which the liquid is drawn by the upward stroke of the piston.

Suction Dredger (Sand-pump Dredger).   A *dredger* using the suction of a centrifugal pump to bring material up a long pipe to discharge it into a barge.

Suction Pump.   A pump which depends on atmospheric pressure for its action and which, in practice, will not lift water from a greater depth than about 7·6 m.   The water is drawn into the barrel by suction, the descent of the piston allows the water to pass through a bucket valve into the upper part of the pump from which it is lifted out and delivered on the upward stroke.

Suction Valve.   *Foot valve.*

Summing Gear.   A *differential gear* which adds or subtracts the motions of two members.

Sun and Planet Wheels.   A gear-wheel (sun wheel) around which one or more planet wheels or planetary pinions rotate in mesh.   See Fig. 61 and **Epicyclic Gearing.**

Sunk Key. See **Key.**

Supercharger. An axial flow or centrifugal compressor which supplies air, or a combustible mixture, to a piston-engine at a pressure greater than atmospheric and is driven either directly by the engine or by a gas turbine motivated by the exhaust gases. The latter is called an 'exhaust-driven supercharger' or a 'turbo-supercharger'.

Compression may be effected in a cylinder—*piston supercharger*, or by separate volumes of air or mixture—*positive-displacement supercharger* or by the relative motion of two specially shaped rotors in a fixed case—*Roots supercharger*—or by the motion of vanes carried in a rotor eccentrically located in a fixed case—*vane supercharger*. See **Roots Blower.**

Multi-speed Supercharger. A gear-driven supercharger with a clutch system to allow the engagement of different ratios to suit changes in altitude.

Supercharging. (*a*) The maintenance of ground-level pressure in the inlet pipe of an aero-engine up to the *rated altitude* by a *supercharger*.

(*b*) Boosting. See **Boost.**

Superfinishing. A process involving short strokes at a very rapid rate with a lighter pressure than in *honing* and *lapping* and with copious amounts of coolant and lubricant. It is applied to both cylindrical and flat surfaces and surface finish is usually in the range 25–100 nm; 10 nm can be obtained.

Superheated Steam. Steam heated at constant pressure to a temperature above that due to saturation and out of contact with the water from which it was formed.

Surface. The boundary of an object separating it from another substance such as the surrounding air.

Surface Texture (Surface Roughness). The surface quality as determined by small departures (roughness) from its general geometrical form.

Surface Texture Roughness Index Number. Numerical assessments of the average height and/or depth of surface irregularities. British Standard index numbers assess the centre-line average values. (See Fig. 65.)

Surface Chuck. *Face chuck.*

Surface Condenser. A *condenser* in which cooling water is circulated in tubes to condense the steam and a vacuum is maintained by an air-pump.

Surface Finish. See **Cleaning, Finishing, Honing, Lapping, Lay, Polishing, Roughness, Waviness, Superfinishing, Tumbling.** Fig. 65.

Surface Grinding Machine. A machine for finishing flat surfaces with a high-speed abrasive wheel mounted above a reciprocating, or rotating, work table on which the work is held, often by a *magnetic chuck*.

Surface Meter. An instrument for measuring the texture of surfaces, using a stylus whose up and down motion is magnified up to 100,000 times to produce a graph of a cross-section of the surface and a number representing the centre-line average height. Cf. **Profilometer.** See also **Surface Texture Index Number** under **Surface.**

Surface Plate (Planometer). A rigid, accurately flat, cast-iron plate for testing the flatness of other surfaces and to provide a plane datum surface for marking off work for machinery.

Surface Texture. The appearance and characteristics of a metal surface after machining. See **Lay** (*a*), **Surface.**

Surging. (*a*) A severe fluctuation or abrupt decrease in the delivery pressure of a centrifugal *supercharger* or of a *compressor*.

(*b*) The coincidence of a harmonic of a *cam's* lift curve with its controlling valve spring's natural frequency of vibration, leading to irregular action.

Surging Drum. *Warping cone.*

Surprise Piece. An extra cam fitted on a *cannon pinion* arbor which advances the *star-wheel* tooth at each hour in a *repeater* mechanism.

Suspension. That portion of a mechanism designed to damp vibration and reduce the effect of external shock loads on the major portion of the assembly. See also **Brocot Suspension, Hydrodynamic Suspension.**

Suspension Links (Vibrating Links). Two parallel flat rods which lift and lower the *slot links* of a steam-engine for reversal. One end of the pair is loosely attached to the tail of the slot link and the other is attached to a short lever keyed on the *weigh shaft.*

Suspension Spring. The thin ribbon of spring steel supporting a pendulum.

Swage. (*a*) To open out (bevel) the end of tubes, etc.

(*b*) A hardened-steel eccentric die and anvil arranged in a suitable holder used for spreading teeth in a saw.

(*c*) A die for shaping a piece of forging; also called 'top and bottom tools '.

Cf. **Dished Plate.**

Swage Saw. A saw that is ground-off in a straight taper and has no parallel portion.

Swaging. Drawing down a piece of wrought metal to a desired form; commonly, the opening out of the ends of tubes to take a threaded coupling.

Swash-plate (Wabbling Disc). A circular plate mounted obliquely on a shaft, as a substitute for a crank mechanism.

Swash-plate Motor or Engine. See **Engines, Servomotor Types.**

Swash-plate Pump. See **Pump.**

Sway Brace. A support, often horizontal, to control vibration and to damp undesirable movements.

Sweat Cooling. The cooling of a component of an engine or mechanism by evaporating fluid through a porous surface layer, such as in rocket-engines and gas-turbine blades. See also **Film Cooling.**

Sweep-seconds Watch. A watch with the seconds hand at the centre of the watch, a movement which requires a supplementary *third wheel* and *fourth pinion.*

Swell. A device with spring control at the back of a shuttle box on a *loom* to hold the shuttle in position.

Swell of Pulley. The curved surface of a pulley rim to prevent a belt from working off while running.

Swifts (or Cylinders). (a) The large rollers in a set of woollen carding engines which, with manual aid, scribble (card) the wool.

(b) Light revolving frames to carry the hanks of wool during unwinding.

(c) The revolving frames used in dark rooms for processing and drying long rolls of film.

Swing Table. The table of a *drilling machine* which swivels around a central pillar.

Swiss Lever. See **Escapement.**

Swiss Machine (Schiffle or Shuttle Machine). An embroidery machine with the shuttles placed diagonally.

Swiss Screw Thread. *Thury screw thread.*

Switch. (a) A mechanism for altering the direction of a moving body.

(b) A device for diverting rolling stock from one rail track to another.

(c) A mechanical device for opening and closing an electric circuit.

Swivel. (*a*) A pin and collar to permit circular motion in a mechanism.

(*b*) A link in a chain consisting of a shank and collar to permit circular motion.

Swivel-head Lathe. A special lathe for boring and turning tapered objects having the mandrel headstock mounted and pivotable on a base plate.

Swivel-pin. *King-pin.*

Swivelling Propeller. A propeller capable of being turned bodily so as to transmit its thrust in any desired direction.

Sylphon Bellows. A thin-walled air-tight cylindrical metal *bellows* like a concertina, which responds to pressure variations.

Synchro. A generic term for a class of electromechanical devices used for data transmission. See also under **Differential.**

Synchro Angle. The angular displacement of a synchro rotor from its electrical zero position.

Synchromesh Gear. A gear in which the driving and driven members are automatically synchronized by small *cone clutches* before engagement.

Synchronizing Gear. A gear to synchronize the firing mechanism of a gun with the rotation of the *airscrew* so that the bullets do not meet the blades.

Synchronous. Occurring at the same time; simultaneous; in step.

Synchronous Vibrations. Vibrations which correspond exactly in period and phase.

T

θ. The symbol for angle of twist.

t. and b. Turned and bored.

t, T. The symbols for time and period respectively. *T* also for kinetic energy and for torque.

T.M. Twisting moment (*torque*).

t.p.i. Threads per inch.

T.S. Tensile strength, ultimate.

Table. The horizontal portion of a machine on which the work is placed for planing and other operations.

Table Feed. A machine which has a stationary cutting tool and a traversing work table is said to have table feed. See **Planing Machine** (*b*).

Tachometer. An instrument for indicating the revolutions per minute of a revolving shaft, operated either by a spring-controlled ring pendulum or by spring-loaded governors, or by magnetic means (see **Governor**). When registering the revolutions of a revolving shaft in cotton spinning it is called an ' indicator '.

Tackle (Lifting Tackle). A combination of pulleys, blocks, etc., for hoisting purposes.

Tail. The back part of a portable crane upon which the balance weights rest.

Tail Pin. The *back centre* pin of a lathe.

Tail Race. The channel conducting water away from a water wheel or other hydraulic machine.

Tail (Auxiliary) Rotor. A small rotor mounted at the tail of a helicopter on a horizontal axis to provide sideways thrust to counteract the torque of a single *main rotor*, and to give directional control; also called ' auxiliary rotor ' and ' anti-torque rotor '. Fig. 85.

Tail Screw. (*a*) The back centre screw of a *headstock*.

(*b*) The screw actuating a *poppet cylinder*.

(*c*) In general, a screw located behind the part of a machine which it actuates.

Tail Stop Screw. The back screw of the *headstock* of a back-geared lathe.

Tailshaft. The shaft driving a marine propeller.

Tailstock (Poppet). The movable head of a lathe which supports the end of the work remote from the driving headstock. (Fig. 102.) It slides on the lathe bed, carries a spindle aligned with the centre of the *fixed headstock* and adjustable, and is coned internally to receive a hardened dead-centre. The term ' tailstock ' likewise applies to milling and grinding machines.

Taintor Gate. (U.S.). See Gate.

Take-off Sprocket Wheel. The *sprocket* which accepts the exposed ciné film from the constant-speed drive sprocket and protects the constant-speed drive from the pull of the *take-up reel*.

Take-up Reel. The reel which accepts the ciné film from the *take-off sprocket wheel* and is driven by a friction drive to maintain adequate tension in the film.

Tandem Engine. An engine with the cylinders arranged axially, or end to end, with a common piston rod.

Tang. Pointed, tapered or narrowed end of a cutting tool, e.g. the narrow outer end of a drill produced by milling two flats on opposite sides of a tapered shank. (See Fig. 114.)

Tangent Screw (Tangent Wheel). A worm screw for making a

Tangent Screw (Tangent Wheel)

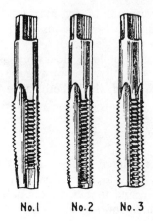

FIG. 177. SCREW TAPS.

No.1   No.2   No.3

fine adjustment in the setting of an instrument about the worm wheel axis to correct its line of sight or to adjust *a vernier*.

Tangential Keys. See Key.

Tank Engine. A locomotive with a tank for carrying its own water, thus dispensing with a tender. The tanks are either on the side (*side tank*) or on the top of the boiler (*saddle tank*).

Tap. A screwed plug used for cutting internal threads in holes and having an accurate thread form and size. It has longitudinal grooves for the clearance of chips whilst cutting. (See also **Boiler Tap, Plug Tap, Second Tap, Stay Tap, Taper Tap, Thoroughfare Tap**). Machinists' taps come in sets of three for each size and are called (1) taper, (2) second, (3) plug. (Fig. 177.)

Machine-working Taps have longer shanks than hand-working taps so that, for example, several nuts can be threaded at the same time in a *screwing machine*.

Tap Grooving. The cutting of the grooves (flutes) in *taps* usually with a milling machine.

Tap Plate. *Screw plate.*

Tap Wrench. A lever with a square hole in the centre into which the square-ended shank of a *tap* can be fitted when using the tap.

Tape Condenser. A mechanism controlling leather tapes and rubbing rollers for converting the web of fibres from the *doffer* of a *carding engine* into a number of *slivers*.

Tape-controlled Machine. Any machine whose operation is in part, or wholly, controlled by a punched-tape system. Usually the operation of feed and speed of a cutting tool or the workpiece

356

position is controlled by electronically reading instructions from the tape. Optical and pneumatic reading coupled with mechanical, pneumatic or hydraulic actuation is sometimes used.

Taper Key. See Key.

Taper Pin. A pin locking member between two mating parts. Fig. 178 shows two applications for a taper pin.

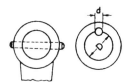

FIG. 178. TAPER PIN APPLICATION.

Taper Pin Drill. A *twist drill* with a straight shank and its flutes tapering from the shank to the beginning of the conical point.

Taper Roller-bearing. A *roller-bearing* with tapered rollers to take end thrust and radial loads using internally and externally coned races.

Taper Screw Chuck (Screw Worm Chuck). A chuck with a taper screw fixed to the centre of a small face-plate to which wood can be screwed for turning.

Taper Screw Thread. A screw thread cut on the face of a cone.

Taper Shank. A *shank* tapered in its length and circular in section as in some *drills* and *reamers*.

Taper Tap. The first *tap* used in threading a hole, with its first few threads ground down to a taper to provide a guide.

Tappet (Tappet Valve). A sliding member working in a guide to operate a *push-rod* or valve system. The tappet's motion is controlled by a cam and the sliding motion eliminates side thrust and distinguishes it from the motion of a pawl or screw. The tappet and cam convert a rotating into a reciprocating motion. In weaving, a tappet mechanism is called a 'wiper'.

Fig. 179 shows a tappet valve in a side-valve engine, *A* being the mushroom head. Fig. 200 shows a tappet in a motorcycle engine.

Tapping. Forming a screw thread in a hole by means of a tap of the correct size.

Tapping Hole. A hole drilled to the same size as the bottom of a screw thread.

Tapping Screw. *Tap.*

T-bolts (Tee-headed Bolts). Bolts with a short cross-piece for a

T-bolts (Tee-headed Bolts)

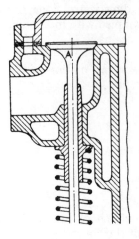

Fig. 179. Tappet Valve.

head which is used for holding work in corresponding slots on a machine table by turning the bolt through 90°.

Tee-rest. See **T-rest.**

Tee-slots. See **T-bolts.**

Tee-valve. A three-way *screw-down stop valve* with the connections in the form of the letter 'T'.

Tee-headed Bolt. *T-bolt.*

Teeth. The tooth-shaped projections on cog-wheels, gear-wheels, ratchet wheels, and many cutting tools. Referring to cutting tools see **Finishing Teeth, Involute Gear-teeth, Roughing Teeth, Semi-finishing Teeth.** See also **Helical Gear-tooth.**

Teleprinter. A telegraph transmitter with a typewriter keyboard, by which characters of a message are transmitted electrically in combinations of five binary digits, being recorded similarly by the receiving instrument, which then translates the message mechanically into printed characters via control movements of an electric typewriter.

Telescopic Slide (Telescopic Shaft). Two or more hollow tubes sliding one within another, providing a long support when extended.

Tell Tale. An indicator to show the amount of movement of a winding engine or of some mechanism, or to indicate the precise time when a series of operations has terminated on an automatic machine.

Telpher Line. A form of *monorail* on which electrically-driven trucks carry suspended loads.

358

Temperature Adjustment. The regulating of a chronometer or watch for temperature changes.

Temperature Compensation. The automatic adjustment of the reading of an instrument to allow for changes in the ambient temperature.

Template (Templet). A thin plate of accurately shaped profile for marking out or checking the result of a machining or other operation. See also **Drill Template, Routing Machine.**

Temple. A *loom* device, consisting of spiked or fluted rollers, which keep the woven cloth at the same width as the spread of the warp threads in the reed.

Tenon. A tongue of one-third the thickness formed on the end of a member to fit into a mortise in a second member. See **Grooving.**

Tenoning Machine. A machine with four saws for cutting *tenons* in wood. Two run vertically together and are adjustable sideways and the other two run horizontally to cut the shoulders. The vertical saws are often replaced by adjustable cutters revolving horizontally.

Tensile Test. A test in which specimens are subjected to an increasing load in tension, usually until they break. A stress-strain curve plotted from the results is used to determine the limit of proportionality, proof stress, yield point, *ultimate tensile stress*; elongation and area reduction of a specimen are also measured.

Tensile-testing Machine. A machine for applying either a tensile or a compressive load to a test-piece, hydraulically or by power-driven screws.

Terotechnology. The installation, commissioning, maintenance, replacement and removal of plant, machinery and equipment; feedback to design and operation thereof, together with the related subjects and practices.

Test Load. *Proof load.*

Test-piece. An accurately made piece of material for a *tensile test, impact test* or other *testing machine.*

Testing Machine. A machine for applying loads to a test-piece to measure extension, etc. See **Haigh Fatigue-testing Machine, Izod Test, Tensile-testing Machine.**

Texture. *Primary.* That component of the *surface texture* remaining after the normal action of a tool during production.

Secondary (*Waviness*). The components of the *surface texture* due to imperfections in the performance of the machine.

Thermal Efficiency. The ratio of the work done by a heat-engine to the mechanical equivalent of the heat supplied in the steam or fuel. See also **Brake Thermal Efficiency, Indicated Thermal Efficiency.**

Thermal Expansion. See **Expansion.**

Thermal Stress. Stress in a structure or mechanism caused by unequal expansion of different parts due to differential heating.

Third Pinion (Watch). The pinion on the same axis as the *third wheel.*

Third Wheel (Watch). The third wheel drives the fourth (seconds) pinion being the wheel between the centre and the fourth wheel of a watch train.

Thompson Indicator. The Thompson Indicator (Fig. 180) enlarges the movement *BB'* to *AA'* through simple levers and is used in *engine indicators.*

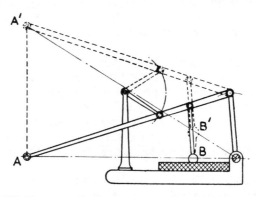

FIG. 180. THOMPSON INDICATOR—STRAIGHT-LINE MOTION.

Thoroughfare Tap. A *tap* with its square head small enough to allow it to pass through the hole which it has tapped.

Thou. See **Mil.**

Thrasher (Thrashing Machine, Thresher). A machine which separates the grain of wheat or other cereals from the straw and chaff. It has beating bars in a rotary cylinder or drum for separating the grain which is winnowed by means of shaking riddles or screens under the action of an air blast and finally delivered as clean wheat into a sack. The machine's name is also spelt ' thresher '.

Thread. *Screw thread.*

Basic Form (q.v.). (Fig. 156.)

Complete Thread. That part of the thread which is fully formed at both crest and root. A chamfer at the start of the thread which does not exceed one inch in length is included within the length of the complete thread. Cf. **Full Thread.**

Depth. The sum of the *addendum* and *dedendum* (see Fig. 172). The radial distance between the *major* and *minor cylinders* or *cones.* See **Screw Thread Diameters.**

Design Form (q.v.) and Fig. 156.

Form (q.v.).

Grade (q.v.).

Incomplete. That part of the thread which is fully formed at the root but truncated at the crest by its intersection with the cylindrical surface of the part.

Thickness. The distance between the flanks of a thread measured at the design *pitch line* parallel to the axis.

Total. The *complete thread*, the *incomplete thread* and the *washout thread.*

Useful. The *complete thread* plus the *incomplete thread.*

Washout. That part of the thread which is not fully formed at the *root.* See also **Full Thread.**

Thread Gauge. *Screw-threaded gauge.* See also **Thread Measurement.**

Thread Grinding. The accurate finishing of screw threads by a specially profiled grinding wheel which is automatically traversed along the revolving work.

Thread Insert (Aero-, Wire-thread Insert). A wire of approximately rhombic cross-section formed into a spring-like helix. It is inserted into tapped holes and retains another threaded member. Insertion is accomplished with a driving tool engaging on the leading end tang. After full insertion the tang is broken off at the notch leaving the insert just above the top of the *countersink* in the tapped hole. Thread inserts are used when a good resistance to wear is required for holes tapped in malleable materials, e.g. threaded holes in light-alloy castings for aircraft where the weight of a more durable material is prohibitive. (See Fig. 181.)

Thread Measurement. A simple method of measuring screw threads is shown in Fig. 182. Three pieces of wire of known diameter are inserted between the measuring contacts.

Thread Rolling. The formation of screw threads by rolling pressure.

Threads, Power Transmission

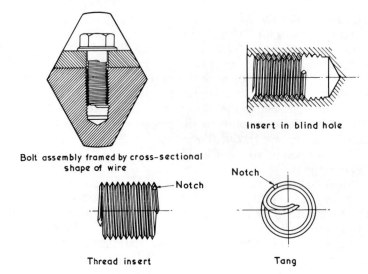

Insert in blind hole

Bolt assembly framed by cross-sectional shape of wire

Thread insert — Notch

Notch — Tang

FIG. 181. THREAD INSERTS.

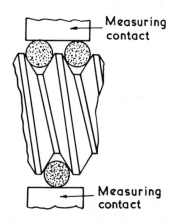

Measuring contact

Measuring contact

FIG. 182. THREE-WIRE SYSTEM OF THREAD MEASUREMENT.

Threads, Power Transmission. Three types of thread suitable for power transmission are illustrated in Fig. 183.

Three-high Mill (Three-high Rolls). A *rolling mill* with three rolls, one above the other, in which bars or plates can be passed forward between the lower pair and backward between the higher pair. Cf. **Two-high Rolls.** (Fig. 150.)

362

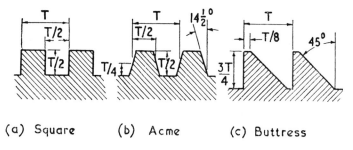

(a) Square   (b) Acme   (c) Buttress

FIG. 183. POWER TRANSMISSION THREADS.

Three-quarter Plate Watch. A watch with the *balance* in the same plane as the upper plate which is cut away for that purpose.

Three-start Thread. See **Multi-threaded Screw.**

Three-throw Crank. A crankshaft often used for pumps, with three cranks at 120° for driving three valves, buckets or pistons.

Three-throw Pump (Treble-barrel Pump). A pump with three working barrels in line, having the piston rods connected to a *three-throw crank*. Cf. **Two-throw Pump.**

Thresher. *Thrasher.*

Thrilling. *Knurling.* See **Knurling Tool.**

Throat. The inner edge of the flange of a railway wheel.

Throat Diameter (of a Worm Wheel). The minimum diameter at the tips of the teeth. See also **Gorge Radius** under **Gorge.**

Throttle. The lever, or switch, for controlling the speed of an engine, by regulating the flow of the fuel-air mixture, of the fuel or of steam in a steam-engine.

Throttle Governing. Governing a steam-engine by the *throttle valve* instead of by the shaft to the *slide valve*.

Throttle Valve. (*a*) The *butterfly valve* of a petrol engine.

(*b*) The *regulator valve* of a gas engine.

(*c*) The governor-controlled steam valve of steam-engines and turbines.

(*d*) The regulating valve controlling the pressure and temperature range of the working fluid in a refrigerator.

(*e*) A flat thin disc valve placed diagonally across a round pipe for partial or total closure of the pipe.

Throttling. Reducing the pressure of a fluid by causing it to pass through minute or tortuous passages.

363

Throw

Throw. The eccentricity of a *crank* or *eccentric*, equal to twice the radius.

Throw Disc. The disc of a *slotting machine* which actuates the ram by a short connecting-rod.

Throw-out Gear. The reversing gear of a marine engine employing only a single *eccentric*.

Thrust. (*a*) The component of the resultant force from a propeller parallel to the propeller axis.

(*b*) The resultant force from a jet-engine or from a rocket-engine.

(*c*) The compressive force in a member of a structure, of an engine or of a mechanism.

Thrust Bearing (Thrust Block). A bearing on a shaft for taking an axial load (thrust) such as a ball-bearing with lateral races, a *Mitchell bearing* or a plain bearing pad.

Thrust Block. *Thrust bearing.*

Thrust Collars. Collars on a shaft or spindle which transmit thrust to a *thrust bearing*.

Thrust Reverser. *Thrust spoiler.*

Thrust Screw. A screw which takes the thrust of a spindle in a drilling or other machine.

Thrust Shaft. A separate length of shafting on which are formed the *collars* for the thrust bearing of a marine engine.

Thrust Spoiler (Thrust Reverser). A controllable device mounted at, or on, a *propelling nozzle* to reduce or to reverse the jet thrust.

Thury Screw Thread (Swiss Screw Thread). A metric thread with a profile angle of $47\frac{1}{2}°$, with half the thread height cut off and rounded to form the crest.

Tie-rods (Tension Rods). (*a*) The rods supporting the *jib* of a crane.

(*b*) A purely tensile structural member.

Tightening Pulley. A pulley hinged on a sliding bearing plate, which, in moving guides, takes up slack in a pulley system.

Time and Motion Study. An analysis of the timing and movements of an operator, or of parts of a machine (or mechanism) to ascertain how time can be saved or movements simplified when undertaking a particular operation on a piece of work.

Time Constant. The interval during which the value of the amplitude of a vibration, that decreases exponentially, decreases to one/e of its initial value.

Timing. (*a*) The setting of the hands of a clock or watch to time.

(*b*) The observing of the rate of a clock or watch.

(c) The setting of the valve-operating mechanism of a piston-engine so that the valves open and close at the correct positions in the operating cycle.

Timing Belt. A belt with transverse teeth on the inner side of the belt which engage in slots in the driving and driven wheels.

Timing Chain. The chain between the sprocket wheels on the crankshaft and the camshaft of an internal-combustion engine.

Timing Gear. The two to one drive connecting the *camshaft* of a single camshaft engine to its crankshaft.

Timing Nuts. The two nuts used for timing purposes, one at each end of the arm on the rim of a chronometer balance.

Timing Screws. *Quarter screws.*

Timing Washers. Thin washers placed under the heads of the screws of a *balance* (Fig. 39) to alter its moment of inertia and thus its time of vibration.

Timing Wheels. Toothed wheels on the crankshaft and a camshaft of a motor-vehicle which are connected by a *timing chain* to give a reduction ratio of one to two for a single camshaft engine.

Tip. The edge where the *tooth flank* meets the *tooth crest*.

Tip Angle. The angle between the axis of a bevel gear and the *tip cone* generator.

Tip Circle. The circle of intersection of the *tip cylinder* by a transverse plane; also called ' addendum circle '.

Tip Cone. The cone containing the tips of the teeth in a bevel gear.

Tip Cylinder. The cylinder containing the tips of the teeth in a helical, spur or worm gear.

Tip Diameter. (a) The diameter of the *tip circle* or *tip cylinder* of a helical, spur or worm gear.

(b) The diameter of the circle of intersection of the *tip cone* and the outer end faces of the teeth of a bevel gear.

Tip-path Plane. The plane of rotation of the blade tips of a rotorcraft in flight; it is higher than the rotor hub.

Tipping Bucket. *Skip.*

Tire (U.S.). *Tyre.*

Titan Crane. See **Crane.**

Toe. (a) The lower end of a vertical spindle working in a *footstep*.

(b) In railway *points* the locked ends of the movable blades.

Toe-in. The small forward convergence of the planes of the front wheels of a motor-vehicle to promote steering stability and to equalize tyre wear.

Toe-out. The outward inclination of the front wheels of an automobile on turns due to setting the steering-arms at an angle.

Toggle-joint (Knee). A lever knuckle-joint consisting of two lever arms forming an angle with each other and hinged at the centre. The free end of one lever is hinged on a fixed pivot and that of the other lever is free to move. Any movement to bring the levers nearly into line causes a very large pressure to be exerted at the ends.

Tolerance. An acceptable range of variation of some dimension.

Limits of Tolerance (Limits). The maximum amounts by which the actual size of a dimension or of a profile, etc., are permitted to depart from the design size or form.

Bilateral Tolerance. A tolerance allowing variation in both directions from the design *size* or form.

Feature Tolerance. A tolerance on size of a particular feature such as a pin diameter or a slot width.

Form Tolerance. The total variation allowed for the form of a feature, shape or profile.

Metal Tolerance. The total dimensional variation, measured normal to the profile, in the amount of metal allowed on the surface of that profile, normally indicated by the maximum positive and/or negative departures from the design form. See also **Limits, Metal.**

Positional Tolerance. The total amount of variation allowed for a positional feature, such as tolerances on distances between centres. These tolerances are normally distributed bilaterally or in all directions round a centre. Cf. **Unilateral Tolerance.**

Unilateral Tolerance. A variation allowed in only one direction from the design size or form, such as in the dimensions of mating features. Cf. **Positional Tolerance.**

Tongs, Lifting. Fig. 184 (*i*) shows lifting tongs carrying a plate in jaws (*a*). Fig. 184 (*ii*) shows the jaws open, held up by *d*, and operated by *a*, through hinge *b* and chain *c*, and pivoted about *e*. This type of tong is used for rough bulky material.

Tongueing. Using cutters set in a revolving block to put an edge on a board, which fits into a groove on the edge of another board. Cf. **Grooving.**

Tool. Any implement by which mechanical operations are performed, whether by hand or machine such as those used for

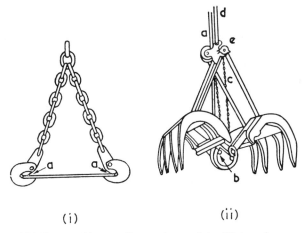

FIG. 184. LIFTING TONGS: (i) carrying a plate, (ii) tong jaws open.

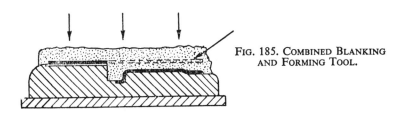

FIG. 185. COMBINED BLANKING AND FORMING TOOL.

forming and cutting. Fig. 185 shows a combined blanking and forming tool See **Blanking, Forming.**

Tool Carriage. The sliding carriage which carries the cutting tool in any automatically operated machine.

Tool Head. *Capstan tool head.*

Tool-holder (Cutter Bar). The bar holding the cutting tools in metal turning and shaping. See **Circular-form Tool in Holder.**

Tool Post. (*a*) A clamp by which a tool is held in the slide rest of a lathe. (Fig. 102.)

(*b*) A clamp by which a tool is held in the ram of a shaping machine.

Tool Rest. See **Rest, Slide Rest.**

Tool Room. That part of the machine shop containing specialized machines, such as jig borers, where the manufacturing tools and jigs are made for the rest of the machine shop.

Tool Stay. A slotted bar held in the socket of the T-rest and with the slot embracing the flattened shank of the boring tool.

Tooling. The cutting of metals by cutters as opposed to the shaping of surfaces by grinding.

Tooth. See **Teeth.**

Tooth Crest. The surface joining the two flanks of a single gear-tooth. See **Tooth Flank.**

Tooth Flank. That portion of a tooth surface which lies within the *working depth*.

Active Tooth Flank. The portion of the tooth flank which makes contact with the mating flank.

Tooth Profile. The line of intersection of the *tooth flank* with a defined surface.

Tooth Space. (*a*) The difference between the radii of the *tip cylinder* and *root cylinder* of a cylindrical gear.

(*b*) The shortest distance between the *tip cone* and the *root cone* measured along the *back cone* generator of a bevel gear.

Tooth Thickness. *Transverse.* The length of arc of the reference circle between opposite flanks of the same tooth of a helical or spur gear.

Normal Chordal. (*a*) The shortest distance between the *tooth traces* of a helical, spur or worm gear.

(*b*) The shortest distance between the intersections of the *tooth traces* of a bevel gear with the *pitch circle*.

(*c*) The shortest distance between the intersection points of the *tooth traces* of a tooth with the *pitch circle*.

Normal. The length of the arc of a normal helix between the two traces forming the gear-tooth.

Tooth Trace. The line of intersection of the *tooth flank* with the *reference cylinder* or *pitch cone*.

Toothed Cylinder. See **Pin Barrel.**

Toothed Wheels. See **Bevel Gear, Helical Gear, Spiral Gear, Spur Gear, Worm Gear.**

Top. Combed *sliver* prepared for spinning into worsted yarn.

Top and Bottom Tools. See **Swage.**

Top Card. An *indicator card* taken from the top of a vertical or oscillating cylinder.

Top Dead-centre. *Inner dead-centre.*

Top Rake. See **Rake.**

Top Steam. The steam entering above the piston in a *double-acting steam hammer*.

Torque. The turning moment about an axis, for example, of the air forces on a propeller, which may be uniform or fluctuating or of the turning moment of an engine crankshaft. Cf. **Moment of a Force.**

Torque Coefficient (of Propeller). The *torque* divided by the product of the density of the fluid, the square of the rotational speed and the fifth power of the diameter.

Torque Converter. (*a*) Any device which acts as an infinitely variable gear, but with varying efficiency such as a *fluid flywheel* or *coupling*. (See Figs. 186, 187.)

(*b*) Any device that multiplies torque, for example, a gear-box.

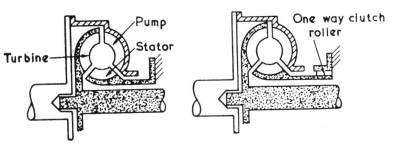

FIG. 186. TORQUE CONVERTER.   FIG. 187. TORQUE-CONVERTER-COUPLING.

Torque Dynamometer. A piece of test equipment for absorbing the power and measuring the *torque* of an engine.

Torque Limiter. A device to prevent the torque of a constant-power *turboprop* from exceeding a certain value.

Torque Link. A scissor type of mechanical linkage to prevent the relative rotation of the telescopic members of an aircraft *shock absorber*.

Torque Meter. A piece of test equipment attached to a rotating shaft which measures the angle of twist of a known length of shaft between two gauge points and thus enables the transmitted power to be calculated.

Torque Tube. A tube which is used to transmit or resist a *torque*.

Torsiograph. An instrument for measuring and recording the amplitude and frequency of the torsional vibrations in a shaft.

Torsion Bar Suspension. A springing system, adopted in some motor-vehicles, in which straight bars anchored at one end are subjected to torsion by the weight of the vehicle.

Torsion Pendulum. A mass suspended by a wire which rotates alternately clockwise and anti-clockwise and executes simple harmonic vibrations with a constant period independent of amplitude.

Torsional Vibration. See **Damper** and **Vibration.**

Torsional Vibration Damper. See **Damper.**

Total Technology. The integration of research, development, design, production, marketing, and plant operation, together with planning and management.

Toughness. A condition intermediate between brittleness and softness, indicated by a high ultimate tensile stress with a low to moderate elongation and area reduction of a test-piece, plus a high value in a *notched-bar test.* See also **Tensile-testing Machine.**

Tourbillon Movement (Watch). A watch escapement on a delicate steel carriage, which revolves round the fourth pinion and carries also the balance; the purpose of this arrangement is to eliminate positional errors. The fourth wheel is a fixture, being screwed to the pillar-plate. The carriage revolves once a minute. Cf. **Karrusel Movement.**

Tower Clock. A clock for mounting in a tall tower usually having exposed dials and hands. See also **Turret Clock.**

Tower Crane. See **Crane.**

Tracing Machines. Wood-working *copying machines.*

Track. (*a*) The rail or rails along which trains travel.

(*b*) The distance between the wheels of a vehicle, the wheels being on the same axle.

(*c*) The distance between the outer points of contact of the main wheels of an aircraft with the ground.

(*d*) The distance between the centre-lines of paired wheels in an aircraft undercarriage.

Track Rod. A rod connecting through *ball joints* the arms carried by the stub axles of a motor-vehicle so as to convey the angular motion from that wheel which is being directly steered to the other.

Tracks. The top faces of the bearers of a lathe bed.

Traction. The propulsion of vehicles by power by virtue of the frictional adhesion of their wheels.

Traction Engine (Road Locomotive). A power-driven vehicle, frequently propelled by steam, with large ribbed wheels for travel on roads, the wheels being gear-driven from a simple or compound engine mounted on top of the boiler. A rope drum is provided for haulage purposes. See also **Road Roller.**

Traction Sheave (of an Elevator). A power-driven grooved pulley over which passes a rope which is driven by the *sheave.* The friction between the rope and pulley constitutes the traction.

Tractive Force. The pull in pounds of a powered vehicle such as a locomotive exerts at the *draw-bar.*

Tractor. (*a*) A vehicle used on a farm especially for drawing wagons and powering agricultural machines.

(*b*) An aeroplane with a power plant producing tension in the propeller shaft.

(*c*) Any mobile prime mover used for towing purposes on land.

Trail. The distance of the point of contact of a steered wheel with the ground behind that where the line of the swivel-pin axis intersects the ground.

Trailing Axle. The rearmost axle of a locomotive.

Trailing Edge. That edge of a wing, aerofoil, strut or propeller blade which last touches the air or water when the craft is in motion. Cf. **Leading Edge.**

Trailing Lengths. The coupling rods of a fully coupled locomotive extending backwards to the *trailing wheels.*

Trailing Points. See **Points.**

Trailing Springs. The springs carrying the axle boxes of the *trailing wheels* of a locomotive.

Trailing Wheels. The wheels on the rearmost axle of a locomotive.

Train. (*a*) The interconnecting wheels and pinions of a mechanism such as a watch or clock and usually designated by a number; for example, a watch train suitable for a *balance* making 20,000 vibrations an hour is called a 20,000 train.

(*b*) A series of connected wheels or parts in machinery.

Train Brake. *Vacuum brake.*

Train of Gearing. A system of wheels and pinions providing an increase in torque with a reduction in speed or vice versa.

Train of Wheels. See **Train of Gearing.**

Tram. (*a*) Silk yarn formed of two or more single threads, slightly twisted and intended for *weft.*

(*b*) A wheeled vehicle running on rails.

Tram Wheel. A flanged wheel.

Transducer. A device for converting a physical quantity (or signal) of one kind into another kind.

Transducer Actuating Mechanism. Four types of transducer actuating mechanisms are illustrated in Fig. 188.

Transfer Line. A series of machines operating automatically on

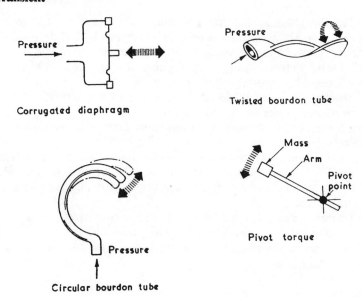

Corrugated diaphragm

Twisted bourdon tube

Circular bourdon tube

Pivot torque

FIG. 188. TRANSDUCER ACTUATING MECHANISMS.

a continuous line of parts.

Transient. A non-steady motion impressed on a system which usually decays to a negligible value, but the lasting effect of the transient may be a different steady state.

Transition Fit. See **Fit.**

Transmissibility. The ratio of the amplitude of vibration of a specified particle within a mechanical system to the amplitude of an applied vibration at a given frequency; normally the system is supported elastically, as by *isolators*.

Transmission. A term applied to various methods of transmitting and transforming power as by (*a*) a line shaft, (*b*) belts and pulleys and (*c*) gears.

Transmission Dynamometer. A device to measure the power transmitted by a shaft either (*a*) by using a *torque meter* to measure the twist over a given length of shaft, or (*b*) by direct measurement of the torque acting on an interposed differential gear. Cf. **Absorption Dynamometer.**

Transpiration Cooling. *Film cooling.*

Transporter Bridge. A bridge consisting of two tall towers connected at their tops by a supporting girder along which a carriage runs, with a small platform suspended from the carriage for transport at ordinary road level.

Transverse Pitch. The distance between the traces of adjacent teeth of a gear measured round the *reference circle*. See also **Pitch,** and Fig. 84.

Transverse Plane. A plane at an angle to the axis of a screw, gear, or body, usually normal to the axis. Cf. **Transverse Section.**

Transverse Pressure Angle. The acute angle between the normal to a *tooth profile* in a *transverse section*, where it intersects the *reference circle* and the tangent to the reference circle at that point.

Working Transverse Pressure Angle. The angle with reference to the *pitch circle* instead of the reference circle.

Transverse Section. A plane section normal to the major axis of a body. Cf. **Transverse Plane.**

Transverse Springs. Laminated springs arranged transversely across a motor-vehicle and parallel to the axles.

Trap Points. *Points* placed in running rails to prevent unauthorized switching of trains on to other tracks.

Travel. The maximum linear movement of a reciprocating part such as a piston, valve, valve rod, etc.

Travelator. A moving platform, constructed on the same principle as an *escalator*, for conveying passengers up or down a long slope, but without steps.

Traveller. A small C-shaped spring clipped upon the ring of a ring spinning frame to act as a thread guide and to assist in the insertion of the yarn twist.

Traveller Gantry. A *gantry* with a movable carriage on rails for movement along or across the gantry; the carriage carries a fixed *crab* or *winch*.

Traveller, Overhead. See Crane.

Travelling Bridge. A movable bridge which can be rolled backwards and forwards across an opening. Fig. 189 shows a bridge carrying granular material which is shipped from a hopper *D* into trucks *E*. The carrier *C* picks material from the trough *B* into which it has been deposited from truck *A*.

Travelling Crane. See Crane.

Travelling Gear. The actuating gear of a *travelling crane*.

Traversing. A longitudinal motion of a cutter on a lathe or of any tool on a machine. Cf. **Facing** (*a*).

Traversing Drill. *Slot drill.*

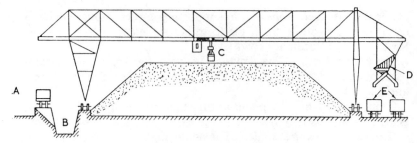

FIG. 189. TRAVELLING BRIDGE.

Traversing Length. The surface length examined in a complete determination of surface *texture*, usually including several *sampling lengths*.

Traversing Mandrel. A *headstock* sliding *mandrel* for cutting ornamental screw threads, with interchangeable guide screws at the tail end.

Traversing Screw. A screw which provides a traversing motion to some part of a machine tool.

Traversing Screw Jack. A *jack* which travels longitudinally upon its base by means of a horizontal traversing screw turning in bearings on the base.

Treadle. (*a*) A foot lever connected by a rod to a crank to give motion to a lathe, sewing machine or other mechanism.

(*b*) A contact operated by the deflection of running rails due to the passage of a train, trucks, etc.

Treadmill. An appliance for producing motion by the stepping of a man or horse, on the steps of a revolving cylinder.

Treble Barrel Pump. *Three-throw pump.*

Treble Clack Box. *Treble valve box.*

Treble Ported Slide Valve. An *exhaust relief valve* in a cylinder with two narrow ports in the body of the valve in addition to the end supply.

Treble Valve Box. A pump casing containing suction, delivery and intermediate check, or retaining valves.

T-rest. A T-shaped rest to support a tool which is clamped to the bed of a wood-turning lathe.

Triblet. (*a*) A tapering mandrel on which rings, nuts, etc., are formed or forged.

(*b*) A steel core upon which tubes are drawn.

374

Tribology. The study of friction, lubrication, and wear.

Trick Valve (Allan Valve). A *slide valve* housing an internal steam passage additional to the exhaust cavity.

Tricycle. (*a*) A vehicle with three wheels, one in front and two behind.

(*b*) An aeroplane's undercarriage system with a single wheel unit under the front of the fuselage and the main undercarriages further back approximately under the centre of gravity.

Trigger. A device, coarser than a *flirt*, for releasing a spring or catch and thus setting a mechanism in motion.

Trim. A term relating to the materials of the disc, body seat ring and stem of valves and stating the percentage of some element of the alloy of which the part is made, such as ' 10 per cent chrome trim '. In ' swing type ' check valves it may also relate to the hinge pin (see **Check Valves**). In ' gate valves ' trim may also refer to the wedge or wedge facing rings.

Trimming Machine (Guillotine). A lever- or treadle-operated machine for cutting, trimming, etc. See also **Guillotine Shears**.

Trimming Press. A press for *trimming* sheet-metal stampings or die forgings.

Trip Dogs. Adjustable *cam followers* which are set to give the correct set, speed, and turret position for each operation of automatic screw-making machines and automatic lathes.

Trip Engines. Engines in which the valves are opened by short levers instead of eccentrics.

Trip Gear. A gear for actuating valves, opening them by a trigger mechanism, which is then tripped out of engagement to allow the valve to close under a heavy spring. Cf. **Corliss Valve, Drop Valve**.

Trip Lever. A *bell-crank lever* for the rapid opening and closing of valves.

Triple-expansion (Steam) Engine. A steam-engine with high pressure, intermediate pressure and low-pressure cylinders working on the same crankshaft. See **Multiple-expansion Engine, Simple Steam-engine**.

Tripping. The running of a tooth of an *escape wheel* past the locking face in an *escapement*.

Trolley (Trolly). (*a*) A low truck running on rails.

(*b*) A small table running on wheels.

(*c*) A truck that can be tilted.

Trolley-bus. An electrically-driven *omnibus* collecting current from an overhead system of wires by means of a pole and *trolley wheel.*

Trolley Wheel. (*a*) A single- or double-flanged wheel on a *trolley.*
(*b*) A small grooved wheel for collecting current from an overhead wire for a *trolley-bus.*

Trunk Air Pump. A marine air pump with a hollow piston rod enclosing the connecting-rod, which is jointed loosely within to its lower end.

Trunk Piston. A piston which is long in relation to its diameter and takes the *connecting-rod* thrust, there being no piston rod or crosshead.

Trunk Plunger. A pump plunger made hollow to admit the connecting-rod which is attached to the lower end.

Trunnion (Trunnion Bearing). A bearing on which a vessel or cylinder can rotate or oscillate.

Trussed Shaft. A long light shaft supported by rods to provide rigidity.

Trying-up Machine. A machine for planing heavy timber in which the cutters are fixed on the face of a circular disc.

Tube Drawing. The manufacture of seamless tubes by drawing a tubular piece of material through *dies* of progressively decreasing size. (See Fig. 56.)

Tube Expander. A boilermaker's tool, consisting of a central tapered mandrel with rollers which are progressively screwed in and rotated to increase the end diameter of a tube.

Tube Extrusion. Fig. 190 shows a ram pushing a mandrel through a billet to form a tube. (See also Fig. 191.)

Tube Mill. A *ball mill* having a cylinder longer than usual; this usually being subdivided internally so that the material to be ground passes from one compartment to the next, the grinding media in successive departments being progressively and appropriately smaller.

Tumbler Bearing. A support bearing for long shafts which, when the carriage, or traveller, comes into contact with it, pivots on an external support and allows the carriage to pass, but returns immediately to its position when released.

Tumbler Drier. A rotating drum with small ledges on the inside which cause clothes, after raising, to tumble from top to bottom as it rotates in a draught of warm dry air. Cf. **Spin Drier.**

Tumbler Lock. A lock in which a tumbler or latch engages with notches in the bolt of the lock thus preventing its motion until the

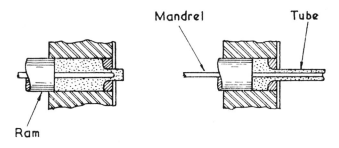

FIG. 190. TUBE EXTRUSION.

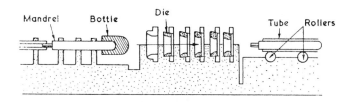

FIG. 191. PUSH-BENCH PROCESS FOR TUBE-MAKING.

tumbler has been lifted or displaced by the key, the key removing the obstacle and then shifting the bolt.

Tumbling. The process of revolving workpieces in a barrel with abrasive and other material for the purpose of deburring, cleaning and improving the surface finish or changing the lustre.

Tup. (*a*) The hammer head of a steam hammer.

(*b*) A *monkey*.

Turbine. A rotary power unit driven by the impact of, or re-action from, a flowing stream of air (*air turbine*), of hot gases (*gas turbine*), of water (*water turbine*), or of steam (*steam turbine*) on the blades, buckets or vanes. See separate terms and **Disc-and-drum Turbine.**

In an ' outward flow ' (Fourneyron) water turbine, the water impinges on the buckets of an outer ring and in the ' inward flow ' or central discharge turbine the method is reversed, the water entering from the periphery and escaping axially.

377

Turbine

In the 'Jonval' or downward axial-flow turbine, the discharge is from above downwards. The water is admitted from the periphery to the 'vortex' turbine and after actuating the vanes passes out at the centre above and below; the turbine may be fixed horizontally or vertically. In the 'Girard' turbine, the water impinges on the curved sides only and it can be used horizontally and vertically.

Turbine Blade. See **Blade.**

Turbine Buckets. (*a*) The *buckets* of an *impulse turbine* or *disc-and-drum turbine*.

(*b*) Turbine blades (U.S.).

Turbine Disc (or Drum). The rotating member on which the blades of the turbine are fixed.

Turbine Pump. A rotary pump with a number of stages which can thus lift higher heads than a *centrifugal pump*.

Turbine Rotor. See **Rotor** and Fig. 142.

Turbine-type Axial Compressor. Fig. 192 shows the stator and rotor of a turbine-type axial compressor.

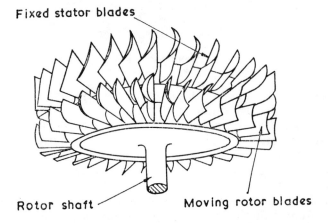

FIG. 192. TURBINE TYPE OF AXIAL COMPRESSOR.

Turbine Wheel, Turbine Rotor. The fundamental Hero type of turbine wheel as used in lawn sprinklers is shown in Fig. 193. See also **Rotor.**

Turbo-generator. A directly coupled steam-turbine and generator.

378

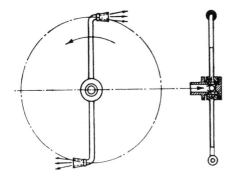

FIG. 193. TURBINE WHEEL.

Turbojet Engine. An air-swallowing engine, composed of a compressor, combustion chambers and a gas turbine, which generates thrust by a jet of hot gases passing down an exhaust cone to a propulsion nozzle. (Fig. 194 shows the by-pass type.) See also **Ducted Fan** and Fig. 59.

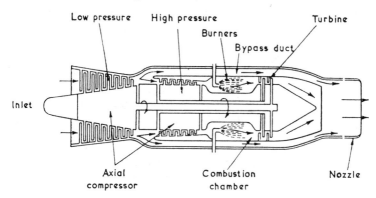

FIG. 194. TURBOJET ENGINE (BY-PASS TYPE).

Turboprop Engine. An air-swallowing *turbojet engine* coupled to a propeller and providing thrust mainly by the propeller and partly by a jet.

Turbo-pump. (*a*) A centrifugal pump with guide vanes at the exit from the *impeller*. Cf. **Turbine Pump.**

(*b*) A combined *ram-air turbine* and hydraulic pump for a guided weapon.

Turbo-starter. An independent turbine, driven by compressed air, a gas source or other means, used for starting an aircraft engine.

Turbo-supercharger. See **Supercharger.**

Turnboot. See **Curb Pins.**

Turnbuckle (Screw Shackle). A long nut screwed at the two ends, one right-handed and the other left-handed, enabling the ends of two rods (or wires) to be joined together and to adjust the total length; also used to adjust the level of plates.

Turning. Using a lathe to produce circular and cylindrical work.

Turning Circle. The circle of minimum radius in which a vehicle can be turned. See also **Steering-lock.**

Turning Cylinders (Slewing Cylinders). The shorter cylinders of an hydraulic crane whose rams cause the crane to slew on its centre. Cf. **Lifting Cylinder.**

Turning Tool. See **Lathe Tool.**

Turns. A small dead-centre watchmaker's lathe used for pivoting, polishing and turning small parts, normally held in a vice and driven by a hand wheel or a bow.

Turntable. (*a*) A circular platform rotating on a centre pivot with wheels under the ends that run on a circular track and with rails across a diameter. Locomotives are driven on to the rails, the table is then rotated either through 180° to allow the locomotive to drive off facing the opposite way or through some other angle so that the locomotive can proceed on to another line.

(*b*) A circular revolving platform for turning road vehicles.

(*c*) A rotating disc for supporting gramophone discs while they are being played.

Turntable Ladder. An extensible ladder mounted on a rotating platform that is carried on the rear end of a motor-vehicle and is supported by feet extended to the ground when in use. By this means, firemen with a hose can be raised on top of the ladder to attack fires at heights of 20 metres or more.

Turret. A revolving tower (or compartment) for gun and gunner(s) in ship, fort or aircraft.

Turret Clock. A large clock in which the movement is quite separate from the dials. The pallets usually span eight teeth. (See also **Tower Clock.**)

Turret Head (Turret) of a Lathe. A device for containing a number of drills, cutting tools, etc., on a lathe to be brought into successive use on the work. See also **Capstan Tool Head** and Fig. 23.

Turret Lathe. A large *capstan lathe* in which the capstan (turret) head and carriage are automatically power-operated in the correct sequence for each particular job. The capstan saddle is on the main slide of the bed which is not the case on a *capstan lathe*. See also **Combination-turret Lathe.**

Turret Rest. See **Capstan Tool Head.**

Turret Saddle. That part of a *turret lathe* which carries the hexagonal capstan tool head. The saddle slides directly on the *lathe bed*.

Twin Hoist. A twin rail-mounted hoist unit with a common hoist cable giving greater height and speed of lift.

Twin Screws. A pair of right- and left-handed propellers on separate parallel shafts

Twin-shaft Turbine. Two similar turbines mounted on the same horizontal shaft but discharging in opposite directions and hence balancing the end thrust. See **Ljungström Turbine, Spool.**

Twist Drill. A hardened-steel drill with cutting edges formed at the periphery of helical flutes and with a conical point backed off to provide clearance for the borings as they are cut off. (See Fig. 114.)

Twist Drill Grinder. A grinding machine set for grinding the constant angle of a *twist drill*.

Twisting Frame. *Doubling frame.*

Two-high Rolls (Two-high Mill). A *rolling mill* with two rolls only, one above the other. If the mill cannot be reversed, a bar, after passing through, must be drawn back over the top, thus losing a pass. Cf. **Three-high Rolls.** (Fig. 150.)

Two-jawed Chuck. A lathe chuck with only two jaws which may be either independent or actuated by a single screw spindle.

Two-start Thread. *Double-threaded screw.*

Two-stroke (2-stroke) Cycle. A piston-engine cycle completed in two strokes, involving one crankshaft revolution, the charge being introduced, compressed, expanded and exhausted through ports in the wall of the cylinder, before and during the entry of the fresh charge. Cf. **Diesel Cycle, Four-stroke Cycle, Otto Cycle.**

Two-throw Crank. An axle or shaft with two cranks, usually at right-angles to each other.

Two-throw Pump. A double-barrelled suction pump operated by two cranks. It is not in equilibrium like a *three-throw pump*.

Tyre. A forged and flanged steel ring shrunk on the rim of a locomotive or other wheel to give added strength and durability.

Tyre Rolling Mills. Vertical or horizontal mills in which the *tyres* are expanded and shaped between inner and outer rolls in a roughing and a finishing pass.

U

u. (*a*) A symbol for velocity. Also v.

(*b*) A symbol for strain energy.

Ultimate Load. (*a*) The maximum load which a structure is designed to withstand without a failure.

(*b*) The product of the *limit load* and the ultimate factor of safety. See **Factor of Safety.**

Ultimate (Tensile) Strength. The breaking load under tension.

Ultimate Tensile Stress. The ratio of the highest load applied to a piece of metal during a tensile test divided by the original cross-sectional area. Also called ' tenacity '.

Ultrasonic Drilling. Drilling with a reciprocating tool which vibrates at an ultrasonic frequency.

Ultrasonics. The study of sound waves with a frequency too high for audibility, that is, greater than about 16 kHz.

Unaflow Engine (Uniflow Engine). A *steam-engine* in which the steam enters through *drop valves* at the ends of the cylinder and exhausts through a piston-controlled belt of ports at the centre.

Uncoiler. A machine for unwinding coils of sheet metal and flattening the sheets by passing through suitably adjusted rollers.

Under Frame. That part of a *truck* containing the axles and their bearing springs, the axle boxes, the buffer and drawbar springs and the wheels.

Undercut. (*a*) A recessed diameter at the end of a thread or similar part, enabling mating parts to lie flush.

(*b*) See **Broach (Rake Angle).** Fig. 17.

Undercut Tooth. See **Saw-tooth Profiles.**

Undershot Wheel. A water-wheel, which receives the water near the bottom of the periphery and its power is obtained almost entirely from the impulse of the water on the vanes. See also **Overshot Wheel, Poncelot Wheel.**

Undulation. A periodic departure, or departures, of the actual tooth surface of a gear from the design surface due to machining or other variations.

Undulation Height. The normal distance between the crests and troughs of tooth undulations.

Undulation Wavelength. The distance between two adjacent crests of an undulation.

Ungeared (Machine). (*a*) A lathe or drilling machine without a *back gear.*

(*b*) A direct drive as from an aircraft engine to a propeller without gearing.

Unified Screw Threads. In 1948, Great Britain, Canada and the U.S.A. agreed to merge the *Whitworth thread* and the American Standard thread into a Unified System of Screw Threads. See also **Screw Thread.**

Uniflow Engine. *Unaflow engine.*

Unilateral Limit System. See Limit System.

Unilateral Tolerance. See Tolerance.

Unit of Work. See Work.

Unit Stress. The stress upon a given sectional area per square inch, square foot or square metre ($1\ Pa = 1\ N/m^2$).

Universal Chuck (Concentric Chuck). Same as *self-centring chuck.*

Universal Joint. A *Hooke's joint* placed at the ends of the propeller shaft in an automobile to allow for the movement of the rear axle relative to the gear-box. (Fig. 195.)

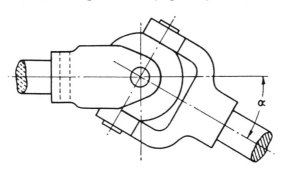

FIG. 195. UNIVERSAL JOINT.

Universal Vice. A *vice* that can be turned on different axes and fixed in almost any desired position. Its movable jaws can be positioned by a hand screw or pneumatically for quick action.

Unsprung Mass. That part of a vehicle *suspension* system which, although cushioned by the springing of a tyre, is not damped by the main vehicle springs.

Unstable Equilibrium. See Equilibrium.

Up-and-down Indicator. A mechanism for indicating when a chronometer or watch needs winding, ' Up ' indicating when fully wound.

Upset, Angle of. See Angle of Upset.

Upset Forging

Upset Forging. A process in which the steel is placed against a stationary die impression, a movable grip die moves in to hold the piece and a plunger advances to form the head on a bar or other piece. Three times the diameter is the maximum obtainable without buckling. The process was originally designed for producing bolt heads but has been developed for other operations.

V

V. The symbol of potential energy.

v. The symbol for velocity—also *u*.

V.H.N. *Vickers hardness number.*

Vacuum. A region in which the gas pressure is very considerably lower than atmospheric pressure.

Vacuum Brake. A brake system on passenger trains. A vacuum, maintained in reservoirs by exhaust pumps, is simultaneously applied to brake cylinders throughout the train. When the vacuum is broken all brakes are automatically applied.

Vacuum Chamber Unit. A chamber, as used in an aneroid barometer, consisting of a number of steel diaphragms formed into a single unit. The chamber is exhausted by using the pipe shown in Fig. 196 and the diaphragms are corrugated to produce greater

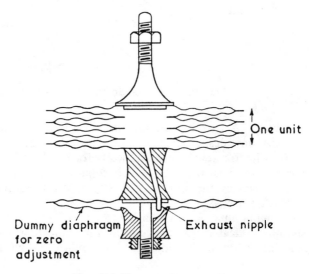

FIG. 196. VACUUM CHAMBER UNIT.

flexibility. Zero adjustment is facilitated by the dummy diaphragm. Pressure changes in the outer atmosphere cause a movement of the centre of the diaphragm which is nearly linear. Temperature changes affect the accuracy and have to be allowed for or compensated. This type of unit has many applications.

Vacuum Gauge. A gauge indicating the amount of *vacuum* in a vessel partially evacuated of its air or in a steam condenser.

Vacuum Pump. See **Air-pump, Pulsometer.**

Valve. A lid or cover to an aperture that opens a communication for a liquid or gas in one direction or closes it in another, or regulates the amount of flow. See **Back-pressure Valve, Bib-valve, Ball Valve, Butterfly Valve, Check Valve, Clack Valve, Conical Valve, Flap-valve, Gate Valve, Needle Valve, Non-return Valve, Poppet Valve, Return Valve, Rotary Valve, Screw-down Stop Valve, Sleeve Valve,** in addition to terms defined below.

Valve Characteristic (Static). The relation between piston displacement and gauge pressures at outlet ports or the pressure difference between them.

Valve Land. The close-fitting part of a piston in a piston valve, covering a port and thus cutting off the flow of fluid through it.

Valve Lap. The distance a piston moves from the central position to the point where the pressure in the port has reached its maximum.

Valve Port. A hole in the valve body of a piston valve through which the fluid flows.

Dropping Valve. A valve which lowers the supply pressure by a constant amount.

Kingston Valve. A wing valve fitted to a ship's side for admitting water to pumps, etc., or for blowing out ballast tanks, flooding, etc.

Pilot Valve. A small valve controlling the motion of a larger valve.

Piston Valve. A device controlling the flow of fluid through holes in the valve body by close-fitting cylindrical pistons.

Reducing Valve. A valve for providing fluid at a low pressure from a supply at higher pressure and independent of the latter.

Reflux Valve. A non-return valve used in pipelines on rising gradients.

Relief Valve. A valve loaded by weights, by springs or hydraulically, that opens when the pressure exceeds a predetermined amount.

Valve-box (Valve Chamber, Valve Chest)

Valve-box (Valve Chamber, Valve Chest). (*a*) The chamber containing the valves or valve in a *force pump* or *steam-engine*.

(*b*) The steam chest containing the *slide valve*.

Valve Chamber. *Valve-box.*

Valve Cock. (*a*) A cock which opens with a lift valve or by means of a slide valve. Cf. **Plug Cock.**

(*b*) A *ball valve.* (See Fig. 197.)

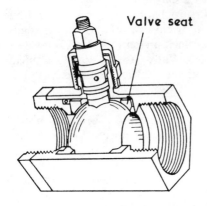

Valve seat

Fɪɢ. 197. Cᴏᴄᴋ Vᴀʟᴠᴇ.

Valve, Cylindrical Slide. Fig. 198 shows the type of cylindrical slide valve used in steam-engines.

Valve Diagram. A graphical method for correlating the movements of the eccentric, of the valve and of the points of admission, cut-off, compression and release for the *slide valve* of a steam-engine.

Valve Face. The sealing surface of a valve which beds or fits on to the seating.

Valve, Four-way Piston. Fig. 199 shows the working of a four-way piston valve with by-pass control.

Valve-gear. The mechanical arrangements for actuating valves. See **Hackworth Valve-gear, Joy's Valve-gear, Marshall Valve-gear, Radial Valve-gear, Valve-gear (Overhead Mushroom), Walschaert's Valve-gear,** and **Link Motion.**

Valve-gear (Overhead Mushroom). Fig. 200 shows the mechanism of an overhead mushroom valve-gear as fitted on motorcycles.

Valve Insert. A *valve seating* of special steel pressed into the heads of high-duty petrol engines.

Valve Land. That part of the piston of a piston valve which closes a port.

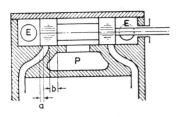

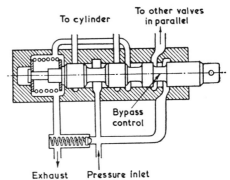

FIG. 198 (*above*). CYLINDRICAL
SLIDE VALVE. (*a*) Exhaust
cap; (*b*) Steam cap; (*E*) Exhaust cap; (*P*) Steam inlet.

FIG. 199 (*above, right*). FOUR-
WAY PISTON VALVE.

FIG. 200 (*below, right*). OVER-
HEAD MUSHROOM VALVE-
GEAR.

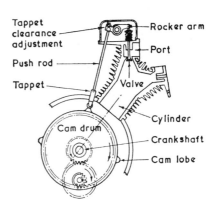

Valve Lap. The distance moved by the piston from the central position to the point where the port pressure is a maximum.

Valve Lifter. A specially made ' G ' clamp used for the removal of internal-combustion engine valves by compressing the valve springs and thus allowing the pins, cotters, etc., to be removed.

Valve Milling Machine. A *milling machine* for milling square or hexagonal portions of valves and cocks.

Valve-opening Diagram. A diagram showing the opening area of a valve plotted against the internal-combustion engine crank angle or piston displacement.

Valve Port. A port in the valve body of a piston valve.

Valve Reseating. Recutting the worn seats of valves.

Valve Ring. *Equilibrium ring.*

Valve Rocker (Rocker Arm). A small pivoted lever which

Valve Rocker (Rocker Arm)

transmits motion from a *cam* or a *push rod* to a valve stem. The rod which oscillates about the spindle *F* in Fig. 137 is a valve rocker. *G* is the locknut for adjusting the *push-rod* clearance. Fig. 200 shows the valve rocker on a motorcycle engine.

Valve Rod (Valve Spindle). The rod to which a valve is attached.

Valve-rod Gland. The *gland* closing the stuffing box of the rod of a slide valve.

Valve Seat (Valve Seating). The bearing surface against which a valve seats when shutting off the flow of gas or liquid. A stop-valve seat is shown in Fig. 201.

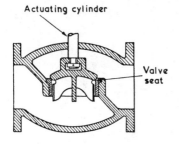

FIG. 201. STOP VALVE.

Valve Sector. (*a*) A *slot link*.

(*b*) The vertical sliding link of an oscillating cylinder engine which communicates the eccentric's motion to the valve rod *weigh shaft*.

Valve Spindle. *Valve rod.*

Valve Spring. (*a*) A helical spring used to close a *poppet valve* after it has been lifted. (See Fig. 137.)

(*b*) Any spring for closing a valve that has been lifted mechanically or by fluid pressure.

(*c*) A spring used to force the packing rings of a *slide valve* against its working face.

Valve-spring Cotters. Various types of cotters for valve springs are illustrated in Fig. 202.

Valve Stem. *Valve rod.*

Valve Yoke. The *bridle* of a *valve rod*.

Vane (Air Vane). A pivoted free surface which turns along the wind direction or a fixed curved surface which changes the direction of flow.

Vane Engine. See **Engine (Servomotor Types).**

Vane Pump. See **Pump.**

388

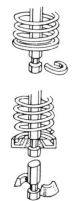

FIG. 202. VALVE-SPRING COTTERS.

Vane. (*a*) A pivoted free surface which turns along the wind direction like a weather cock.

(*b*) A fixed curved surface for changing the direction of flow of a liquid or gas.

(*c*) A blade of a fan.

(*d*) The bucket of a turbine.

Vanish Cone. The cone whose surface passes through the roots of the *washout thread*. See **Thread.**

Vapour Blast Cleaning. Blasting the surface of a workpiece with a high-velocity jet of vapour to give a smooth clean surface.

Variable-area Propelling Nozzle. A turbojet *propelling nozzle* with variable outlet area to match the different operating conditions especially with an after-burner (see **Reheat**). It may be actuated mechanically or aerodynamically, called ' eyelid ' or ' petal ' type, these names being self-descriptive.

Variable Cut-off. A *cut-off* actuated from a governor which is brought into action according to the load on the engine.

Variable Expansion. The expansion of steam in a steam-engine where the amount alters under the varying conditions of working, automatic or otherwise. See **Varying Travel.**

Variable Gears. Toothed wheels which transmit varying velocity ratios during the course of a single revolution. See also **Sector Gears.**

Variable-pitch Propeller. A propeller, the *pitch* of whose blades can be varied in flight.

Variable-speed Gear. A device, consisting of smooth speed cones or *expansion pulleys,* whereby the speed ratio between two shafts can be varied without shifting the belt.

Variable Stroke (Variable Capacity Engine). See **Engine (Servomotor Types).**

Varying Travel. The varying travel of a *slide valve* to alter the amount of *lap* for variable expansion. Cf. **Constant Travel.**

Vee-belts. Belts with a cross-section of vee shape for use with *expansion pulleys,* etc. A typical cross-section is shown in Fig. 203.

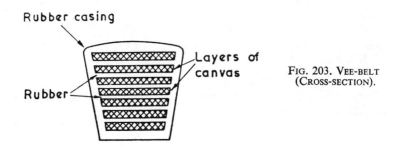

Rubber casing

Layers of canvas

Rubber

FIG. 203. VEE-BELT (CROSS-SECTION).

Vee-strips. *Adjustment strips.*

Vee-thread (Angular Thread). A screw thread with a vee-shaped profile.

Velocity. The rate of change of position, or rate of displacement of a point with respect to a specified reference frame of co-ordinates and expressed in feet (or centimetres) per second. The direction as well as the magnitude are needed for the complete specification of a velocity.

Velocity Ratio. (*a*) The ratio of the distance moved through by the point of application to the corresponding distance moved by the load. See also **Efficiency, Mechanical Advantage.**

(*b*) The ratio of the velocities of bodies which are mutually connected by gearing, etc.

Verge Escapement. See **Crown-wheel Escapement.**

Verge. *Pallet arbor.*

Vernier. A movable auxiliary scale, sliding in contact with the main scale of graduation, to enable readings to be made to a fraction (e.g. a tenth) of a division in the main scale.

A vernier calliper for measuring internal and external dimensions is illustrated in Fig. 204 (*a*).

Vernier Depth Gauge. This gauge is illustrated in Fig. 204 (*b*).

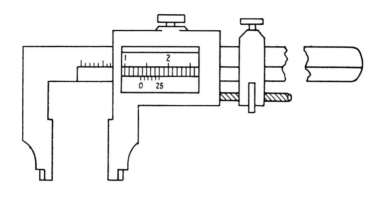

(a)

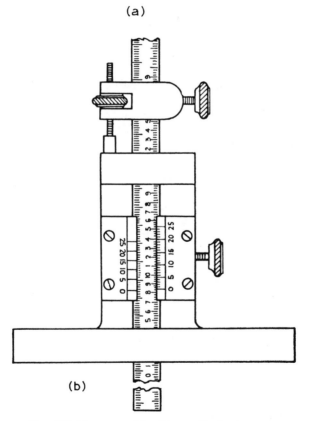

(b)

FIG. 204. VERNIERS: (*a*) callipers, (*b*) depth gauge.

391

Vertical Boring Mill. A mill in which the work is held on a revolving horizontal table and the tools held in a turret on a single upright head or on a cross-rail spanning two uprights.

Vertical Crane. See **Crane.**

Vertical Engine. An engine with the cylinders arranged vertically above the crankshaft. Cf. **Inverted Engine.**

Vertical Escapement. See **Escapement.**

Vertical Milling Machine. A *milling machine* with a vertical spindle, as found in *end mills* and often used for *profiling.*

Vertical Shaping Machine. *Slotting machine.*

Vibrating Conveyor. A trough attached to *vibrators* which impart an upward and reciprocating movement to propel material upward and forward during the upward movement and which draw back the trough underneath the material in the subsequent movement.

Vibrating Links. *Suspension links.*

Vibration. A repetition of some value of velocity in an arbitrarily selected particle in a *mechanism.* See also **Mounting** and **Damper.**

Circular Vibration. The motion of a body, of which any point moves in a circle and in which any line maintains a constant direction; the motion is resolvable into two components of equal magnitude in space and time quadrature.

Forced Vibration. A vibration maintained by periodic force(s) and having frequencies related to the frequencies of the force(s).

Free Vibration. A vibration with periods depending solely on the properties of the disturbed system.

Natural Mode of Vibration. The manner of vibration associated with each natural frequency. The fundamental natural frequency.

Normal Mode of Vibration. An undamped natural mode with each particle in a simple harmonic motion at the same frequency in phase or antiphase. The decrements are zero and the phase differences are either 0 or π. See also **Mode** and **Phase.**

Self-induced (Self-excited) Vibration. The state of steady or growing vibration of a mechanical system caused by the extraction of energy from an external source.

Steady-state Vibration. That state when the velocity of each material particle in a mechanical system is a periodic quantity or the sum of periodic quantities.

Torsional Vibration. Variations in the torque transmitted by a crankshaft.

Vibration Isolator. A resilient mounting used to absorb the vibration of a fixed mechanical installation. See also **Isolator.**

Vibration-testing Machine. A machine for subjecting specimens to desired frequencies and amplitudes of vibration to determine their operational characteristics. These machines are of three main types: (*a*) wholly mechanical using cams or cranks, (*b*) electromagnetic using a moving-coil transducer to give linear vibrations, (*c*) hydraulic linear actuator incorporating a servo valve of which a number can be easily linked together.

Vibrator. (*a*) An instrument or mechanism which causes rapid vibrations in rectilinear or rotary motion and is usually controlled either by an electromechanical device such as a tuning fork or the stretched wires in a magnetic field carrying a mirror to make a sound track.

(*b*) Pneumatic equipment such as that used for stirring a concrete mix after laying or a rotary immersion vibrator for settling laid concrete.

(*c*) An instrument for checking the time of vibration of a watch or clock *balance* and its spring against a master balance with whose vibrations it is compared.

Vibratory Feeder. A feeder delivering material in accurate quantities down an inclined surface, controlled magnetically or by vibrating the surface.

Vibrometer. An instrument for indicating variations from the correct balancing of revolving machinery.

Vice (Vise). A workshop tool for holding or gripping work firmly, consisting of a pair of steel-faced jaws, one of which is movable by a screw or by a lever; the other jaw is fixed. See **Machine Vice.**

Vice Chuck. *Machine vice.*

Vickers Hardness Number (V.H.N.). A number equal to $0.927p$, where p is the yield pressure, the area on which it acts being 0.927 times that of the surface area of the contacting faces. The pyramidal indenter is square-based and the opposite faces contain an angle of $136°$. The Vickers and Brinell hardness numbers for a given load are nearly equal. Cf. **Brinell Hardness Test.**

Virgule Escapement. See **Escapement.**

Virtual Centre. *Instantaneous centre.*

Viscosity. The internal friction due to molecular cohesion in a fluid; the resistance to the sliding motion of adjacent layers of a fluid when in motion.

Viscosity

The ' coefficient of viscosity ' is the tangential force per unit area necessary to maintain unit relative velocity between two parallel planes which are unit distance apart.

The ' kinematic viscosity ' is the viscosity of the fluid divided by its density.

Vitrified Bond. See **Bond.**

Volute Spring. A spring formed in a helix of a flat ribbon plate, rod or wire extensible in the direction of the axis of the coil; used as a buffer spring or to carry the axle boxes of the driving axles of locomotives.

Vortex Chamber. *Whirlpool chamber.*

Vortex Turbine. See **Turbine.**

W

W. The symbol for weight and for work.

ω. The symbol for angular velocity.

Wabblers. The coupling boxes connecting the *breaking pieces* with the necks of the puddling *rolls.*

Wabbling Disc. *Swash-plate.*

Wagon Retarder. *Retarder.*

Waldegg Valve-gear. *Walschaert's valve-gear.*

Wallow Wheel. A *bevel wheel* on a vertical shaft with the teeth facing downwards (see **Crown Wheel** (*a*)).

Walschaert's Valve-gear (Waldegg Valve-gear). A radial-type gear driving the valve of a steam locomotive through a combination lever whose oscillation is the resultant of sine and cosine components of the piston motion. See also **Radius Rod.** Cf. **Joy's Valve-gear.**

Wankel Engine. An internal-combustion engine with a rotary piston, which is a three-lobe rotor that rotates in the combustion chamber to provide the same cycle as in a reciprocating piston engine.

Warning. The partial unlocking of the striking train of a clock. See **Warning Piece** and **Warning Wheel.**

Warning Lever. The lever operated by the hour *warning wheel.*

Warning Piece. A projection on the *lifting piece* projecting through a slot in the dial plate of a striking clock and against which a pin on the *warning wheel* butts and holds up the train. Exactly at the hour, or other period, the warning piece drops clear of the pin and frees the striking train.

394

Warning Wheel. The last wheel in the striking train of a clock, that is held up by the *warning piece*.

Warp. (*a*) To pull a load by the winding of a rope, or chain, upon a drum.

(*b*) The threads spread out on the beam of a *loom* which will run lengthwise in the woven fabric.

Warp Machine. A straight-bar lace frame with *spring needles* in which individual threads pass to individual needles to form the fabric. The warp comes from the warp beam and from independent beams. A bar warp machine makes warp net from beams only.

Warping Cone (Surging Drum). A conical or capstan-shaped drum for receiving a coil of rope or chain.

Warping Mill. (*a*) A cylindrical cage for winding threads from jack *bobbins* to fixed lengths for use in a lace machine. See also **Jack** (*b*).

(*b*) A large-diameter wooden reel upon which threads are wound during the making of a *warp*.

Washer. An annular piece of metal, leather, rubber, etc., for distributing pressure under a nut or making a tight joint between surfaces. Two typical types of tab lock-washers, which prevent the nut rotating, are shown in Fig. 205.

Washing Down. Thinning down to a feather edge.

Washout Thread. That part of the profile of a screw thread not fully formed at the root.

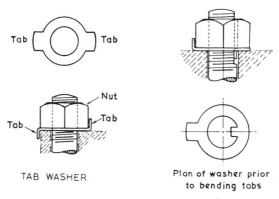

TAB WASHER

Plan of washer prior to bending tabs

INTERNAL TAB WASHER

FIG. 205. WASHERS.

Watch. A small portable timekeeper. Hence watch *movement*.

Watch and Clock Mechanisms. See **Barrel** and **Barrel Arbor, Balance Spring, Centre Pinion, Centre Wheel, Escape Pinion, Escape Wheel, Escapements, Fourth Pinion, Fourth Wheel, Fusee Arbor, Going-barrel, Great Wheel, Index, Keyless Mechanisms, Staff** and **Wheel, Third Pinion, Third Wheel;** Figs. 31, 36, 39, 62, 161.

Water Cylinder. The pump barrel of a steam pump.

Water Jacketing. The casing of engine cylinders with a jacket through which water flows to keep them cool.

Water Turbine. See **Francis Water Turbine, Kaplan Water Turbine, Mixed-flow Water Turbine, Pelton Wheel, Propeller-type Water Turbine, Turbine.**

Water Wheel. A wheel with a horizontal axis and buckets or *floats* on its rim, the water flowing onto or into the buckets to provide power. See **Breast Wheel, Overshot Wheel, Turbine, Undershot Wheel.**

Water-pressure Engines. Turbines, hydraulic rams, etc., driven by water pressure or by a head of water.

Watt Governor. A *pendulum governor*. A pair of links, pivoted to a vertical spindle, terminate in heavy balls; shorter links are pivoted to the mid-points of these links and to the sleeve operating the engine throttle.

Watt's Straight-line Motion. A type of motion used by Watt to guide the piston rod in many of his early steam-engines. The point *C* in Fig. 206 oscillates along a straight line as levers *A* and *B* rotate about their fixed pivots.

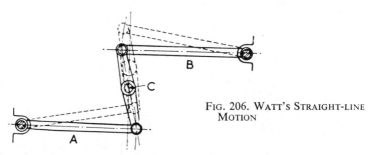

FIG. 206. WATT'S STRAIGHT-LINE MOTION

Wave-form. The average shape of each recurring outline in a profile that repeats regularly, such as in a screw thread.

Wavelength. The distance from crest to crest in a *wave-form*, such as a screw thread, or in an *undulation*.

Wavelength Cut-off. The maximum or minimum wavelength which an electrical (or similar) measuring instrument is adapted to register.

Waviness. Surface irregularities that have greater spacing than *roughness*. The height is peak-to-valley distance and the width is the spacing of adjacent waves.

Way Shaft. *Weigh shaft.*

Ways. The smooth top surface of a *lathe bed* on which the carriage and tailstock slide.

Wedge Gate Valve. A *gate valve* in which closure is effected by the wedge action between the gate and the body seats, the gate being either solid or cored in one piece or in two pieces. Cf. **Double Disc Gate Valve.**

Wedge Facing Rings. Rings of different material from the wedges and secured to them, on which the wedge faces are machined.

Wedge Gearing. Gearing composed of wheels with circumferential grooves which fit into each other and thus provide a friction drive.

Weft, Pick, Shot. The threads across the width of a fabric. A single thread is called a ' pick ', or a ' shot ' of ' weft '.

Weft Fork. A short lever, forked at one end and hooked at the other, with the fulcrum in the centre, which stops a *loom* when the *weft* breaks or runs out.

Weigh Shaft (Way Shaft, Reversing Shaft). The shaft for moving the *slot links* to put a steam-engine into forward or backward gear.

Weighbridge. A table carried on a system of levers with arms so proportioned that a large weight on the weighbridge table can be balanced by a small weight moved along a steelyard.

Weighing Machines. See **Balance, Beranger Balance, Roberval Balance, Spring Balance, Steelyard** and **Weighing Machine, Self-indicating.**

Weighing Machine, Self-indicating. A weighing machine with a counterpoise in which the load is indicated by a movable pointer on a graduated scale as shown in Fig. 207. The graduation is based on the equality $P.l = Wx$, where P is the weight to be measured, and W is the counterpoise weight. In the simple form shown in the diagram the scale is not uniform; by using a flexible strip, moving on a cam, for P, the scale can be made uniform.

Welding. The joining or fusion of pieces of metal by raising

Welding

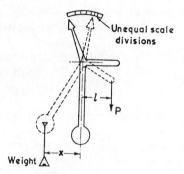

Fig. 207. Self-indicating
Weighing Machine.

the temperature at the joint to make the metal plastic so that the pieces can be joined or fused together.

Arc Welding employs the heat of an electric arc to bring the metal to a molten state for joining by fusion.

Braze Welding utilizes a filler rod that melts at a temperature greater than 700 K, but lower than the melting point of the base metal; it can be applied to all metals that melt above 800 K, except aluminium and magnesium. See also **Brazing.**

Electron Beam Welding uses an electron beam for fusing the metal to make a joint.

Gas Welding uses an oxy-acetylene or oxy-hydrogen flame to obtain the desired temperature.

Resistance Welding (Fig. 208) joins metals by the simultaneous application of pressure and heat.

Well Crane. See **Crane.**

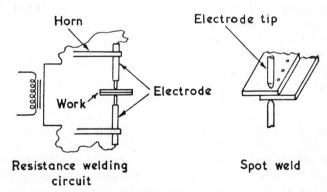

Fig. 208. Resistance Welding.

398

Westinghouse Brake. An air brake on railway rolling stock controlled by a reservoir of compressed air underneath the engine and connected to a cylinder and piston under each carriage.

Wharf Crane. See **Crane.**

Wheel. (*a*) A solid disc or a structure made up of a *rim* and *hub* united sometimes by spokes, and capable of rotation on its own axis, such as a car wheel or a pulley.

(*b*) The larger of an unequal pair of gears. Equal gears are usually called wheels if the diameter is greater than the width of the *rim*. Cf. **Pinion** (*d*).

Wheel Base. The distance between the leading and trailing axles of a vehicle.

Wheel-cutting Engine. A machine for cutting the teeth of wheels for clocks and watches.

Wheel-and-disc Drive (Gear). A *ball-and-disc gear* (see **Integrator**), in which the roller and balls are replaced by a wheel.

Wheel Dresser. Alternate plain and star-shaped hardened-steel discs mounted and freely revolving on the end of a long handle. The discs are applied under pressure to the abrasive wheel that has to be dressed.

Wheel Plate. *Quadrant plate.*

Wheel-quartering Machine. A horizontal drilling machine with opposed spindles at opposite ends of the bed for drilling simultaneously the crank-pin holes in both wheels on a locomotive coupled-axle to ensure the precise angular relationship.

Wheel Teeth. See **Teeth.**

Wheel Wobble. See **Shimmy.**

Whelps. The equidistant longitudinal strips around the barrel of a *capstan* or *warping cone* to increase the bite of the rope.

Whip. (*a*) A *whip crane*. See **Crane.**

(*b*) An arm of a *windmill* carrying the cross-pieces and sails.

Whip Crane. See **Crane.**

Whip Gin. *Gin block.*

Whipping Drum. The winding barrel of a *whip crane*.

Whirling Arm. An apparatus consisting of a long horizontally rotating arm for carrying models for aerodynamic tests or for subjecting apparatus and men to high accelerations.

Whirling of Shaft. The deflection from straightness of a shaft at certain critical rotational (whirling) speeds.

Whirlpool Chamber (Vortex Chamber). The space surrounding the *impeller* of a centrifugal pump into which it discharges.

White Metal. An easily fusible alloy with a tin base (over 50%) containing lead, antimony and copper, used for lining bearings. See also **Babbitt's Metal**.

Whitworth Quick-return Motion. Fig. 209 shows a simple quick-return motion, the point *B* rotating about *O* causes the

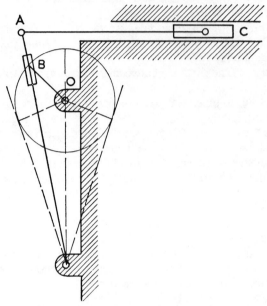

FIG. 209. WHITWORTH QUICK-RETURN MOTION (1).

slide *C* connected through *A* to move slowly from left to right and return quickly as indicated by the relative lengths of the arcs of the circle (*a* and *b*, Fig. 210).

Fig. 210 shows a practical example of the motion suitable for a *shaping machine*. The distance *BO* can be set to vary the reciprocating distances between *F* and *G*. The main drive is supplied through *A*. See also **Quick Return**.

Whitworth Screw Thread. A symmetrical vee-thread of 55° included angle; one-sixth of the sharp vee is truncated at top and bottom, the thread being rounded equally at crests and roots by circular arcs blending tangentially with the flanks, the theoretical depth of thread being thus 0·640327 times the nominal pitch. B.S.W. is a 'coarse thread' series from ⅛ in. to 6 in. diameter.

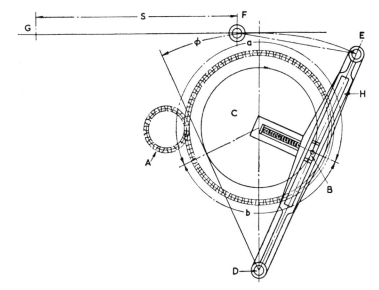

FIG. 210. WHITWORTH QUICK-RETURN MOTION (2).

B.S.F. is a ' fine thread ' series from $\frac{3}{16}$ in. to $4\frac{1}{4}$ in. diameter. A truncated Whitworth thread has flat crests.

Whizzer (Hydro-extractor). A perforated cylinder inside which loose material is dried when flung outwards by revolving paddles to remove the water by centrifugal force. Cf. **Spin Drier, Tumbler Drier.**

Wicket Gate. A regulator of the supply of water to *water turbines* having guide vanes pivoted at their centres which are automatically adjustable to the rotation speed of the turbine wheel and vane angles regulated by a *pendulum governor* acting through a servo-motor.

Wig-wag. A machine which vibrates a polisher used for polishing watch and clock pivots, etc.

Winch. (*a*) A steam-engine on a ship's deck used for hoisting cargo, etc., with various drums and barrels driven by gearing at various speeds.

(*b*) A hoisting machine operated by hand or driven by power.

Wind Pump. A pump which is operated by the force of the wind rotating a multi-bladed propeller, often used for raising water from below ground level. Sometimes called a *windmill*.

401

Wind-driven Turbine. *Ram-air turbine.*

Windage. The loss of energy of rotating machinery due to air resistance.

Winding Drum. The drum on which a haulage rope is wound when raising and lowering mine cages or moving trucks on inclines.

Winding Engine. An engine for hoisting a load up a shaft. See **Hoist.**

Winding Gear. The mechanical gear associated with *lifts* and *hoists.*

Winding Square. The square end of a *fusee* arbor on which the key fits.

Windlass. (*a*) A horizontal drum for hauling or for hoisting, using the wheel-and-axle principle. See also **Chinese Windlass.**

(*b*) An apparatus for hoisting an anchor by means of its cable and for lowering it with a brake to regulate the paying out of the cable.

Windmill. A mill worked by the action of wind on large sails mounted on *whips*, the whole apparatus being rotatable on top of the building and the angle of the sails adjustable. Cf. **Wind Pump.**

Windmilling. (*a*) An aircraft's propeller rotating but delivering no power to the propeller shaft.

(*b*) A compressor rotating by the pressure of the airstream.

(*c*) The rotor of an autogiro (because it is not power driven) is windmilling; likewise, the rotor of a helicopter in case of engine failure.

Wing Rail. *Check-rail.*

Wing Valve. A conical-seated valve guided by radial vanes or ribs when fitting inside a circular port.

Wiper. (*a*) An oscillating bar which cleans the face of a car's windscreen.

(*b*) The cam teeth on the wheel of a tilt hammer.

(*c*) A mechanism used in weaving to convert a rotating into a reciprocating motion.

Wire-drawing. (*a*) The manufacture of wire by pulling rod or wire through successively smaller holes in hard steel or ceramic die-blocks.

(*b*) The throttling of a fluid, such as steam, by passing it through a small orifice or restricted valve-opening.

Wire Gauges. See **British Standard Wire Gauge, Brown and Sharpe Wire Gauge, Gauges Commonly Used, Paris Wire Gauge.**

Wire Thread Insert. See **Thread Insert.**

Wobble Crank. A short-throw crank for giving an elliptical motion to a sleeve valve by a short connecting-rod and ball-joint.

Wobble-plate Engine. *Swash-plate engine.* See **Engines (Servo-motor Types).**

Wohler Test. A fatigue test of specimens of materials held with one end in a rotating chuck and carrying a weight on a ball-bearing at the other end.

Woodruff Key. See **Key.**

Woodruff-keyway Mill. A milling cutter for standard *Woodruff keyways*, each size of key needing a separate cutter. See **Woodruff Key.**

Woof. *Weft.*

Work. The overcoming of resistance (force) through a certain distance. The units of work are the *erg* and the *foot-pound.* See also **Energy.**

Working Barrel. A pump *barrel* containing the piston and bored portion and sometimes the *clack* valves.

Working Cylinder. The exploding cylinder in a gas-engine which has a separate *compressing cylinder.*

Working Depth. The depth on the tooth face to which the tooth of the mating gear extends, being less than the total length of tooth from point to root by the amount of the bottom clearance.

Working Load. The mean ordinary load to which a structure or mechanism is subjected. Cf. **Proof Load.**

Working Pressure. The pressure at which an apparatus or engine works, as distinct from its test pressure.

Working Stress. The safe stress for a structure or mechanism, based on experience and distinct from any proof stress.

Work-piece (Work). That piece of material on which work is to be done by a machine.

Workshop Gauge. A gauge used for checking during the actual production of the work on a machine.

Workshop Microscope. A machine-mounted optical microscope for observing threads, cutting tools and grinding wheels *in situ.*

Worm (Worm Gear, Worm Wheel). A *helical gear* that meshes with a *worm wheel* in sliding contact. See also **Worm Gearing** and Fig. 211.

Hour-glass Worm. A worm cut to provide a very low ratio in the steering system of a motor-car, corresponding to about four turns lock-to-lock, in the straight-ahead position.

Worm (Worm Gear, Worm Wheel)

Multi-start Worm. A worm in which two or more helical threads are used to obtain a larger pitch and a higher velocity ratio than with a single-start worm.

Worm Gearing. Gearing composed of *worms* and *worm wheels.* The worms normally drive the wheels and provide gears of high reduction ratio connecting shafts with axes at right angles, but whose axes do not intersect. See also **Worm.** (See Fig. 211.)

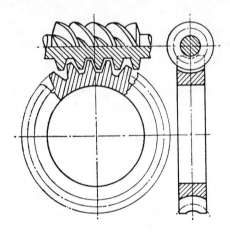

FIG. 211. WORM AND WORM WHEEL.

Worm Thread. That portion of a *worm* bounded by the *root cylinder, tip cylinder* and the two *helicoid* surfaces.

Worm Thread Flank. The surface lying within the *working depth.*

Worm Thread Profile. The line of intersection of a flank with a defined plane.

Worm(screw) Conveyor. A *conveyor* in which a revolving worm continuously propels loose material along or through a tube. Fig. 41 shows various types. Cf. **Archimedean Screw.**

Worm Wheel. A concave face gear-wheel with teeth capable of line contact with a *worm.* See also **Worm Gearing** and Fig. 211.

Worm Wheel Hobbing Machine. A machine for cutting teeth using a *hob* which is the exact counterpart of the *worm.*

Worm-and-wheel Steering-gear. A gear in which the worm on a steering column meshes with a worm wheel or sector, the latter being attached to the spindle of a *drop arm.*

Wrap-up. A means for providing relative displacement between a driving and a driven member when a threshold value is exceeded, subsequently annulling it and restoring a positive drive when below that value, such as in the case of a given torque or of a given number of turns, etc.

Wrenching Allowance. The length of screw thread, for accommodating the relative movement between the pipe end and the coupling and needed for wrenching beyond the position of hand engagement.

Wringing. The process of sliding one *slip gauge* on another by a wiping motion to remove all air and dirt between the mating surfaces. When properly wrung together, the gauges cling to each other with negligible error.

Wrist Pins. (*a*) Pins, on the big end of a *master connecting-rod* of a *radial engine*, which form the *crank pins* of the *link rods*.

(*b*) A pin held by the *crosshead* on a steam-engine and by which the movement of the piston rod is imparted to the connecting-rod.

Wrist Plate (Motion or Rocking Disc). The plate, attached to the side of the cylinder of a Corliss engine and worked by levers, which transmits motion by connecting-rods to the valve spindle. See **Corliss Valve.**

X

X Spring. Two superimposed laminated springs forming a letter X, as used on carriages and some balances.

Y

Y.P. Yield point.

Y Lever. The longest lever of a *weighbridge* to which the steel-yard's rod is attached.

Yale Lock. A cylinder lock for doors in which a key raises a number of springs to different heights for its release. See Fig. 212.

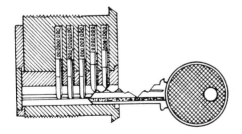

Fig. 212. Yale Lock.

Yankee Machine. *Single-cylinder machine.*

Yard Traveller. Outdoor overhead traveller. See **Crane.**

Yield Point. That point in the loading test on a test-piece when the deformation increases suddenly and a substantial amount of *plastic deformation* takes place under constant (or reduced) load; for example in the stretching of a test-piece in tension.

Yielding Attachment. A special method by which the outer end of a *mainspring* is attached to the *barrel* to permit concentric uncoiling of the spring.

Yoke. The frame formed in a slide-valve spindle to embrace the box-like portion of the valve and thus form the connection.

Yoke (of a Valve). The exterior part of an outside screw valve in which the actuating thread of the stem engages either directly or through a bush or through a yoke sleeve. The yoke may be integral with or separate from the *bonnet.* Cf. **Bridge.**

Young's Modulus. The ratio of the stretching force, as on a test specimen, per unit cross-sectional area to the elongation per unit length. Its value is of the order of 10^5 Pascals or 10^7 lb/in.^{2} for metals.

Z

Zero (Setting). The setting of the zero of an instrument before measurements are started.

Zerol* Bevel Gear. A *spiral bevel gear* with curved teeth and having a zero degree mean *spiral angle.* [*Registered Trade Mark.]

Zinc and Steel Pendulum. See **Pendulum.**

APPENDIX

DEFINITIONS OF UNITS

Base Units

Metre (m)—The metre is the length equal to 1 650 763·73 wavelengths in vacuum of the radiation corresponding to the transition between the levels $2p_{10}$ and $5d_5$ of the krypton-86 atom.

Kilogramme (kg)—The kilogramme is the unit of mass; it is equal to the mass of the international prototype of the kilogramme.

Second (s)—The second is the duration of 9 192 631 770 periods of the radiation corresponding to the transition between the two hyperfine levels of the ground state of the caesium-133 atom.

Ampere (A)—The ampere is that constant current which, if maintained in two straight parallel conductors of infinite length, of negligible circular cross-section, and placed 1 m apart in vacuum, would produce between these conductors a force equal to 2×10^{-7} newton per metre of length.

Kelvin (K)—The kelvin of thermodynamic temperature is the fraction 1/273·16 of the thermodynamic temperature of the triple point of water.

Candela (cd)—The candela is the luminous intensity, in the perpendicular direction, of a surface of 1/600 000 m^2 of a black body at the temperature of freezing platinum under a pressure of 101 325 N/m^2.

Mole (mol)—The mole is the amount of substance of a system which contains as many elementary entities as there are atoms in 0·012 kg of carbon 12. The elementary entities must be specified and can be atoms, molecules, ions, electrons, other particles or specified groups of such particles.

Other units

Bar: a unit of pressure $= 10^5$ N/m^2.

Coulomb (C): the unit of electric charge; it is the quantity of electricity transported in one second by a current of one ampere.

Electron volt (eV): a unit of energy used in nuclear physics $= 1·60206 \times 10^{-19}$ J.

Degree (°): a unit of plane angle $= \pi/180$ rad.

Farad (F): the unit of electric capacitance; it is the capacitance of a capacitor between the plates of which there is a potential difference of one volt charged with one coulomb.

Appendix

Henry (H): the unit of electric inductance; it is the inductance of a closed electrical circuit in which an e.m.f. of one volt is produced when the current varies uniformly at the rate of one ampere per second.

Hertz (Hz): the unit of frequency; it is equal to one cycle per second.

Joule (J): the unit of energy or work; it is the work done when a force of one newton acts over a distance of one metre.

Kilowatt-hour (kWh): a unit of energy; it is the energy expended when a power of 1000 watts is supplied for one hour.

Knot (kn): a unit of speed; it is equal to one nautical mile per hour.

Litre (l): a unit of volume equal to one cubic decimetre.

Lumen (lm): the luminous flux emitted within a unit solid angle of one steradian by a point source having a uniform intensity of one candela.

Lux (lx): the unit of illumination $= 1 \text{ lm/m}^2$.

Newton (N): the force required to accelerate a mass of one kilogramme at one metre per second squared.

Ohm (Ω): the resistance between two points on a conductor at a potential difference of one volt when a current of one ampere is flowing.

Pascal (Pa): a unit of pressure $= 1 \text{ N/m}^2$.

Poise (P): a unit of dynamic viscosity, usually quoted in centipoise (cP); $1 \text{ cP} = 10^{-3} \text{ Ns/m}^2$.

Radian (rad): the angle subtended at the centre of a circle by an arc whose length is equal to the radius of the circle.

Siemens (S): the unit of conductance $= 1/\Omega$.

Steradian (sr): the solid angle subtended at the centre of a sphere by a cap with an area equal to the radius squared.

Stoke (St): a unit of kinematic viscosity, usually quoted in centistokes (cSt); $1 \text{ cSt} = 10^{-6} \text{ m}^2/\text{s}$.

Tesla (T): the unit of flux density $= 1 \text{ Wb/m}^2$.

Tonne (t): a unit of mass $= 10^3 \text{ kg}$.

Volt (V): the unit of e.m.f. and potential difference; it is the difference in potential between two points on a conductor carrying a constant current of one ampere when the power dissipated is one watt.

Watt (W): the unit of power $= 1 \text{ J/s}$.

Weber (Wb): the unit of magnetic flux; it is the flux which, linking a coil of one turn, produced in it an e.m.f. of one volt as it is reduced to zero at a uniform rate in one second.

DERIVED UNITS

Some of the derived units are given special names and these are listed below:

Quantity	Special Name	Units
Electric capacitance	farad (F)	As/V
Electric resistance	ohm (Ω)	V/A
Electromotive force, potential difference	volt (V)	W/A
Energy, work	joule (J)	Nm
Force	newton (N)	kgm/s^2
Frequency	hertz (Hz)	1/s
Illumination	lux (lx)	lm/m^2
Inductance	henry (H)	Vs/A
Luminous flux	lumen (lm)	cd sr
Magnetic flux	weber (Wb)	Vs
Magnetic flux density	tesla (T)	Wb/m^2
Power	watt (W)	J/s
Quantity of electricity	coulomb (C)	As

Other derived units are:

Quantity	Units
Acceleration	m/s^2
Angular acceleration	rad/s^2
Angular velocity	rad/s
Area	m^2
Density	kg/m^3
Diffusion coefficient	m^2/s
Dynamic viscosity	Ns/m^2
Electric field strength	V/m
Frequency	1/s
Kinematic viscosity	m^2/s
Luminance	cd/m^2
Magnetic field strength	A/m
Pressure or stress	N/m^2
Surface tension	N/m
Thermal conductivity	W/mK
Velocity	m/s
Volume	m^2

Appendix

MULTIPLES

Multiples of units may be denoted by various prefixes as shown in the following table.

Factor	Prefix and Symbol
10^{12}	tera (T)
10^9	giga (G)
10^6	mega (M)
10^3	kilo (k)
10^2	hecto (h)
10	deca (da)
10^{-1}	deci (d)
10^{-2}	centi (c)
10^{-3}	milli (m)
10^{-6}	micro (μ)
10^{-9}	nano (n)
10^{-12}	pico (p)
10^{-15}	femto (f)
10^{-18}	atto (a)

Examples

$$km = kilometre = 10^3 \text{ metres}$$
$$mm = millimetre = 10^{-3} \text{ metres}$$
$$mA = milliampere = 10^{-3} \text{ amperes}$$
$$\mu s = microsecond = 10^{-6} \text{ seconds}$$

Indices not included in the table, e.g. 10^{-8}, are not recommended for use.